Recruiting and Educating Future Physics Teachers:
Case Studies and Effective Practices

Cody Sandifer, Towson University, Editor

Eric Brewe, Florida International University, Editor

Sponsored by

Physics Teacher Education Coalition (PhysTEC)
www.PhysTEC.org

July 2015

Recruiting and Educating Future Physics Teachers:
Case Studies and Effective Practices

Published by:

American Physical Society
One Physics Ellipse
College Park, MD 20740-3845
U.S.A.
www.PhysTEC.org

This book is an outcome of the PhysTEC project, which is supported by:
American Physical Society
American Association of Physics Teachers
APS 21st Century Campaign
National Science Foundation

Funding
This book is funded by the National Science Foundation through the PhysTEC project. PhysTEC is a project of the American Physical Society and the American Association of Physics Teachers.

This material is based upon work supported by the National Science Foundation under Grant Nos. PHY-0108787 and PHY-0808790. Any opinions, findings, and conclusions or recommendations expressed in this material are those of the authors and do not necessarily reflect the views of the National Science Foundation.

Left cover photo by Ken Vickers
Center cover photo by Cornell University Photography Services
Right cover photo by Brent Jones

Cover design by Krystal Ferguson

ISBN: 978-0-9848110-5-2

Table of Contents

Articles

Preparing Future Physics Teachers: Overview and Past History

Case Studies of Successful Physics Teacher Education Programs

Recruiting and Retaining Future Physics Teachers

Structuring Effective Early Teaching Experiences

Book design by Krystal Ferguson

PHYSTEC PREFACE

Scientific research has helped humankind in countless ways, and its pursuit at universities across the country has redefined academic culture. University faculty reward systems have evolved to strongly favor research publications and external grant funding. While academic research has been extraordinarily productive, and has contributed substantially to the tremendous advances in science and technology that have taken place over the past 60 years, it has also diminished another role traditionally played by universities and colleges in the United States: preparing a corps of K-12 professionals to educate the next generation.

In addition to an increased focus on research at universities, our society has also seen a significant devaluation of the teaching profession. While there remain many dedicated, energetic, and innovative classroom leaders, finding and preparing the next generation of qualified teachers is a significant concern. The current shortage of well-prepared science and mathematics teachers threatens our nation's competitiveness and core strengths [1].

These challenges come at a time when the need for physics teachers is greater than ever. The number of students taking high school physics has doubled in the past 20 years [2] and reports continue to indicate that physics as a discipline has the most severe shortage of well-educated teachers in American high schools [3]. Ensuring that these teachers have an adequate understanding of physics and knowledge of how to help students learn this challenging subject is key. Physics departments play a critical role in the education of these teachers.

Helping physics departments reclaim the mission of preparing physics teachers has been the goal of the Physics Teacher Education Coalition (PhysTEC) for the past 14 years. The project's successes provide tangible evidence that real progress is possible. While we may have hope, there is still much work to be done to bring excellence to all of the nation's physics classrooms.

University faculty committed to improving the education of future teachers must also consider the rapid evolution in our understanding of how students learn. Standing and reciting facts fails to either engage or provide substantive learning to the vast majority of students. The exciting message is that physics education research is making substantial progress in improving our understanding of both how students learn and how to inspire them.

This book and its predecessor, *Teacher Education in Physics: Research, Curriculum, and Practice*, are part of the PhysTEC project's attempt to raise the stature of physics teacher education within academia, and to inspire and collect scholarship that helps practitioners in the field better understand, and ultimately improve, teacher preparation. The first book, published in 2011, was a collection of peer-reviewed articles that had already been published or accepted for publication in either the *American Journal of Physics* or *Physical Review Special Topics – Physics Education Research*. The book's editors, David Meltzer and Peter Shaffer, were seeking to inspire or collect research that met the specific standards of these journals, and they used the peer-review process of the journals to act as a filter for the book's content. They also worked tirelessly to encourage and work with authors to bring scholarship into final form so that it could be published by the journals, and, ultimately, in the book. In this way, the first book inspired the publication of five new manuscripts, and reprinted six other previously published articles.

The present book, by contrast, is composed entirely of new invited and contributed works. PhysTEC project leaders realized that there were many lessons that had been learned by dedicated individuals and groups that might not otherwise be shared with the broader community unless a suitable impetus was provided. This book supplied that motivation by providing a venue for articles that explore current best practice in physics teacher education with a broader perspective than would be allowed for publications in the journals mentioned above. *Recruiting and Educating Future Physics Teachers: Case Studies and Effective Practices* provides a peer-reviewed collection of work that explores practical considerations necessary to build and sustain teacher education programs in physics. The 21 articles brought together here cover a wide range of topics that the community has grown to understand as fundamental to building thriving teacher education efforts.

While the book draws almost entirely from lessons learned in the United States, and applied to the education system in this country, there are many lessons that also apply more broadly. The PhysTEC project hopes that this collection will serve as a starting point for newly developing programs to gauge where critical actions should be focused, and as a resource to improve existing programs, by enabling physics teacher educators to learn about effective practices across the spectrum of activities necessary to build great programs.

Finally, the project would like to thank the books' editors, editorial board members, reviewers, and authors for their significant contributions in developing this work. We hope that the results will prove useful and inspirational to the many individuals who work tirelessly to help prepare the next generation of physics teachers.

Theodore Hodapp
APS Director of Education and Diversity
PhysTEC Project Director (2004–2014)

[1] Committee on Prospering in the Global Economy of the 21st Century and Committee on Science, Engineering, and Public Policy, *Rising Above the Gathering Storm: Energizing and Employing America for a Brighter Economic Future* (National Academies Press, Washington, D.C., 2007). http://www.nap.edu/catalog/11463.html

[2] S. White and C. L. Tesfaye, High school physics courses & enrollments, AIP focus on (June 2014). http://www.aip.org/sites/default/files/statistics/highschool/hs-courses-enroll-13.pdf

[3] American Association for Employment in Education, *Educator Supply and Demand Report 2014-2015* (American Association for Employment in Education, Columbus, OH, 2015).

EDITORS' PREFACE

This collected set of 21 invited and contributed papers focuses on the recruitment and preparation of future high school physics teachers. The book's intended audience includes physics faculty, physics department chairs, and secondary education faculty who are interested in improving their institutions' secondary physics education programs, or who desire to create new programs where such programs do not currently exist.

All manuscripts in this book, with the exception of one historical survey, are written as "how-to" guides that describe in-class and out-of-class practices that can lead to an increase in the numbers of physics majors who enroll in teacher preparation programs, and preservice physics teachers who stay in the major, graduate, and become effective high school teachers. This book focuses primarily on undergraduate teacher education programs, though in a few cases these are interwoven with graduate programs.

The intent of this volume is twofold. First, we hope to give readers a reference guide to the tricks of the trade in physics teacher education, which we hope will in turn guide the revision and implementation of physics teacher education programs. Second, we hope to provide authors an outlet to share their knowledge of effective physics teacher education practices, as this knowledge is often not well adapted for standard research journal publication.

Submission and Review Procedures

The call for proposals for *Recruiting and Educating Future Physics Teachers: Case Studies and Effective Practices* was released in August, 2012, and described the invited and contributed sections of the book. Potential authors were invited to submit 1,500-word manuscript proposals; 29 contributed proposals were ultimately submitted. Each proposal was reviewed by an editorial board member and an editor (Brewe or Sandifer). Proposals were evaluated in terms of overall quality, relevance to the book topic, potential to help practitioners improve physics teacher education at their institutions, the degree to which evidence was provided to support the effectiveness of the described activities, and adherence to the proposal guidelines. The board member assigned to each proposal made a recommendation to the editors as to whether the authors should: (1) be encouraged to submit a full manuscript, with no suggestions necessary; (2) be encouraged to submit a full manuscript that incorporates minor suggestions from the board member; (3) be encouraged to submit a full manuscript that incorporates major suggestions from the board member; or (4) not be encouraged to submit a full manuscript. Nineteen of the 29 proposal author teams were encouraged to submit full manuscripts.

Every author who was encouraged to submit a full manuscript did so, and consequently 19 contributed manuscripts were submitted. We used a review process similar in structure and rigor to that of a journal such as *Physical Review Special Topics – Physics Education Research*. Each manuscript was reviewed by three reviewers: one editorial board member and two reviewers selected by the board member or the editors. The same evaluation criteria applied to the proposals were also applied to the full manuscripts, along with two additional criteria: whether the manuscript was written for a general audience of physics faculty, and whether the institutional context, design principles, and underlying philosophies (if appropriate) were adequately described. Each reviewer recommended whether the manuscript should be: (1) accepted for publication with no suggestions; (2) accepted for publication with minor suggestions; (3) revised and resubmitted, anticipating acceptance but contingent on re-review to address major suggestions; or (4) not be accepted for publication. Sixteen of the 19 full manuscripts submitted to the contributed sections of the book were ultimately accepted.

How to Use this Book

Faculty who are new to physics teacher preparation, or who are experienced teacher educators interested in an overview of state-of-the-art practices, should begin their exploration of this book with "Characteristics of thriving teacher education programs," by Vokos and Hodapp. In this comprehensive summary, the authors argue that physics departments play a critical role in teacher recruitment and preparation. The authors draw on their combined experiences as chair of the Task Force on Physics Teacher Preparation and director of the PhysTEC project (2004–2014), respectively, to outline eight key components of highly successful programs. Detailed examples from universities across the United States are provided to exemplify each component.

A logical next step for the reader is to carefully digest the book's invited case studies, in which faculty from Seattle Pacific University (a private university), the University of Arkansas (a research-intensive public university), and Middle Tennessee State University (a comprehensive public university) describe the approaches they used to significantly increase the recruitment, retention, preparation, and graduation of future teachers in their physics departments.

Finally, certain readers may be at a stage where they are revisiting particular aspects of their teacher education programs. Readers in this category might consult the contributed book sections that pertain to their interests: Recruiting and Retaining Future Physics Teachers; Structuring Effective Early Teaching Experiences; Preparation in the Knowledge and Practice of Physics and Physics Teaching; or Mentoring, Collaboration, and Community Building. Manuscripts in these sections highlight how different departments and institutions have successfully handled very specific, and potentially problematic, aspects of their programs.

Book Sections and Common Themes

This book is organized into six different sections, based on the original request for manuscripts. The first two sections primarily contain invited manuscripts (five invited manuscripts and one contributed manuscript), whereas the remainder of the book consists of contributed manuscripts.

Preparing Future Physics Teachers: Overview and Past History

The invited Vokos and Hodapp article, as described above, presents an overview of eight key components that are characteristic of thriving physics teacher education programs. These components emerged from 15 years of ongoing site visits to successful secondary physics education programs throughout the nation.

Many of the program revisions chronicled in this book are grounded in the authors' participation in the Physics Teacher Education Coalition (PhysTEC) project. Rather than have all authors dedicate portions of their manuscripts to lengthy descriptions of PhysTEC, we decided that it would be more useful and efficient to invite the current PhysTEC project director, Monica Plisch, to write a paper describing her insights into the purposes, structure, and necessary activities of the project. That paper, "The Physics Teacher Education Coalition," is the second article in this section.

The third and final paper in this section is a contributed paper, "The roots of physics teaching: The early history of physics teacher education in the United States," by Gunning and Sheppard. This is an enlightening synopsis of physics teacher education from the post-Revolutionary period to the Second World War.

Case Studies of Successful Physics Teacher Education Programs

It is common practice in the education field to select an institution ("case study") and thoroughly document its successes, failures, key personnel, program structure, and institutional context. In reading such articles, members of the community can scrutinize the case study and decide whether the institution's programs and activities might be adapted for use at their home institutions.

We took such an approach in this book by inviting authors from respected physics departments at Seattle Pacific University, the University of Arkansas, and Middle Tennessee State University to publish the case studies of their teacher preparation successes. Healthy enrollments in their preservice teaching programs and impressive graduation rates provide convincing evidence of the success of their reform efforts.

The case studies are strikingly similar in that all three institutions experienced long droughts of student interest in their respective secondary physics education programs, after which programmatic changes, attitudinal shifts, increased communication, and concentrated recruitment efforts led to dramatic upswings in physics majors choosing high school teaching as a career.

Recruiting and Retaining Future Physics Teachers

A secondary physics education program cannot thrive without active recruitment and retention, both of physics majors and physics education majors. The articles in this section serve to emphasize this point. It is not sufficient to hope that students will find their way into a preservice teaching program, or be retained in such a program, without help or guidance. Strong recruitment and retention measures must be put in place, or an institution's preservice physics teacher programs will fall short of faculty and administrator expectations.

The contributors to this section suggest that recruitment and retention activities should include expected advertising avenues, such as posting flyers in hallways and classrooms, but also novel approaches that include physics course reform, the creation of physics-specific teaching methods courses, early teaching opportunities, collaborations with area schools and teachers, an increased focus on teaching-oriented advising and mentoring, and the hiring of faculty and staff—including current or retired high school teachers—whose primary responsibility is to support the teacher education program.

Structuring Effective Early Teaching Experiences

This section reinforces the idea that early teaching experiences (ETEs) can be used to recruit and retain future high school physics teachers, and can benefit the sponsoring physics department in terms of increased student learning and incremental course reform. ETEs can be structured in a variety of ways, and no single structure works for all physics departments. ETEs can be embedded in both classroom and non-classroom contexts, for instance, and can be organized as Learning Assistant programs, teaching institutes, and early teaching courses that place interns (potential teacher candidates) in different K-12 teaching environments.

Common themes include the observation that the ETEs should be led by physics department personnel, rather than led by outside departments or organizations, and that successful ETEs tend to share common characteristics: careful and in-depth lesson planning, intern participation in "active learning" lessons, explicit reflections on teaching and learning, the involvement of practicing or retired master K-12 teachers, and close faculty or staff supervision.

Preparation in the Knowledge and Practice of Physics and Physics Teaching

It is essential to avoid a common pitfall of teacher education reform, which is to assume that the extent of the physics department's involvement in teacher preparation begins and ends with ensuring that a sufficient number of physics course credits are included in the secondary education track or concentration. While content preparation is a crucial component in the preparation of preservice teachers, the authors in this section share a common vision that other factors must also be considered. These factors include the manner in which core physics courses are taught, student participation in physics research and scientific practices, the connections between content understanding and effective instruction, and the respecting and fostering of genuine scientific curiosity and creativity.

Mentoring, Collaboration, and Community Building

This section begins with a compelling summary of the powerful roles that Teachers-in-Residence (expert high school teachers) can play in reinvigorating teacher education programs, and continues with persuasive descriptions of how far-reaching mentoring practices, faculty-sponsored learning communities, and a series of physics-specific teaching methods courses can positively impact preservice physics teacher education. Also suggested are steps that research-intensive institutions might take to enrich and expand their physics departments' teacher education efforts.

Final Thoughts

Beyond descriptions of teacher education programs and activities, there is a strong emphasis throughout the book on implementation advice, ongoing challenges, and lessons learned. Authors were strongly encouraged to share their insights into these issues since, while it is critical to grasp the final structure of an institution's thriving physics teacher education program, it is just as important to learn about the authors' flashes of insight, formative trials and tribulations, ongoing problems, and regrettable mistakes. The only thing worse than reinventing the wheel, as they say, is reinventing the broken wheel.

Some information in this book will be surprising, and some will be more or less applicable to your physics department depending on your institutional supports and constraints—but hopefully each article will be thought-provoking and useful in your mission to address your institution's physics teacher education needs.

Acknowledgments

Sincere thanks are due to Theodore Hodapp and Monica Plisch from PhysTEC and the American Physical Society, who proposed the idea for the book, offered us our editorial positions, and supported the book's production. The editorial board members provided critical guidance early in the process and also served as proposal reviewers, for which they have our gratitude. The 44 full manuscript reviewers did a stellar job of communicating their thoughtful comments to the authors, and they deserve special thanks for submitting their reviews by the deadline. Gabriel Popkin worked tirelessly to copyedit all 21 manuscripts, Alyssum Pohl joined us as a meticulous reference editor at a time of great need, and Krystal Ferguson skillfully crafted the book's design and layout, for which they all have our heartfelt appreciation. Lastly, we thank the authors for sharing the wonderful teacher education activities occurring at their home institutions. It is their dedication to the recruitment and education of future physics teachers that made this book possible.

Cody Sandifer, Editor
Department of Physics, Astronomy and Geosciences
Towson University

Eric Brewe, Editor
Department of Teaching and Learning
Florida International University

Characteristics of thriving physics teacher education programs

Stamatis Vokos
Department of Physics, Seattle Pacific University, 3307 Third Avenue West, Seattle, WA 98119

Theodore Hodapp
American Physical Society, One Physics Ellipse, College Park, MD 20740

Physics departments have a critical role to play in the preparation of high school physics teachers. Yet many physics departments in the United States are not actively engaged in physics teacher education. The Physics Teacher Education Coalition and the National Task Force on Teacher Education in Physics have observed some of the best physics teacher preparation programs in the country over many years. These observations have been collected to give examples of practical ways to transform teacher education in physics. This chapter describes eight key components needed to build thriving programs and gives specific examples of institutions at which these efforts have been developed and refined. The key components are (1) recognition and support for the departmental teacher education champion, (2) targeted recruitment of preservice physics teachers, (3) active collaboration between physics departments and schools of education, (4) a sequence of courses focused on the learning and teaching of physics, (5) early teaching experiences, (6) individualized advising of teacher candidates by knowledgeable faculty, (7) mentoring by expert physics teachers, and (8) a rich intellectual community for graduates. The intent of this article is to provide physicists with a framework they can use to take appropriate actions at their own universities, to provide a set of good-practice ideas, and to identify transformational people and established programs.

I. INTRODUCTION

A. National need

The need for qualified physics teachers—a need greater now than at any previous time in the nation's history—is well established [1]. Each year, approximately 3,100 teachers find themselves in front of a high school physics classroom for the first time. Of these teachers, fewer than 10% are newly-minted graduates of a physics teacher education program. Figure 1 indicates the frequency of graduation rates of high school physics teachers at U.S. universities [2].

More than half (about 1,700) of these new teachers of physics are experienced teachers who have not taught physics before. Compounding this, as seen in Figure 2, only 47% of high school physics classes are taught by a teacher with a degree in the discipline [3]. This is consistent with the fact that fewer than half of the teachers of high school physics have a major or minor in physics or physics education, or any preparation specifically devoted to the teaching of physics.

A vision that every U.S. high school student will have the opportunity to learn physics with a qualified teacher is currently far from being realized. The negative consequences of the status quo for the science enterprise writ large are outlined elsewhere, as is the indispensable role that physics departments must play in physics teacher education to ameliorate the situation [4,5]. To change the precarious state of pre-university physics education, physics departments can look to a small number of excellent programs that stud the otherwise gloomy national landscape. In this chapter we describe key characteristics of such thriving programs, as instantiated in local contexts.

B. Guiding efforts

The ideas that we discuss in this chapter are guided by the report of the Task Force on Teacher Education in Physics (T-TEP), *Transforming the Preparation of Physics Teachers: A Call to Action* [2], and by the experience gained over more than a dozen years through the efforts of the national Physics Teacher Education Coalition (PhysTEC). This includes experience honed from numerous site visits to programs in all stages of development. The chapter is also informed by the compendium of research articles *Teacher Education in Physics: Research, Curriculum, and Practice* [6] and grounded in the teacher education literature. We seek to operationalize the T-TEP recommendations and the findings from the PhysTEC project through the presentation of specific examples, in order to illustrate general principles.

The American Physical Society (APS), the American Association of Physics Teachers (AAPT), and the American Institute of Physics (AIP) constituted T-TEP to document the current state of physics teacher education in the nation and to make recommendations for how this state could be improved. Over a period of four years, T-TEP collected and analyzed data through surveys, site visits, literature reviews, and formal and informal input from many individuals and organizations. Members of T-TEP brought together knowledge and experience from critical pieces of the complex system of physics teacher education. Physics faculty, education faculty, university administrators, high school teachers, and professional organization representatives produced a report that captured the findings of T-TEP's four-year investigation and issued recommendations, which were endorsed unanimously by the Task Force [2].

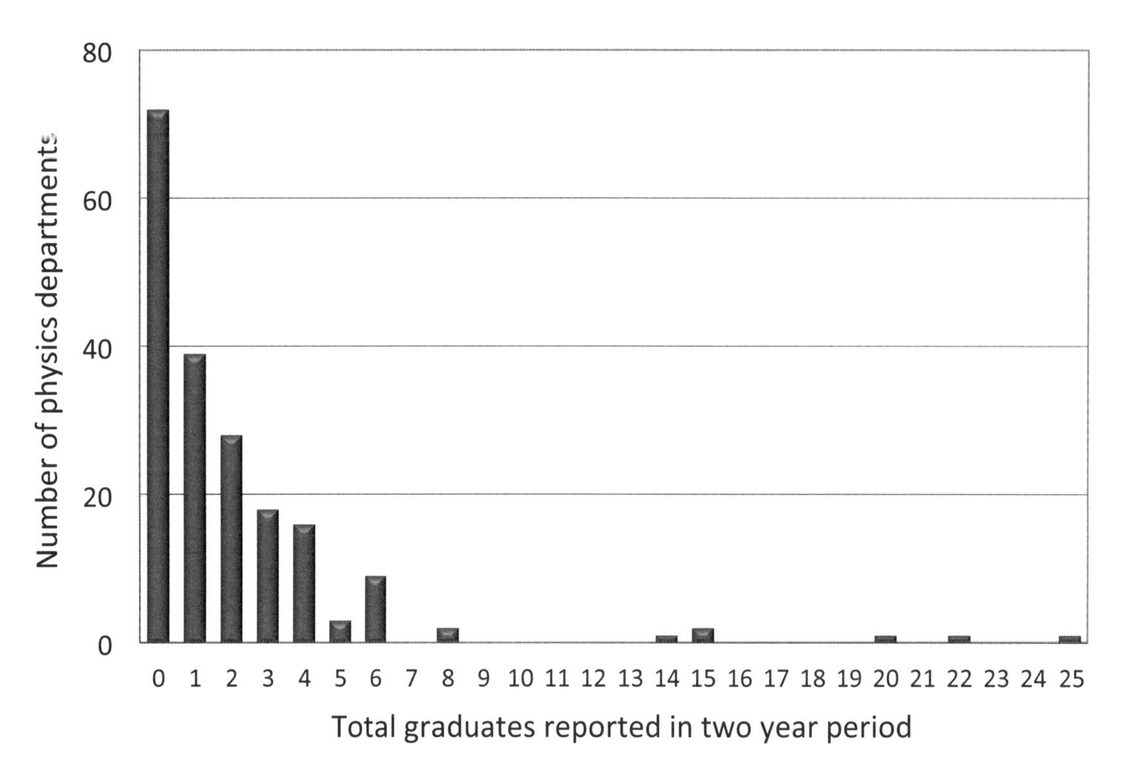

FIG. 1. Distribution of teacher education programs in U.S. physics departments by the total number of graduates in a two-year period [2].

The PhysTEC project began in the late 1990s as a result of conversations among leaders of APS, AAPT, and AIP, who sought to identify and address the most pressing problems facing physics education at the close of the last century. The significant lack of qualified high school physics teachers stood out as *the* critical issue facing not only physics education but also the physics enterprise itself. In 2000 a pilot project was launched to explore ideas, and in 2001 the project began in earnest with six funded sites. As of 2014, PhysTEC has funded more than 30 physics departments, raised upwards of $20 million in external funding, and recruited over 300 universities to join a coalition of institutions committed to improving teacher education in physics. PhysTEC's annual conference draws leaders in physics teacher education from across the country and around the world to discuss good practices, interrelated pedagogical and intellectual themes, and practical methods of building and improving programs. The project's leaders have, over the past decade, visited numerous institutions to probe underlying structures that sustain productive and powerful teacher education programs [7].

C. Chapter description

This chapter provides a framework of key elements and considerations for starting, strengthening, and sustaining a vibrant teacher education program in physics. Analysis of extant programs indicates that there is no single model—no one-size-fits-all solution—for implementing these elements successfully. Rather, the intrinsic institutional diversity in higher education, including differing missions and the contexts of specific physics departments, demands a variety of models. Through the articulation of unifying ideas and programmatic elements, this chapter seeks to communicate a coherent vision for transforming the professional preparation of physics teachers from the present "inefficient [and] incoherent" state to one in which the national system of preparing physics teachers is equipped to "deal with the current and future needs of the nation's students" [2].

In the PhysTEC project, and during site visits and investigations conducted by the Task Force, it became clear that developing and sustaining effective teacher education programs depends critically on the local context [8]. The presence of an engineering school, research and teaching expectations of faculty members, reputation of the university, and communities served by the university all play important roles in "fitting" program components to a particular institution. Differences in these variables did not preclude success; however, they did alter the shape and ultimate expression of the components of the universities' teacher preparation programs. Our selection of cases to illustrate various characteristics was designed to provide as wide a representation of institutional contexts as possible. It is our hope that you, the reader, may see the potential at your university for

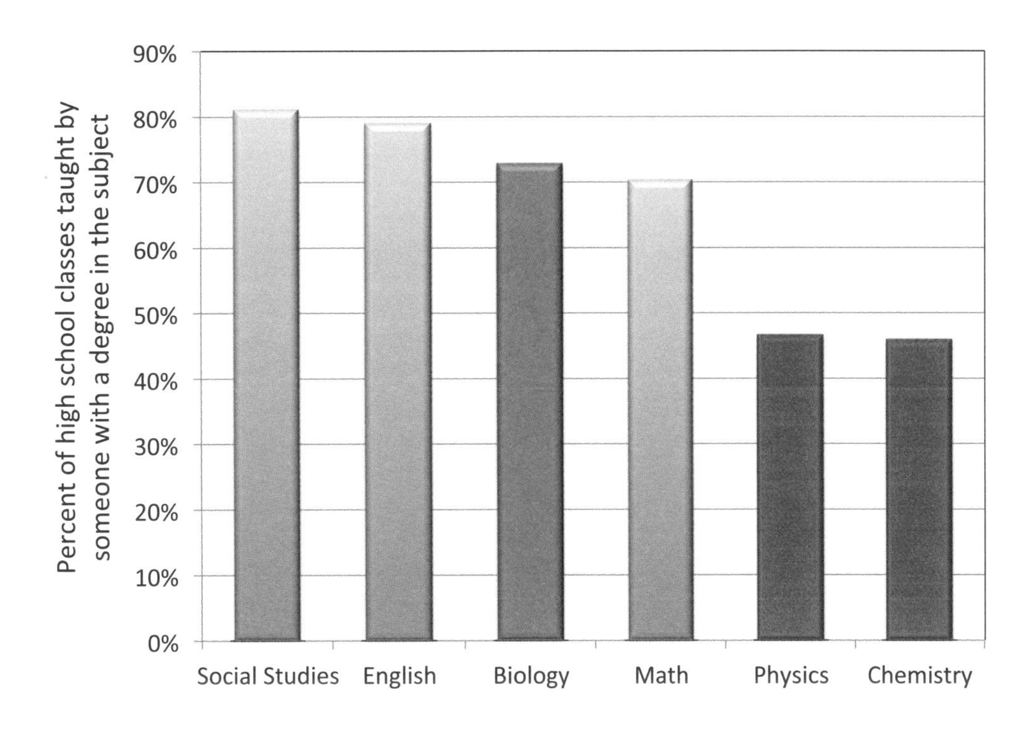

FIG. 2. Percent of U.S. high school classes taught by a teacher with a degree in the subject [3].

implementing one or more of these components based on your own context, and we encourage you to contact the leaders named here to learn more about the details of their successes in creating, adopting, and adapting ideas to fit them to their local settings.

This chapter is organized around the *characteristics* of thriving physics teacher education programs as articulated by the T-TEP report. These are (1) recognition and support for the teacher education *champion*, (2) targeted recruitment of preservice physics teachers, (3) active collaboration between physics departments and schools of education, (4) a sequence of courses focused on the learning and teaching of physics, (5) early teaching experiences, (6) individualized advising of teacher candidates by knowledgeable faculty, (7) mentoring by expert physics teachers, and (8) a rich intellectual community for graduates. Characteristics (6) and (7) are combined because of their complementary nature.

The discussion of each characteristic comprises two parts. The first part consists of a general description of the characteristic, expanded along several dimensions that illustrate that topic. Second, we provide specific examples of the characteristic, either from the sites identified in the T-TEP report or from institutions that the PhysTEC project has noted as having implemented or developed effective practices. These examples are selected to reflect a diverse range of institutional types and settings. It is hoped that the specificity included here and in the following chapters will inspire the nation's physics departments and schools of education to unleash their inherent creativity to improve physics teacher education in ways that are guided by the research base and the accumulated knowledge of the field.

II. CHARACTERISTICS

A. Recognition and support for the champion

It should come as no surprise that champions—individuals with passion and acumen—are key to building successful programs of teacher education. Champions are critical influences in building and growing programs of almost any kind. Champions can be a single individual or a small group, but their ability to build sustainable programs depends ultimately on how well they are supported. PhysTEC and T-TEP have identified these individuals as critical and have found a variety of ways in which support for them has been configured and backed by the department and the institution. To the best of the field's collective knowledge, no thriving physics teacher education program functions without a champion.

Physics departments seeking to build a thriving physics teacher education program should create an environment in which faculty who are professionally committed to the preparation of physics teachers may step up (or be hired) to help the department engage in this effort or to extend already existing efforts. These potential teacher education champions, who are at the nexus of intellectual activity and administrative leadership of the program, should have specific strengths in several areas as described below.

Perhaps the most common trait of successful champions is their political and social acumen for negotiating paths through the maze of institutional support. Central to the task of securing support for new academic programs is the art of using these programs to help chairs, deans, and provosts solve their problems. To accomplish this, a champion must understand the problems these administrators face and communicate how the teacher education program presents a solution. Communication with administrators should not stop once the initial support for a program has been secured, however. Providing periodic updates to chairs, deans, and provosts is also critical because this steady stream of information helps ensure that these administrators are armed with the tools they need to help the champion build on the program's initial successes. Great administrators know that investing in a high-functioning program is a better bet than trying to fix a poorly operating unit.

Navigating these paths is almost always more easily accomplished by an individual with seniority, or by a group that includes someone (e.g., a chair or former administrator) who understands how doors are opened and appointments are secured. Placing an untenured faculty member or a staff person in the champion role without senior support can marginalize the champion, endanger his or her career, and potentially place the teacher education program at risk.

Interpersonal skills are critical, too, as the art of understanding the social dynamic of relationships helps the champion recognize who can help the project succeed. Building a strong collection of collaborators and allies enables opportunities in unforeseen ways. Interpersonal skills are critical in an additional way: Given the small numbers of prospective physics teachers at a typical institution, the champion's role in engendering a community atmosphere among these prospective teachers is crucial. The champion often provides the connective tissue in the program.

Of course all of this is moot if the department's or the institution's mission and direction are not aligned with the goals of teacher education. Changing this alignment is difficult at best; however, concerted and aligned individual efforts can provide seeds for change. One opening for change was described at a PhysTEC site visit: "Never waste the opportunities offered by a good crisis" [9]. A number of successful programs have done just this, seeking to address low enrollment or national and regional calls to increase the number of science teachers, or to improve perceived decline in student quality. Having a ready solution to these problems can alter the future direction of a department or operating unit. For most departments it will not become the primary *raison d'être*, but it can often be supported alongside complementary goals. However, we have never seen this occur without individuals who bring energy, organizational skill, and passion for accomplishing this activity—in other words, champions.

Example 1. University of Arkansas, Fayetteville: Gay Stewart

Professor Gay Stewart was recruited in 1994 by the physics department at the University of Arkansas, Fayetteville, the flagship institution in the state system, to help the department increase undergraduate graduation rates in physics by reforming the teaching of physics at the introductory level. At the time of Gay's hire, the University of Arkansas was a doctoral-degree granting research institution with high research activity (RU/H in the Carnegie Classification scheme); it was upgraded to the highest category of research activity, RU/VH, in 2011. Thanks to Gay's extensive efforts, and with a second faculty member hired in 2001 with a similar set of professional expectations, the department experienced a dramatic increase in the numbers of physics majors and prospective physics teachers [10]. The number of physics majors graduating per year leapt from fewer than five in the mid-1990s to more than 25 in 2012. During her time in Arkansas, which lasted until 2014, Gay was a tireless advocate for the program at the local, state, and national levels. She secured significant intramural and extramural funding (see Figure 3) to support and expand the program and to serve physics and physical science teachers in the state and surrounding regions.

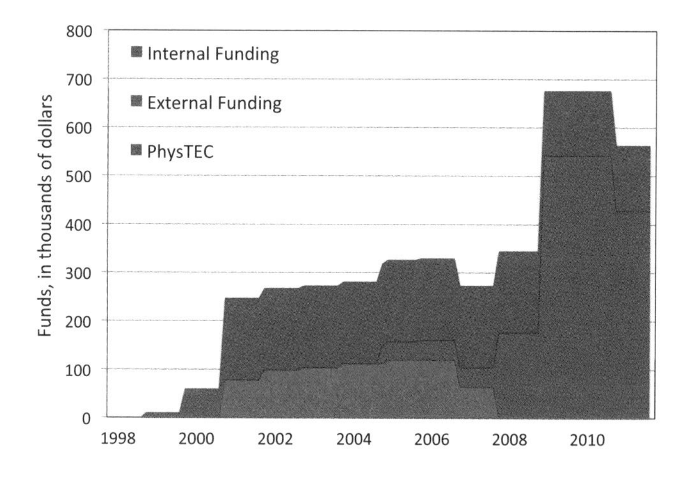

FIG. 3. Cumulative internal and external funding for teacher education at the University of Arkansas [8]. PhysTEC funds are shown specifically, to indicate the catalytic effect such funding can have on changing institutional culture.

Students received individualized attention and mentoring even after graduation, at which time they were tapped as potential cooperating teachers to prepare the next generation of student teachers. Stewart often referred to her students as "my kids," demonstrating her deep dedication toward their personal and professional success. She and her husband, physics professor John Stewart, regularly opened their home to practicing and prospective physics teachers. Gay's efforts to support physics teacher education, including her work on the redesign of the AP Physics curriculum and her election to serve as AAPT president, brought the department national recognition. Gay was viewed as a trusted partner on campus—one with deep institutional knowledge—and as a faithful collaborator off campus, for school system administrators and other members of the national physics education community alike. She adapted national programs like the AAPT Physics Teacher Resource Agent program to local conditions, making the department and the university active partners in state efforts to improve science education.

Is there a faculty member in your department with interest and drive who can pursue teacher education? Will that person receive professional recognition for his or her actions?

Example 2. Middle Tennessee State University: Ron Henderson

Middle Tennessee State University (MTSU) is a large (about 21,000 FTE students), comprehensive regional university, with a physics department of 11 regular faculty members and associated staff and adjuncts. In 2008, MTSU was graduating only three physics majors per year, or 0.13% of all undergraduate degrees at the university. This number compared poorly to the national average of 0.32% of all undergraduate degrees going to physics majors. As one strategy for increasing enrollment, Professor Ron Henderson, the new physics chair at the time, decided to focus some of the department's energy on preparing new high school physics teachers [11]. Ron Henderson invited prominent physics education faculty to give colloquia and held discussions with department faculty on embracing teacher education as an important strategic direction of the department. Under Ron's leadership, the department also revised its course sequence and reoriented its overall educational philosophy to embrace a more holistic view of physics education within the institution's regional context. Preparing students for graduate work was now no longer the only option. Unbeknownst to Ron, the university administration was engaging in similar initiatives and simultaneously secured funding to have MTSU become a UTeach replication site. The synergy of departmental and university goals created a budding collaboration between physics—the only science department to jump in on teacher education—and the College of Education.

MTSU's physics department subsequently went from being considered for elimination to being recognized by the university's president as one of the university's top 10 programs, winning the inaugural President's Award for Exceptional Departmental Initiatives for Student Academic Success in 2013. Ron also argued successfully for a new faculty line in physics education research and ultimately hired Dr. Brian Frank, whose research interests align well with the program and whose expertise has helped the department improve undergraduate courses with an eye toward increasing retention and graduation rates. Recent graduating classes have more than tripled in size, and the department, which had not graduated any physics teachers in the preceding 15 years, now has more than 15 secondary education students who either recently graduated or are currently enrolled. These individuals form the department's personal army of recruiters for future entering classes. Ron did not waste his crisis.

Are you attracting enough well-prepared students to become undergraduate majors? Can you support an individual to build teacher education as a critical component of attracting students to the physics program?

Example 3. University of North Carolina at Chapel Hill: Laurie McNeil

Professor Laurie McNeil knew that North Carolina needed many more high school physics teachers than were currently being educated in the state. As chair of the physics department of a state flagship university, University of North Carolina at Chapel Hill, she brought resources to bear to establish a program that was aligned carefully between the physics department and the School of Education [12]. Along with securing external funds for program startup, Laurie was able to fund a lecturer position in physics education (currently held by Dr. Alice Churukian) that would serve as the nexus of activities. Funding for this position came from several sources: the department itself, the office of the provost, and the office of the dean. McNeil's role in a Carnegie-classified RU/VH university (research university, very high research activity) cannot easily deviate from the standard framework of promotion and tenure at an institution of this ranking; to maintain the confidence of her department she must keep an active research program with doctoral graduate students. Consequently, her role as a champion has been to secure funding, promote the program to the tenured and tenure-track faculty, and establish and maintain communications both horizontally (among the faculty) and vertically (among various levels of the administration). Funding is required to maintain operational aspects of a teacher education program through the lecturer position, a part-time master teacher, and other supportive activities (e.g., Learning Assistants). One testament to Laurie's effectiveness is the degree to which various groups (including the physics department) have been willing to continue support for these components in tight budgetary times.

> *Does your champion have the right connections with the administration and within the department to secure the support necessary for success?*

B. Targeted recruitment of preservice physics teachers

"If you build it, they will come." This may work in the movies [13], but reality is a little different. Any new or potentially growing program has to be built with at least two components: a marketing effort to attract students and a high-quality program to keep them. The most successful teacher education efforts in the country have significant and targeted efforts to recruit students into their programs. For developed programs, word of mouth can be sufficient, but newer efforts require something to jumpstart this process.

Having seen a number of institutions go from having no future physics teachers to boasting vibrant programs, we can testify that there are some fairly basic techniques that anyone wanting to increase the number of physics teacher candidates in their program can apply. Flyers or postcards for student events and posters in physics department hallways not only inform students of the program but also send the message that the physics faculty have endorsed this academic pathway. We see posters giving specific messages, such as, "Come to this opening event," and also ones that proclaim, "Teach—Make a difference." The latter type of message resonates with many students who are trying to do just that with their lives. Stories in the student newspaper documenting how alumni have entered into challenging careers in teaching or returned to help build a community of teachers through courses or professional development activities at the university can be great advertisements as well.

Faculty at PhysTEC Supported Sites and at institutions in the broader PhysTEC coalition often make a point of using in-class announcements in each introductory physics course to advertise to majors the possibility of becoming a teacher. The faculty also explicitly identify the person whom a student can contact about pursuing such a course of study and share this contact's name with students. One of the rules of marketing any program is to advertise it using someone with whom students can identify—age does matter—so the ideal person to communicate the message is a student who already is in the secondary education program or a teacher who assists with the effort. Even if the person making the announcement in class is older, the act of announcing makes it clear that the professor considers high school teaching important and has agreed to sacrifice class time to communicate the message. And while we are thinking about the role of professors, ask yourself: Are we trying to shape students to be like us or are we trying to open opportunities for them as they explore careers and callings in physics? Make sure that all faculty members, especially those who advise students, understand how to appropriately inform students how to pursue teaching, or at least know with whom students should speak to get that information.

Another powerful recruiting tool is to give physics majors the opportunity to experience teaching firsthand. Undergraduates often have no idea what it is like to see students learn difficult concepts or to feel the pride in helping students along the path of learning; they are unsure if they will be good at teaching, and do not understand the rich intellectual challenge involved in understanding and improving student learning. One program that has been particularly successful at providing early teaching experiences is the Learning Assistant (LA) program, which has been championed significantly by faculty at the University of Colorado Boulder. LAs are undergraduates who facilitate the learning of their peers in group-intensive instructional settings (most often in introductory courses). To become an LA, a student must enroll in a course or seminar in learning theory, with examples drawn from the discipline. Some programs allow LAs to continue for a second semester only if they begin a course of study that will lead to teacher certification. This first exposure to teaching can be a basic introduction to assuming some of the roles of a teacher, with students often reporting that they had no idea how challenging and how rewarding teaching can be.

Since the first LA program began, several hundred faculty have participated in national and regional LA workshops. Materials for the learning theory course and supporting documents to build such programs are available [14]. In addition to recruiting students into teaching, LA programs have won recognition from faculty for the impressive student learning gains achieved in the courses the LAs support. (See also the discussion in Section II.E of this chapter.)

Finally, perhaps the most important recruiting tool is the simple act of a faculty member telling a student, "You know, I think you would make a great teacher." Hearing this from a person who knows the student's intellectual and personal potential can have tremendous impact.

Example 1. Illinois State University

Illinois State University is perhaps the quintessential "normal" school. Located in the town of Normal, Illinois, the university was the first public university in the state—Abraham Lincoln had a part in founding the school—and it has a substantial history of preparing teachers. Despite this auspicious beginning, in the 1990s the physics department educated few high school physics teachers. In 1994 the department made a commitment to improve things, and Professor Carl Wenning was asked to lead the department's efforts [15]. The department built a physics teacher education program that eventually included a sequence of six courses in the teaching of high school physics. Carl created a variety of materials including web pages [16] and brochures that could be distributed to undergraduates [17], encouraging them to consider becoming physics teachers. Carl worked with science teacher organizations around the state to distribute flyers at various events. Subsequently, his graduates became ambassadors for the program and were encouraged to return to take part in professional development activities. Many of the program's current students first heard about the program from their high school physics teachers.

The program direction continues today under Ken Wester, a former high school physics teacher who has received the following commitment from the department chair: "If we need to cut positions, the physics teacher preparation program staff will be the last to go."

Can encouraging students to consider teacher education at your institution be part of your recruiting efforts for the physics major? Are you reaching students outside of the physics department (e.g., engineering students, pre-health majors, etc.)?

Example 2. Seattle Pacific University

Seattle Pacific University (SPU) is a small, private, Christian liberal arts university with 4,300 students, of whom 3,400 are undergraduates. The physics department has an extensive commitment to the professional preparation and development of teachers. This deep commitment is due in part to the fact that all current physics faculty were hired in the last 12 years and all place significant value in physics teacher education. A physics education research agenda targeting teacher learning, funded continuously for more than 10 years at a cumulative level of $7 million, has enabled the department to grow from three full-time faculty in 2002 to seven today (four tenure-track faculty, one resident master teacher, and two education research scientists). With close collaboration between the physics department and the school of education, SPU is now a national leader in recruiting prospective teachers of physics.

Two main avenues of recruitment are noteworthy. First, the Learning Assistant program in physics is a well-established program, with strong support from the provost and from the dean of the College of Arts and Sciences. Prospective LAs are identified and personally recruited by faculty members. Requirements for participation are high: Students must commit to taking the LA pedagogy course *every* quarter they serve as LAs. This requirement exposes them to topics that go beyond those in one-quarter or one-semester versions of such courses. LAs are pictured prominently in the department hallways. A powerful LA community organized by Professor Amy Robertson, SPU's LA program coordinator and LA pedagogy instructor, supports newcomers and fosters innovation, thereby promoting the development of science teacher identity among fledgling teachers [18].

Second, the department capitalizes on the SPU students' strong commitment to serving social justice causes. Many students who are interested in science and who want to use that calling in service of the community arrive at the university with narrow models for how to achieve their goals. For many students, medicine is the only candidate for such a vocation. By presenting physics teaching as a science-rich, intellectually demanding career option with social impact similar to their original aspirations, the department recruits prospective teachers by redirecting students who would have pursued biology instead, and at the same time increases the total number of majors and minors. This strategy has worked at other institutions at which students have an orientation toward service. Brigham Young University is the most prominent example in physics teacher education, graduating 10 to 15 new high school teachers every year.

Have you considered adding an LA program? If teacher education is aligned with your institution's mission or strategic plan, how have you shaped recruiting efforts to capitalize on this alignment?

C. Active collaboration between physics departments and schools of education

At many universities, one of the most significant obstacles that has to be overcome in the establishment or improvement of a physics teacher education program is the absence of substantive collaboration between the physics department and the school or college of education. In certain cases, there is even active mistrust and ill will between the two. Although one could enumerate several reasons that would go a long way to *account* for these attitudes, we prefer to take a more positive stance by *understanding* differences in the ethos of these two university structures and *capitalizing* on commonalities.

Despite differences in the research paradigms to which physics and education faculty adhere, it is important to recognize the specialized expertise residing with faculty from both parts of campus. Mutual respect is a prerequisite in any attempt to build on expertise in areas of common interest. Physics faculty are steeped in their long scientific tradition and the habits of mind of the physics community. Education faculty have sophisticated and systemic understandings of schools, students, and educational issues. Physics teachers need both forces to push in the same direction. Physics faculty have much to gain from active collaboration with education colleagues. In our experience, physics faculty who have achieved a high level of authentic collaboration with their education counterparts have been able to creatively tackle institutional and state challenges that would otherwise have been intractable. In addition, education faculty with expertise in assessment or program evaluation can assist individual or departmental efforts to secure internal and external funding from sources that presuppose sophisticated knowledge of educational issues. This collaboration often starts with immediate, local issues in the physics teacher education program and extends well beyond it into collaborative work in other areas, including common work on research projects that require savvy education and outreach activities.

Since student credit hours are often the coin of the realm for designing undergraduate degree programs, a collaboration built on well-deserved trust can enable physics courses (appropriately guided by state teacher certification requirements and the university's education department) to double-count in the major and the teacher education program. This is not just about university turf; having a streamlined program is essential to stewarding scarce student financial resources at a time when education costs are coming under increased scrutiny. Equally important is a thoughtful commitment to internal programmatic coherence as a way of improving the quality of the preparation of prospective physics teachers.

Joint faculty appointments shared between physics and education departments can serve as a mechanism for achieving such collaboration. Several institutions have such positions, which require expertise in both physics and education, and, often, formal precollege teaching experience. Professional success of such faculty requires careful articulation of the reward structures in both departments, as well as specificity in the institutional expectations for promotion and tenure.

That said, a joint appointment does not fit all institutional contexts. Even without it, substantive collaboration is still possible. A first step for a physics faculty member might be meeting with an education colleague or inviting an education colleague to present a physics department colloquium, with the goal of better understanding that colleague's professional perspective on teacher education.

To paraphrase the apocryphal quote attributed to the legendary bank robber Willie Sutton, a physics department needs to collaborate with education faculty "because that is where the teachers are" [19]. Engagement of physics faculty with prospective and practicing physics teachers can open up significant opportunities for the department.

Example 1. City College of New York

Part of the City University of New York system, which comprises almost two dozen colleges, The City College of New York (CCNY) serves science teachers through a very close collaboration between the Division of Science and the School of Education. Joint faculty appointments in both administrative entities facilitate this collaboration. This complex collaboration provides student teachers with coordinated perspectives from subject matter specialists who are also steeped in K-12 contexts. The physics-education collaboration at CCNY allows coordination of resources such as space, supplies, scheduling, and pursuit of external funding. It also enables the envisioning and implementation of new initiatives.

Dr. Richard Steinberg is an excellent example of this type of collaboration. A professor of physics and secondary education, Richard has a deep knowledge of physics, expertise in physics education research, and experience with precollege physics teaching in New York, where he recently spent a sabbatical year teaching in a public school in a low-income neighborhood. His education students receive the benefit of participating in science methods and curriculum courses that are closely aligned with the topics and instructional approaches used in the content courses and that are informed by recent personal experiences in the classroom. Complementing this, in his physics courses, Steinberg sends the explicit message that teaching is a worthwhile career option to be considered alongside other options for majors.

How can collaboration with the school of education help your institution improve preservice physics teacher education?

Example 2. Florida International University

Florida International University (FIU) is Miami's first and only four-year public research university. This rapidly growing campus, which had 53,000 students enrolled in Fall 2013, claims the top spot in the nation in awarding bachelor's and master's degrees to Hispanic students. Half of all teachers in the Miami-Dade County Public Schools are FIU alumni. Research is a big part of the identity of FIU and of the physics department—FIU is designated as RU/H in the Carnegie Classification scheme.

The FIU physics department also has a rich collaboration with the university's College of Education. Professors Laird Kramer (physics) and Eric Brewe (education)—both active in physics education research—have brought national recognition to the research portfolio achieved through this collaboration, which they have parlayed into significant administrative support for the university's physics teacher education program. At a very difficult fiscal time, Laird and

Eric managed to institutionalize the resident master teacher position in the physics department; this position plays a crucial role in the quality of the teacher education program. The trust developed as a result of the collaboration has enabled the program to navigate two challenging transitions: the relocation of the secondary teacher education program from education to physics, and the establishment of a vibrant course of study for a teacher preparation program that over the previous decade had graduated zero physics teachers.

The physics department supports physics teachers locally and regionally by providing Modeling Instruction (MI) workshops [20]. With a strong focus on discipline-specific pedagogical preparation, FIU prospective physics teachers participate in both the Learning Assistant program and an MI workshop, in which they work side-by-side with practicing teachers. The MI workshop experience, in turn, is consistent with the pedagogical approaches taken in the university-level introductory physics course that future teachers take as students and then support as Learning Assistants.

The detailed structure of degree requirements for future physics teachers would have been impossible without a strong collaborative spirit between physics and education. The collaboration has benefited students in additional ways. For instance, the physics education research group at FIU has used funding from the National Science Foundation's (NSF's) Robert Noyce Teacher Scholarship Program to provide research experiences for undergraduate students at FIU and other PhysTEC sites.

FIU serves as a model for institutions that aspire to have a strong physics education research group. Groups of this nature can tackle a coordinated agenda that both promotes physics-heavy learning and addresses important broader educational questions.

> *How can your collaboration with the school of education help you build institutional support for improving physics teacher education? Are there challenges within the institution that will make this collaboration a win-win situation for both entities?*

D. A sequence of courses focused on the learning and teaching of physics

In site visits conducted by PhysTEC and T-TEP, brand new teachers invariably expressed the wish that they had been better prepared for the actual daily tasks of teaching the subject matter to students. On the one hand, new teachers said that their physics courses were often taught with a focus on the rigor of mathematical formalism rather than on the depth of conceptual understanding—an approach that was not conducive to learning to teach physics at the precollege level. On the other hand, new teachers found that exposure in education courses to general strategies for increasing student engagement or assessing student learning was not very useful either. New teachers indicated that they needed specific preparation for teaching physics topics and for assessing student learning of those topics. Such preparation might include, for example, an effective teaching sequence to help students come to recognize that forces arise in pairs that have equal magnitudes in all cases.

This sentiment is reflected in T-TEP's findings. The development of specialized professional knowledge for teaching specific topics is not a typical outcome of the traditional preparation of physics teachers. In the usual model, it is assumed that physics teachers learn physics in the physics department and then learn how to teach in their certification program. Any additional pedagogical training that is specific to physics is left to the student teaching experience, which is supervised infrequently by a university representative for whom physics is typically not an area of expertise.

An apprenticeship model is not *a priori* deficient. On the contrary, some successful international models of teacher preparation build in long-term (one- or even two-year) apprenticeships. In the United States this is a problematic model, because the majority of high school physics classes are taught by teachers without a strong grounding in physics and even less in the teaching of physics. When new teachers apprentice to experienced teachers who lack deep subject matter understanding or self-identification with the discipline-specific values and habits of mind of the physics community [21], the development of a new teacher's specialized content knowledge for teaching physics is neglected, and such individuals are forced to invent or scrounge for tools and ideas to help them teach difficult topics. Furthermore, unlike apprentice biology or math teachers, the physics student teacher cannot easily benefit from the expertise of his or her collaborating teacher's disciplinary colleagues, because, according to AIP, approximately 80% of all physics teachers are their schools' only physics teacher [22]. As in natural selection, an organism living in an isolated environment adapts to the local situation, not to the broader biosphere. Our teacher may learn to teach in ways that are acceptable in his or her school, but may not go outside his or her comfort zone to learn to use evidence-based methods to improve student learning.

We found a few important exceptions to the sentiment expressed by novice teachers at most locations. At a handful of institutions, new teachers felt that their extensive preparation to teach specific topics in physics served them very well in their actual classrooms. This preparation involved special courses on the learning and teaching of physics. These courses typically include components that improve future teachers' content knowledge and knowledge of topic-specific pedagogy, and provide an integration of these two areas—i.e., a knowledge of ways in which physics content is taught and students learn this content. This final knowledge integration is sometimes termed pedagogical content knowledge (PCK) and includes an understanding of how students learn physics, the ideas students bring to the learning environment, effective teaching strategies, and appropriate ways to assess students' evolving understanding of physics.

A significant issue with offering courses of this nature is the small number of potential enrollees. Low enrollments often dissuade chairs and deans from allowing these courses to be offered. This is a vicious cycle that prevents most institutions (including many institutions that have active physics teacher preparation programs) from creating a dedicated sequence of courses and experiences that can provide this specialized knowledge for teaching physics. As a result, without courses that specifically target learning in the discipline, there is functionally little difference in how future physics teachers are educated compared to how future biology, chemistry, or earth science teachers are educated. Consequently, program graduates feel unprepared to teach physics and are unlikely to recommend the institution from which they graduated as a valued resource for prospective teachers. To compound matters, many institutions lack faculty with appropriate expertise to offer physics-specific pedagogy courses. Collaborative hires between physics and education offer one such route to overcome this difficulty.

A winning strategy employed by institutions to reach minimum class sizes is to combine, in the same physics-specific course, preservice teachers and inservice teachers who seek continuing education for professional advancement or improved effectiveness in the classroom. It is important to caution that these two groups have overlapping but different needs. Inservice teachers are professionals with specialized expertise about students and schools—not to mention more life experience. In turn, preservice teachers often have stronger physics preparation. A successful course recognizes and leverages these differences to create activities that are designed to serve both populations. Additional categories of students who can be included in such courses are Learning Assistants and undergraduate and graduate students who are interested in educational issues and who may have an interest in teaching at some future time. Such courses then help participants begin to learn how to teach physics in a manner similar to how physics courses should help students learn how to learn physics.

Example 1. Rutgers University

Of all of the physics teacher education programs we have visited, the program at Rutgers, the State University of New Jersey, stands out as one of the most ambitious and comprehensive. The acknowledged intellectual leader of this program is Professor Eugenia Etkina, a Russian-educated physics teacher educator and currently professor of science education at the Rutgers Graduate School of Education. Beginning with a reform of the university's physics teacher education program in 2001, Eugenia crafted a carefully sequenced set of six courses [23,24] coupled with early teaching experiences to prepare future secondary physics and physical science teachers.

The sequence of courses required for a master's of education degree incorporates the learning of physics into the context of how students approach the discipline, and develops teachers' understanding. In these courses, future teachers are immersed in the nature of science, in scientific thinking, and in appropriate ways to construct a learning environment that will help their students eventually engage in expert behavior in physics. Each course in the program is accompanied by an early teaching experience that allows future teachers to immediately implement the knowledge they build in class and to reflect on this implementation. What stands out in this course sequence is the carefully planned progression of involvement of teachers from observation to authentic practice in a *physics* classroom. In each stage, the future teacher is engaged in reflection and progressively asked to take on more complex tasks.

The number of courses—six—devoted to this purpose at Rutgers University is probably unrealistically high for most institutions. The actual number is less important than the existence of a coherent set of learning and practice opportunities that focus on the teaching and learning of specific physics topics—an opportunity that general science methods courses do not typically provide.

While many teacher education programs aim to construct similar environments, few take such a comprehensive approach toward directly learning and teaching physics. The reason is clear: Few programs graduate enough secondary physics education students to allow a dedicated sequence of courses like this, although this is not the case in a number of foreign countries (e.g., Germany and Finland). One potential way to address this is at the heart of a systemic recommendation of the T-TEP report: The nation should create regional centers in physics education. The Rutgers program would make an excellent model for such centers.

Does your university see itself as a leader in science teacher education? Do you offer a cohesive sequence of courses that focuses on the teaching and learning of different topical areas of physics at the high school level?

Example 2. California State University, Long Beach

What if you cannot afford to staff or populate an integrated sequence of courses with students? California State University, Long Beach (CSULB) has developed one innovative solution to this question with a set of courses: Physics 390, Exploring Physics Teaching; and Physics 491, Pedagogical Content Knowledge in Physics. Physics 390 introduces students to the teaching profession through readings, observations, and discussions. It serves as the required pedagogy course for students wanting to serve as Learning Assistants, and gives many students their first look inside a high school classroom from the teacher's perspective. The exact focus of Physics 491 changes from semester to semester, though it always explores the teaching of fundamental content topics (e.g., energy and momentum, waves and optics, etc.). The course investigates teaching and learning concepts fundamental to physics at the high school level and awards credit to preservice as well as inservice teachers. The powerful ideas here are that (1) the course can reach enrollment minimums by involving both current and future teachers; (2) the content of the course changes each semester, allowing a more complete look at the different techniques and ideas students bring to the range of material; and (3) preservice and inservice teachers work together and consequently build relationships and form networks. This last point is a critical first step in building a learning community that supports new and existing teachers; recall that nearly every person who teaches high school physics works in professional isolation as the only person in his or her school who teaches physics. An added benefit is that rotating the curriculum encourages inservice teachers to take the course several times, allowing them to thoughtfully engage in the breadth of physics topics. In addition, the course is offered at a time when practicing teachers can attend. Teachers enrolled in the course, especially those whose initial licensure is not in physics, praise the experience as critical in enabling them to feel ready for the classroom, and claim that the course is more helpful than any other methods course they have taken.

Physics 491 is taught by Professor Laura Henriques at CSULB, who resides professionally in the department of science education. This allows her to straddle the fence between physics and education departments without her tenure or administrative home being tied to either one. More information on these courses can be found on the department website [25].

Can you offer a physics-specific methods course for prospective high school teachers that is also available to practicing teachers? Are there places in the physics curriculum that would allow physics majors to study learning and teaching physics?

E. Early teaching experiences

Many students have impoverished ideas about physics teaching. According to such ideas, a physics teacher drones on and on explaining laws that were developed centuries ago, helps students memorize relevant formulas, grades their algebraic manipulations of said formulas, and penalizes them for forgetting to make the standard assumptions about friction. Early teaching experiences provide an important counterbalance to these narratives by immersing a student in the complex workings of the classroom and offering exposure to what teaching actually entails. Although undergraduates are not far removed in age from pre-university students, their lingering student perspectives on their high school classroom experiences are not necessarily accurate representations of the work of a teacher. New teachers often feel the need to explain everything themselves, rather than to listen to student explanations. Careful listening is an important skill, as it helps the teacher learn how to effectively guide students toward developing their own understanding of the subject matter. A common aphorism in teacher education is, "A teacher has one mouth and two ears and should use them in that proportion." High-quality early teaching experiences can help students learn to heed this injunction.

There are several kinds of early teaching experiences. A teaching experience at the university level (as a Learning Assistant, for instance) has different characteristics than an experience at the precollege level. Both have value, as they serve complementary priorities. Ultimately, it is essential for prospective teachers to experience the K-12 classroom, for that is where they will eventually spend a great deal of time.

One example of early teaching experiences in the precollege setting is provided by the UTeach model, which was developed at the University of Texas at Austin and which has been replicated, as of this book's press date, at 44 universities in 21 states plus the District of Columbia [26]. The UTeach courses—Step 1, Inquiry Approaches to Teaching, and Step 2, Inquiry-Based Lesson Design—provide the context for university students to engage in short teaching stints in elementary (Step 1) and middle school (Step 2) classrooms and are low-barrier recruitment mechanisms for prospective science, technology, engineering, and mathematics (STEM) teachers. The courses serve an additional and important purpose: They give aspiring teachers early feedback about whether or not such an endeavor fits their dispositions. Another example of an early teaching experience is the Teacher Immersion Institute at Chicago State University and its emulators [27].

In addition to serving as recruitment tools, these early teaching experiences can also be designed to fulfill requirements. If the experience has sufficient intellectual content to qualify for course credit, it can get a student closer to a major or count toward teaching certification requirements.

Example 1. University of Colorado Boulder

The University of Colorado Boulder (CU-Boulder) has been a pioneer of a certain type of early teaching experience, namely the Learning Assistant program [28]. LA programs following the CU-Boulder model are currently in place at 36 institutions. Such programs (1) introduce students to authentic teaching experiences and the possibility of becoming a teacher, (2) improve student learning in university courses, and (3) infuse a scholarly approach to teaching and learning into departmental culture.

Learning Assistants lead small groups of their peers in structured cooperative activities in university courses. To prepare for their roles, they enroll in special pedagogy courses that address larger issues in learning. In addition, LAs carefully go through the material they are expected to help students learn each week. This trio of weekly activities—teaching in group work-intensive settings, subject matter preparation, and pedagogical education—provides LAs with a meaningful yet enjoyable first experience with teaching. At CU-Boulder, Professor Valerie Otero and LA program co-director Dr. Laurie Langdon provide intellectual and programmatic leadership for the university-wide program that, as of Spring 2013, included 93 LA-supported courses in 12 departments.

The LA program in the physics department had its genesis during Professor Steve Pollock's initial implementation of the University of Washington's *Tutorials in Introductory Physics* [29]. Since then Pollock, Professor Noah Finkelstein, and other physics colleagues have expanded the program to sections of several introductory physics courses and introduced LAs into upper-division physics courses to improve student learning.

Has anyone from your department attended one of the regional LA alliance workshops?

Example 2. Towson University

There are various ways to introduce students to physics teaching, including a staged entry into the profession. Towson University (TU), a large regional university in Maryland with historic beginnings as a teachers' college, has designed such a staged entry via a series of three one-credit courses that begins with informal learning environments (e.g., nature centers) in semester one, progresses to teaching in elementary and middle schools in semester two, and then to high school in semester three [30].

In each early teaching course, interns (i.e., prospective teachers) make observations and develop and teach lessons under supervision. As in many of the best programs, the choice of who supervises this experience is critical. Professor Ron Hermann, whose main responsibilities lie in high school teacher education, leads these programs, along with former Teacher-in-Residence [31] Jim Selway. They have compiled a list of local teachers they can trust to provide appropriate mentoring and feedback in this critical phase of preservice teacher preparation. Ron's position at TU is unusual, as his position sits in the Department of Physics, Astronomy and Geosciences, but his responsibilities lie in teacher education—and remarkably, he is not alone. In his department a full six faculty members address science education at a variety of K-12 grade levels, reflecting the college of science's commitment to teacher education. Hermann and his colleagues continue to refine early teaching courses and other aspects of the program as TU becomes a UTeach replication site.

On a historical note, the TU program originally consisted of a two-semester sequence that placed interns solely in informal, elementary school, and middle school contexts. The sequence was intended to keep the stress factor low, paralleling the model used by UTeach. The third-semester course, which utilizes high school physics and physical science placements, was created to meet the needs of interns interested in assisting their favorite high school instructors. The high school level is where these interns ultimately wanted to work, and the new course allowed them to experience the realities of high school teaching and, with appropriate mentoring, see real success in instruction. Using elementary and middle school environments to introduce undergraduates to teaching is extremely useful, but in our experience the passions of most prospective physics teachers do not lie in the lower grades, which is why early teaching opportunities in high school can be so desirable.

Have you asked your students if they had excellent high school physics teachers? Can you establish links with these teachers to provide your students with authentic classroom experiences?

F. Individualized advising of teacher candidates by knowledgeable faculty

and

G. Mentoring by expert physics teachers

Advising and mentoring are overlapping activities that reflect the important roles that university representatives play in helping students navigate state-mandated mazes of certification requirements and in helping to transform prospective teachers into budding teaching professionals. Practically, the issue often arises as follows.

Suppose a physics student in your program has indicated an interest in becoming a teacher. What now? For some students, the fickle finger of fate points away from the physics department and toward the school of education. "Go ask them," a physics advisor may say; while for others the answer is, "Not sure; we never had anyone ask before." Retaining this student is the critical challenge: He or she has already indicated an interest in teaching; now the responsibility lies with faculty or staff to help the student navigate this path. As it turns out, there is almost always a variety of paths to a teaching career and—just to keep you on your toes—they change from time to time with new course requirements and administrative hoops, which fluctuate enough to dissuade some students from pursuing teaching.

The best programs have a go-to person who understands how to "work" the system so that a student can graduate in a reasonable amount of time without having to take coursework or paths that put him or her in jeopardy of not finishing. Flexibility is key. So is understanding the values of the school of education, as that unit is a critical gatekeeper at many of the intersections along the road to certification. It's also a good idea to have someone in the physics department with knowledge of how the state licenses teachers—information that an advisor must know to effectively guide prospective teachers.

Helping a student meet requirements for classroom observation time before enrolling in a course of study is an example of programmatic flexibility. For instance, can the university count any of a student's time spent as an LA toward fulfilling this requirement? If a physics faculty member has developed relationships with colleagues in the school of education, a quick phone call or email can resolve problems with program requirements that might keep such a student from advancing.

Knowing which connections to tap into is critical, but so is knowing the student. An advisor's job is to know the multitude of certification paths, and to provide advisees with appropriate navigation down the best roads for their particular situations. Students approach teaching from a variety of backgrounds, experiences, and personal challenges, and certain routes to certification are better for some students than for others. Understanding a student's background helps an advisor provide the proper guidance, and the advisor probably won't learn this background in a 10-minute time block allocated for advising. More time may be necessary.

Along the road toward certification, students also need good mentoring from someone who knows the teaching ropes and who can help them understand what it will really be like when they start their first jobs. The PhysTEC project has relied for this critical mentoring role on master teachers, or Teachers-in-Residence. This role is one that is difficult for a physics faculty member to play, as it involves critical knowledge of the classroom environment, the realities of day-to-day teaching assignments, and a good-sized "bag of tricks" that can help deal with different classroom challenges. Experienced high school teachers who provide this guidance and support can be resident, either temporarily or permanently, in the physics department or in the school of education. Mentors can also be practicing high school teachers ("cooperating teachers") who allow student teachers or observers in their classrooms, or be part of a group of teachers who have participated in professional development activities provided by the department. No matter where these teachers reside, it is imperative to help students make connections with them. These connections will form the basis for developing ongoing mentoring relationships that will build self-confidence and hone skills of novice teachers.

Choosing savvy academic advisors and knowledgeable and supportive teaching mentors is critical. Although many education schools do a good job at this, the typically low number of preservice physics teachers in their programs often leaves prospective teachers without developed professional networks to which they can turn for advice or mentorship. Placing a teacher in the wrong classroom can reinforce bad strategies through the repeated rehearsal of ineffective teaching approaches: Practice doesn't make perfect, it makes permanent! Getting this initial exposure in physics teaching right is critical, and the physics department is in a unique position to assist.

Example 1. Brigham Young University

Brigham Young University (BYU) in Provo, Utah, has emerged in the last decade as the institution educating the largest numbers of qualified physics teachers in the nation. The physics department graduates about 55 physics majors and an average of 14 physics teachers annually. These large numbers are due in part to the successful alignment of physics department programs with the institutional mission. The department taps the students' close affiliation with the Church of Latter-day Saints to help them combine their interest in physics with their outlook toward service. The BYU program has improved precollege physics education for students in the Provo area. Surrounding school districts have increased the number of physics classes they offer, taking advantage of the steady supply of qualified physics teachers that is available to them. Following the example of the physics department, the biology, geology, and chemistry departments have also capitalized on the opportunity to provide large numbers of well-qualified teachers to their community as well.

Under the leadership of Professor Robert Beck Clark (now retired), the physics department in this research institution hired Duane Merrell, a former physics teacher, into a special tenure-track teaching position to guide the departmental efforts in teacher preparation. In making such a bold move, BYU has elevated the importance of preparing teachers to a level above that at nearly every other institution in the country. Assisted by a master teacher hired by the School of Education on a two-year rotating basis as a clinical faculty associate, Duane recruits and advises students, teaches courses, and supervises student teaching placements. Duane is a magnet for students who are interested in teaching. In him they find someone deeply knowledgeable about flexible entry points to the program, an ally in securing employment (95% of program graduates find jobs), and a critical friend who provides them with physics-specific feedback based on weekly visits to their classrooms. Duane also serves as a resource for the department at large. The authentic collaboration between physics and education that he has engendered has enabled him to secure intramural and extramural funding for the program.

Who among the department faculty or staff understands the details of the pathways students may take in pursuing a teacher education track? Do all academic advisors know to refer students to this person for advice on becoming a physics teacher?

Example 2. Virginia Tech

Virginia Polytechnic Institute and State University, or Virginia Tech, is a large state-supported institution with about 30 full-time faculty members in the physics department plus six instructors to augment its teaching staff. Virginia Tech made a commitment to increase the number of high school physics teachers and became a PhysTEC Supported Site in 2011 under the leadership of Professors Beate Schmittmann and John Simonetti, then the chair and associate chair of the physics department. Having the departmental leadership running such a program—and we do mean that they assumed personal involvement in all aspects of the project—communicated a meaningful message to faculty and, most significantly, to students about the elevated importance of such an effort. When Beate left to take a deanship at another university, John assumed the leadership role of the project, with the new chair serving as a partner in promoting teacher education at the university. John either advises all potential undergraduates considering high school teaching or makes sure they find the right advisor who knows the ins and outs of certification pathways.

In 2011 the department hired Alma Robinson, a master teacher from northern Virginia, as its Teacher-in-Residence. Not only does Alma know the campus well, having completed her bachelor's degree in physics at Virginia Tech, she is also a gifted instructor. When she taught a section of introductory physics, Alma got higher learning gains and better student evaluations from students at Virginia Tech than did any other professor teaching that semester. She is now helping the department improve its introductory courses and is providing experienced mentoring to students who are pursuing high school physics teaching. A sign of the institution's commitment to teacher education was its decision to hire Alma as an instructor after PhysTEC funding for her position ended. John reports that she is indispensable in the department's effort to improve the undergraduate physics major. (The number of majors in the department recently doubled in size.)

Have you considered what roles a full- or part-time master teacher might play in your department? Can you use such an individual to help teach selected courses as well as to broadly promote teacher education?

H. A rich intellectual community for graduates

Junior physics researchers come to understand the ethos of the discipline through guidance and enculturation by a larger community that includes collaborating faculty in one or more institutions, support staff, postdocs, senior graduate students, junior graduate students, and undergraduates. Similarly, a rich intellectual community is necessary for novice teachers to find support, for veteran teachers to mentor others, for physics faculty to work side-by-side with teachers on pedagogical issues, for prospective teachers to find role models, for teachers to build networks that help in securing employment, and for everyone to benefit from and contribute to the community's intellectual and physical resources.

This characteristic of thriving physics teacher education programs is rare among universities, as the very small number of prospective teachers leads individuals down solitary paths that require solving the dilemmas of physics teaching practice on their own. The cost of this isolation to the profession of physics teaching is significant. Novice teachers cannot rise above isolation without a community. Communities offer new ideas, intellectual challenges, support during professional obstacles, opportunities for reflection on pedagogical practices, and access to developments in physics research. Given the solitary nature of typical physics teaching assignments, the absence of a university physics department that can nucleate these interactions leaves a void that typically goes unfilled.

Post-secondary institutions can benefit in several ways when such communities are present. For example, the community—consisting of physics and education faculty and veteran and prospective physics teachers—provides a ready-made network of professionally active cooperating teachers for the university's student teachers. In addition, the physics department gains access to high school students who are interested in physics and might be convinced to study at the university.

The institutional rewards for the department can be significant. At a time of national crisis in precollege STEM education and calls for universities to prepare significantly more science and mathematics teachers, deans, provosts, and presidents can use the physics department's activities as *prima facie* evidence of their institution's efforts in this area. Furthermore, research findings into changes that occur to participants of such communities suggest that physics faculty members are as likely to benefit from their participation as are the program graduates. Benefits to faculty include greater sophistication around issues of teaching and learning and a deeper appreciation of precollege teacher contributions to the continuum of physics education.

This community transcends institutional borders. Participation in AAPT and PhysTEC meetings allows members of physics and education departments the opportunity to join coalitions of similarly minded faculty, get access to novel ideas, and be professionally strengthened through engagement in these larger communities.

Example 1. Modeling Instruction

The graduates of the Modeling Instruction (MI) program at Arizona State University form the largest intellectual community of physics teachers in the United States; this program now also enjoys a significant international following. Beginning in 1990, the MI teaching methods were developed during a decade of physics education research efforts led by Professor David Hestenes. This research, along with dedicated efforts by Dr. Jane Jackson (the program's administrator) and a group of high school physics teachers, propelled Modeling Instruction into a national movement. In 2005, the MI teachers created the American Modeling Teachers Association (AMTA) [32] to expand the program in collaboration with supportive physics faculty at other universities. As of 2013, MI boasted more than 6,000 graduates of summer Modeling workshops, seven active listservs, and an extensive trove of official curriculum resources. Summer 2013 was typical, with 51 workshops held throughout the country for some 800 teachers, most of whom were physics teachers.

Although there are a small number of longstanding national workshops for inservice physics teachers, the MI community is unique in the loyalty of its participants, many of whom fiercely identify themselves as modelers. One reason for this loyalty is that David promoted the vision of science education reform as driven primarily by an organized community of teachers. Other programs do an excellent job in facilitating leadership development by individual participating teachers, but the AMTA goes further, striving to create a cohesive community of teachers in schools, high school science departments, local districts, groupings of neighboring districts, and so on. Some university physics departments that have promoted MI workshops as an important outreach activity (e.g., Buffalo State College in New York) have used the community of modelers as an invaluable resource for their preservice program. Placement of student teachers in modelers' classrooms extends the face-to-face and virtual relationships developed as a result of preservice teachers taking MI workshops as part of a certification program.

The MI program fosters leadership development from among the teaching ranks. Through an apprenticeship model, the community promotes talented participants into becoming workshop leaders in local and national workshops. Supporting these leaders is also critical, and organizations like AMTA can help MI teachers and universities that would like to become involved in offering MI workshops.

> *There is probably a Modeling Instruction group near your campus. Do you know who they are? Would you host a Modeling gathering?*

Example 2. University of Washington

The University of Washington (UW) is a large public institution in Seattle that is strongly focused on research. In this environment, the Physics Education Group at the University of Washington (UWPEG) [33] has been for almost 40 years the center of a rich intellectual community involving physics faculty, physics education researchers, curriculum developers, graduate students, postdocs, scores of short- and long-term visitors, master teachers, undergraduate students who aspire to become physics teachers, and inservice teachers of science at the elementary and secondary levels. The accumulated research and practical experience in this teaching and learning constellation is aptly recognized as a unique treasure. Professors Lillian C. McDermott, Paula Heron, and Peter Shaffer, together with present and former members of the group, have collectively been the engine that combines research on the learning and teaching of physics, research-validated curriculum development, and effective instruction at dozens of institutions. Most notably, in recognition of her seminal contributions to the establishment of physics education research as a field for scholarly inquiry by physicists, Lillian has received some of the highest awards that professional societies can bestow.

The flagship of the UWPEG physics teacher education program is an academic-year course sequence for prospective physics and math teachers. The group also offers an extensive professional development program, a five-week NSF Summer Institute in Physics and Physical Science for Inservice K-12 Teachers, and a full academic-year seminar. Participants in the inservice program meet once per week for two hours during a tuition-free academic-year continuation seminar. This seminar has been offered for decades on the same day of the week and at the same time. Teachers who have participated in the group's programs know that this course is a constant in their lives. Teachers may drop out for an academic quarter or year in response to professional and personal demands, but many rejoin over and over as they are pulled back by a core supportive community that seeks to improve physics learning in the precollege classroom.

These courses serve as the laboratory for cumulative, research-validated improvements to UWPEG's curriculum development projects, most notably *Physics by Inquiry* [34] and *Tutorials in Introductory Physics* [29]. In addition,

Tutorials for Teachers is a recent effort to deepen teachers' knowledge of more advanced physics topics and expose them to research results on students' conceptual understanding of these topics. The local, national, and international activities of all the group's members over 40 years have created at the University of Washington a Center for Physics Education, whose reach and impact can be felt most notably in a powerful community dedicated to the professional success of teachers of physics at all levels. Most physics teacher educators in the country can trace major intellectual influences on their work to the work of this community.

Efforts like those developed at UW do not start with a quantum jump, but rather are built over time. Departments considering building professional communities of people who share a passion for physics teaching and for learning how to do it better should investigate the work of programs like those developed at UW, to see where first steps can be taken or which components might be adopted or adapted to the local situation. Building the rapport of local teachers takes time but can ultimately lead to more majors, better-prepared students, and a healthier physics community.

III. CONCLUSION

Most people trace significant events in their life to a teacher who provided encouragement and advice and sparked their interest in a particular vocation. Physics, like other subjects, relies on the strength of its teacher corps to prepare the next generation of individuals who will tackle the significant problems of the coming decades. Universities that at one time saw their primary missions as helping to educate teachers have now taken different paths and adopted different priorities. Physics has suffered in this transformation, and will continue to suffer as long as we neglect teacher education as a critical component of the discipline.

The good news is that recapturing this mission is relatively easy. Still, it does require intellectual and political effort to reestablish this priority and to direct resources toward ensuring that every student in the country has access to high-quality physics instruction from an engaged and supported teacher. Our vision is that by adapting the crucial components described above to local contexts in institutions throughout the country, we can form a national *system* of thriving physics teacher education programs that will address the compelling national need that we now face.

The eight characteristics fleshed out here represent the collective understanding of excellent programs, but building a program where none exists today does not require a massive shift on all fronts. Rather, choosing appropriate first steps such as supporting (or becoming) the champion, understanding the role of advising, and stepping up recruitment of physics majors and prospective teachers is a good way to begin. As your department begins to attract more students to teaching, other components will be needed, and it will be time to visit some of the acknowledged leaders mentioned here to understand how things have progressed and where powerful ideas have helped shape and improve teacher education. Physics departments are critical to advancing this educational mission, and no one else in any sphere has as much to lose as we do. High-quality physics teacher education cannot be someone else's problem—it must be ours.

[1] T. Hodapp, J. Hehn, and W. Hein, Preparing high school physics teachers, Phys. Today **62** (2), 40 (2009). doi:10.1063/1.3086101

[2] D. E. Meltzer, M. Plisch, and S. Vokos, eds., *Transforming the Preparation of Physics Teachers: A Call to Action. A Report by the Task Force on Teacher Education in Physics* (T-TEP) (American Physical Society, College Park, MD, 2012). http://www.phystec.org/webdocs/2013TTEP.pdf

[3] http://nces.ed.gov/surveys/sass/, retrieved Feb. 12, 2015.

[4] D. E. Meltzer, M. Plisch, and S. Vokos, The role of physics departments in high school teacher education, APS News **22** (8), August/September 2013. http://www.aps.org/publications/apsnews/201308/backpage.cfm

[5] D. E. Melzer and V. K. Otero, Transforming the preparation of physics teachers, Am. J. Phys. **82**, 633 (2014). doi:10.1119/1.4868023

[6] D. E. Meltzer, ed., and P. Shaffer, associate ed., *Teacher Education in Physics: Research, Curriculum, and Practice* (American Physical Society, College Park, MD, 2011). http://www.phystec.org/webdocs/TeacherEducationBook.cfm

[7] See Plisch, this volume.

[8] R. Scherr, M. Plisch, and R. M. Goertzen, Sustaining programs in physics teacher education: A study of PhysTEC Supported Sites, in *Proceedings of Noyce Conference* (National Science Foundation, Washington, DC, 2014). http://www.phystec.org/items/detail.cfm?ID=13289

[9] Attributed to Niccolo Machiavelli.

[10] See Stewart and Stewart, this volume.

[11] See Henderson and Frank, this volume.

[12] See Churukian and McNeil, this volume.

[13] Adapted from a phrase in the 1989 film *Field of Dreams*. In the movie, a novice farmer hears a voice whisper, "If you build it, he will come," in his remote cornfield and interprets the whisper as a call to build a baseball field to entice the ghosts of legendary baseball players to return to play.

[14] For materials associated with the Learning Assistant program at the University of Colorado Boulder, see http://laprogram.colorado.edu/

[15] See Wenning, this volume.

[16] See http://www2.phy.ilstu.edu/ptefiles/

[17] Illinois Section of the American Association of Physics Teachers, *A Career in Science Teaching? Think About It!* http://helios.augustana.edu/isaapt/teach/brochure3.pdf

[18] See Close, Seeley, Robertson, DeWater, and Close, this volume.

[19] Willie Sutton is supposed to have said that he robbed banks "because that's where the money is."

[20] Modeling Instruction is discussed under Characteristic H in this chapter.

[21] See, for instance, M. Marder, A problem with STEM, CBE Life Sci. Educ. **12** (2), 148 (2013). doi:10.1187/cbe.12-12-0209

[22] C. L. Tesfaye and S. White, Challenges high school teachers face, AIP focus on (April 2012). http://www.aip.org/sites/default/files/statistics/highschool/hs-teacherchall-09.pdf

[23] For a more complete description of the course sequence, see: E. Etkina, Pedagogical content knowledge and preparation of high school physics teachers, Phys. Rev. ST Phys. Educ. Res. **6** (2), 020110 (2010). doi:10.1103/PhysRevSTPER.6.020110

[24] See Etkina, this volume.

[25] For information about the PhysTEC project at CSU Long Beach, see http://www.csulb.edu/colleges/cnsm/sas/phystec/

[26] For information about the UTeach program, see http://www.uteach-institute.org/

[27] See Sabella, Robertson, and Van Duzor, this volume.

[28] See Otero, this volume.

[29] L. C. McDermott, P. S. Shaffer, and the Physics Education Group at the University of Washington, *Tutorials in Introductory Physics* (Pearson, Reading, MA, 2002).

[30] See Sandifer, Hermann, Cimino, and Selway, this volume.

[31] For a discussion of the multiple roles of Teachers-in-Residence in physics teacher education, see Plisch, Blickenstaff, and Anderson, this volume.

[32] The AMTA currently has 1500 members. For information, see http://modelinginstruction.org/

[33] For information about the activities of the Physics Education Group, see http://depts.washington.edu/uwpeg/

[34] L. C. McDermott and the Physics Education Group at the University of Washington, *Physics by Inquiry: An Introduction to the Physical Sciences* (John Wiley & Sons, New York, NY, 1996).

The Physics Teacher Education Coalition

Monica Plisch
American Physical Society, One Physics Ellipse, College Park, Maryland 20740

In response to the severe shortage of qualified high school physics teachers, a consortium of professional societies launched the Physics Teacher Education Coalition (PhysTEC) project. PhysTEC has supported more than 40 institutions to build model physics teacher preparation programs. Collectively these sites have more than doubled the number of graduates who are highly qualified to teach physics. The project also established a national coalition of more than 300 institutions committed to PhysTEC's mission, to improve and promote the education of future physics teachers. PhysTEC has developed many avenues for dissemination, and has partnered with organizations that have a common interest in teacher preparation.

I. MISSION AND GOALS

The United States is facing a severe shortage of qualified high school physics teachers. Fewer than half of all high school physics classrooms have a teacher with a degree in the subject [1]. Every year, more than 3,000 new and inservice teachers find themselves at the front of a physics classroom for the first time, and only 35% of these teachers have a degree in physics or physics education [2]. The shortage of qualified teachers is exacerbated by rapidly growing enrollments in high school physics; the number of students completing a high school physics class has more than doubled in the past two decades [3].

A study of physics teacher preparation in the U.S. found a low level of engagement among physics departments [4]. Only about 20% of physics departments have a track, concentration, or minor for future physics teachers that has recently graduated even one teacher. Very few physics departments graduate more than two students per year from physics teacher education programs, and most programs lack training in physics-specific pedagogy. Multiple data sources show that the number of graduates from physics teacher education programs is much smaller than the national need.

In 2001, a consortium of professional societies including the American Physical Society (APS), the American Association of Physics Teachers, and the American Institute of Physics, launched the Physics Teacher Education Coalition (PhysTEC) to address the severe shortage of qualified physics teachers [5]. The mission of PhysTEC is *to improve and promote the education of future physics teachers*. The project's stated goals are to:

- Transform physics departments to engage in preparing physics teachers;
- Demonstrate successful models for increasing the number of highly qualified teachers of physics prepared at colleges and universities [6]; and
- Spread best-practice ideas throughout the physics teacher preparation community.

II. PHYSTEC INSTITUTIONS

Since its inception, PhysTEC has funded more than 40 institutions to build model physics teacher education programs. These institutions were selected through a competitive award process to receive funding and other support to implement PhysTEC key components (described below), with the overall goal of increasing the number of well-prepared physics teachers graduating from each institution [6]. PhysTEC Supported Sites represent a wide variety of institution types, including primarily undergraduate institutions, regional comprehensive universities, and research-focused universities, as well as public and private institutions and minority-serving institutions. Figure 1 shows the locations of all current and previously supported sites, indicated by red squares [7].

Nearly all supported sites have been successful at increasing the number of well-prepared physics teachers that they graduate. Collectively these sites have more than doubled the annual number of graduates who are highly qualified to teach physics. Figure 2 shows the increases in the annual numbers of physics teachers graduating from several of the earliest supported sites; in some cases, these numbers have increased by a factor of 10 or more. Follow-up surveys of PhysTEC graduates show that close to 90% of program graduates become secondary teachers, and of those about 90% teach physics or physical science in a given year [8]. The five-year classroom retention rate for PhysTEC graduates is above 70%, which exceeds the national average for public school teachers. Moreover, PhysTEC graduates are more diverse than the physics teaching workforce as a whole in terms of race and ethnicity.

Another key success of PhysTEC Supported Sites involves sustaining programs after project funding ends. For example, nearly all the sites featured in Figure 2 graduated an elevated number of physics teachers in post-funding years, compared to baseline numbers. An in-depth study showed that most PhysTEC Supported Sites increased their annual graduation rates of physics teachers as well as their funding for physics

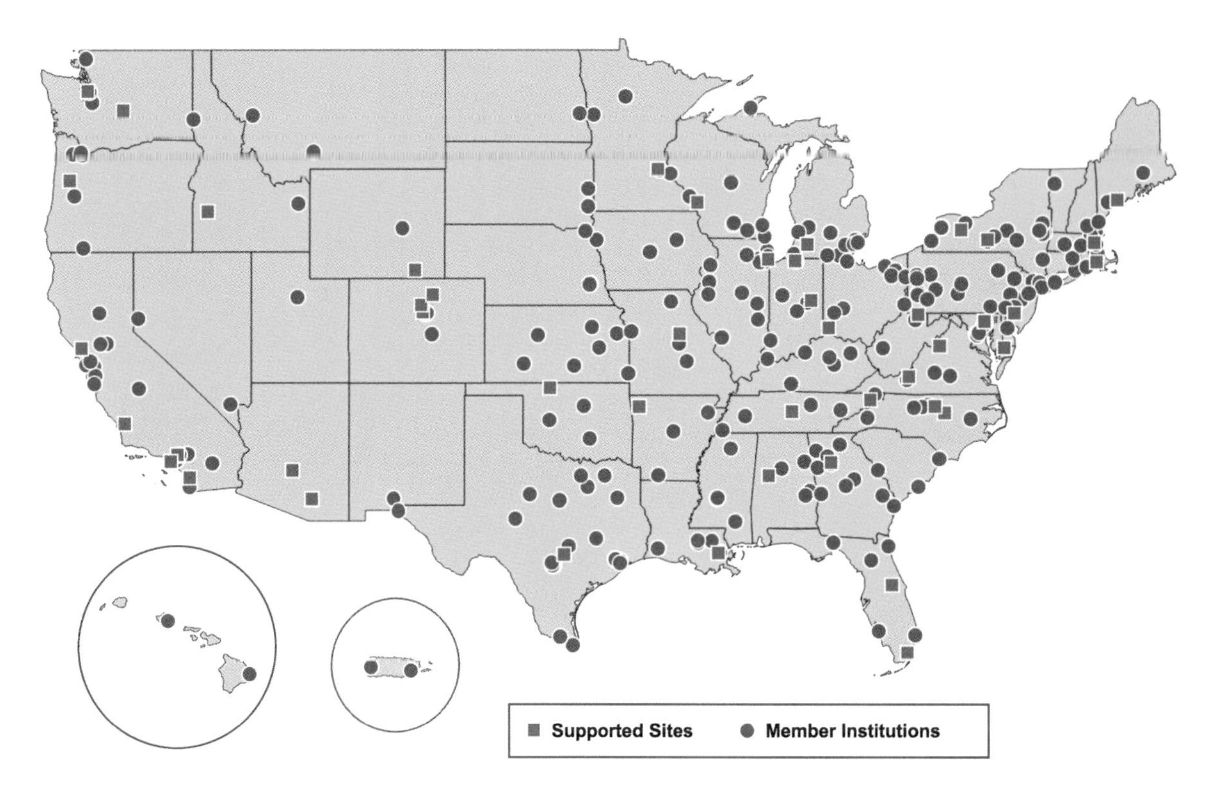

FIG. 1. The map shows more than 300 PhysTEC Member Institutions, which represent about one third of all U.S. physics departments. Supported sites receive funding to build model programs, and member institutions are committed to the PhysTEC mission.

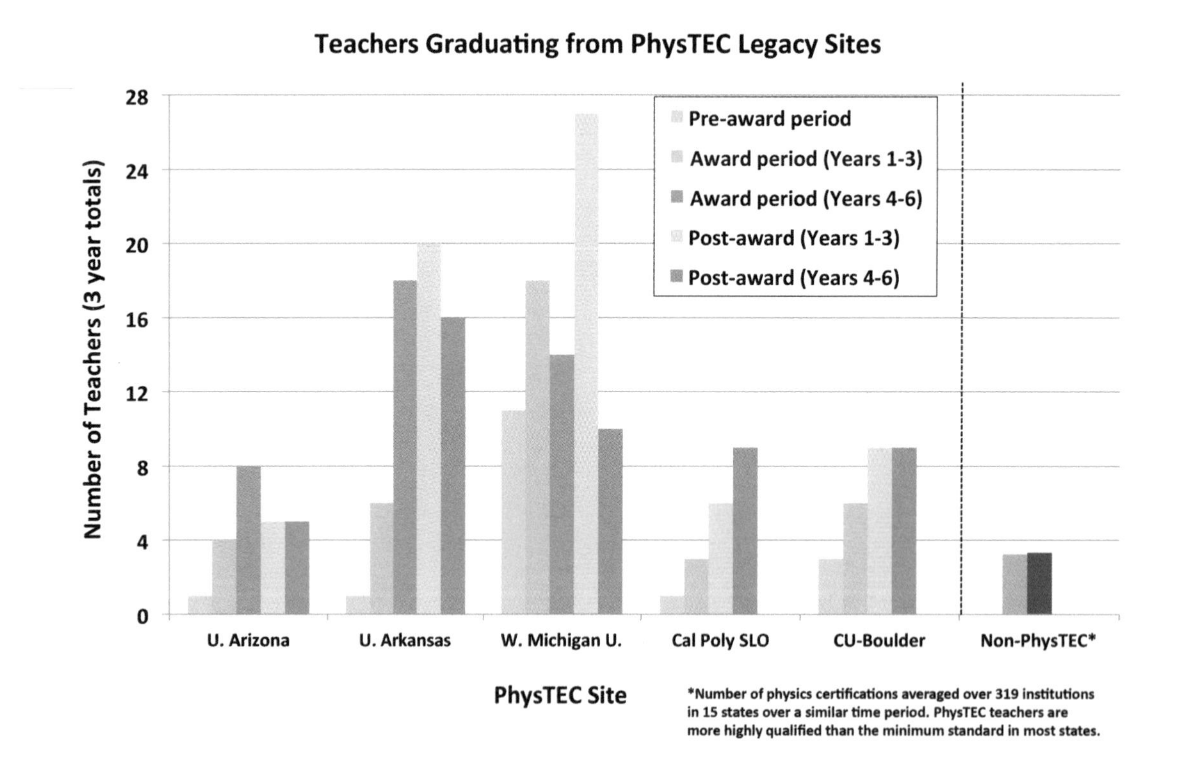

FIG. 2. The graph shows the numbers of graduates from PhysTEC programs at five institutions that are among the earliest sites supported by the project. All institutions show significant increases in physics teacher graduation rates during the funded period, and nearly all institutions sustained these increases after project funding ended. By comparison, institutions that were not PhysTEC Supported Sites showed negligible increases in the numbers of graduates with physics certifications during a comparable time period.

teacher education; some sites even continued to increase funding and graduation rates in the post-award period [9]. Many of the individual program components also continued at these sites after PhysTEC funding ended.

PhysTEC has also established a national coalition of more than 300 institutions committed to the project's mission of improving the education of future physics teachers. Figure 1 shows the location of PhysTEC Member Institutions as blue dots; member institutions have a geographical distribution across the U.S. that approximately correlates with population density. Each member institution is listed on the PhysTEC website [5] and has a page with a description of activities and a list of contacts. The project actively maintains the list of contacts and keeps members informed and up-to-date with an email list. PhysTEC Member Institutions are actively engaged in the project; for example, many have applied for project funding and/or have sent one or more faculty members to a PhysTEC conference or workshop.

III. KEY COMPONENTS OF PHYSICS TEACHER EDUCATION PROGRAMS

The PhysTEC key components are the 10 elements that should be present in a high-quality physics teacher education program [10]. Each PhysTEC Supported Site is expected to implement these components consistently yet flexibly according to the resources and expertise available at the institution and the local context. The list of key components has evolved over time as the project has gained experience with building programs. PhysTEC key components are listed here in bold with brief descriptions:

- The **champion** [4] is a change agent at the university. He or she most often resides in the physics department, and ensures the success of the physics teacher education program by negotiating for staff and funding and by advocating for policy changes that are favorable to the program.
- The **Teacher-in-Residence (TIR)** is a highly experienced high school physics teacher, working full- or part-time at the university, who serves as a recruiter, teacher, and mentor for future and new physics teachers.
- **Collaboration** between physics departments and schools of education is critical to building high-quality, streamlined pathways through which students can earn a degree plus teaching certification.
- **Institutional commitment** [9] includes financial and other support needed to sustain physics teacher education programs at an institution.
- **Program assessment** is necessary for measuring program success, identifying areas for improvement, and ensuring that project activities support stated goals.
- **Recruitment** of future physics teachers involves advertising opportunities in physics education, providing excellent advising on teaching careers, establishing a positive climate for teaching within the physics department, and other activities aimed at increasing enrollment in the physics teacher education program.
- **Early teaching experiences** help students decide to become physics teachers and begin developing the many skills these students need to become effective teachers.
- **Pedagogical content knowledge** [11] is knowledge that is specific to teaching physics. It can be learned through physics-specific methods courses, one-on-one mentoring, and teaching experiences.
- **Learning Assistants** (LAs) [12] are undergraduate students who serve as peer instructors and help improve student learning in large-enrollment courses. LA programs provide LAs with well-structured early teaching experiences, exposure to effective instructional methods, and an introduction to the scholarship of teaching and learning.
- **Induction and mentoring** provide critical support to preservice and recently graduated teachers. Mentors can include TIRs, the cooperating teacher supervising the student teaching experience, and others. Induction support for new teachers can include professional development workshops and involvement in a community of practicing physics teachers.

IV. ENGAGING THE PHYSICS COMMUNITY AND BEYOND

PhysTEC holds an annual conference that is the only national conference dedicated solely to physics teacher preparation. The PhysTEC conference regularly attracts about 120 participants. Most attendees are physics faculty members engaged in teacher preparation; other attendees are education faculty members, Teachers-in-Residence, and future physics teachers. PhysTEC also organizes topical and regional workshops that attract a significant number of participants. Such workshops have focused on effective implementation of Learning Assistant programs and how to engage state university systems. An example of the latter is when PhysTEC brought together representatives from the University of California and California State University systems for a conference on teacher education in the physical sciences. PhysTEC conference evaluation reports indicate that participants strongly value the opportunity to network with each other at events that are strongly focused on physics teacher education.

In an effort to increase the number of scholarly papers on physics teacher education and support faculty engaged in this work, PhysTEC published the first collection of peer-reviewed papers on physics teacher education [13]. PhysTEC also sponsored the Task Force on Teacher Education in Physics (T-TEP), which produced a comprehensive report on the state and needs of the physics teacher education community in the United States [4]. The present volume provides another venue for faculty engaged in physics teacher education to publish in a peer-reviewed publication and disseminate practical knowledge for implementing programs. The members of the PhysTEC leadership team also publish articles in a wide variety of publications, from newsletters such as *APS News* and the *APS Forum on Education* newsletter to journals and magazines such as the *American Journal of Physics* and *Physics Today*, to help raise the profile of physics teacher education.

APS and PhysTEC collaborate with other organizations that have a goal of improving science and mathematics teacher education. APS has helped the American Chemical Society initiate the Chemistry Teacher Education Coalition [14], which is closely modeled after PhysTEC. PhysTEC has collaborated with the UTeach Institute [15], which is replicating a science and mathematics teacher education program that began at the University of Texas at Austin. In addition, the Association of Public and Land-grant Universities has launched the Science and Math Teacher Imperative, which has been a partner to PhysTEC. These partnerships have resulted in joint conferences and, in some cases, jointly funded sites.

V. CONCLUSIONS

PhysTEC is a project led by professional societies that addresses the severe shortage of qualified physics teachers by engaging physics departments in teacher education. By implementing PhysTEC's 10 key components of teacher education programs, PhysTEC Supported Sites have collectively more than doubled the number of teachers they graduate each year who are highly qualified to teach physics. In the 2013–2014 academic year, supported sites graduated nearly 80 well-prepared physics teachers, or about one in eight of the nation's new, well-prepared physics teachers. The project has established a coalition of more than one third of all U.S. physics departments (over 300 institutions) committed to the PhysTEC mission. In addition, the project has organized many conferences and workshops, generated numerous publications and reports, and initiated collaborations with aligned organizations. The PhysTEC project is having a national impact and is poised to end the shortage of well-prepared physics teachers.

ACKNOWLEDGMENTS

The PhysTEC project would not be possible without the contributions of many people, including those at professional societies involved in the project management team, the site leaders and others at their institutions, project consultants, and many others. The full list of project personnel, present and past, is available at www.phystec.org. This material is based upon work supported by the National Science Foundation under Grant Nos. 0108787, 0808790 and 0833210. Any opinions, findings, and conclusions or recommendations expressed in this material are those of the authors and do not necessarily reflect the views of the National Science Foundation. Additional project funding comes from the APS Campaign for the 21st Century.

[1] J. G. Hill and K. J. Gruber, *Education and Certification Qualifications of Departmentalized Public High School-Level Teachers of Core Subjects: Evidence from the 2007-08 Schools and Staffing Survey, Statistical Analysis Report. NCES 2011-317* (National Center For Education Statistics, U.S. Department of Education, Washington, DC, 2011). http://nces.ed.gov/pubs2011/2011317.pdf

[2] S. White and C. L. Tesfaye, *Turnover Among High School Physics Teachers* (American Institute of Physics, College Park, MD, 2011). http://www.aip.org/statistics/trends/reports/hsturnover.pdf

[3] S. White and C. L. Tesfaye, *High School Physics Courses & Enrollments: Results from the 2008-09 Nationwide Survey of High School Physics Teachers* (American Institute of Physics, College Park, MD, 2010). http://www.aip.org/statistics/trends/reports/highschool3.pdf

[4] D. E. Meltzer, M. Plisch, and S. Vokos, eds., *Transforming the Preparation of Physics Teachers: A Call to Action. A Report by the Task Force on Teacher Education in Physics* (T-TEP) (American Physical Society, College Park, MD, 2012). http://www.ptec.org/webdocs/2013TTEP.pdf

[5] See www.phystec.org

[6] The PhysTEC project defines a highly qualified (i.e., well-prepared) physics teacher as having a major or minor in physics (or equivalent coursework), and having completed a teacher education program leading to certification.

[7] For a full list of PhysTEC supported sites as well as access to reports on program activities, see http://www.phystec.org/institutions

[8] For more on outcomes of PhysTEC graduates, see http://www.phystec.org/webdocs/Outcomes.cfm

[9] R. E. Scherr, M. Plisch, and R. M. Goertzen, *Sustaining Programs in Physics Teacher Education: A Study of PhysTEC Supported Sites* (American Physical Society, College Park, MD, 2014). http://www.phystec.org/sustainability, retrieved Sep. 10, 2013.

[10] For more on PhysTEC key components, see www.phystec.org

[11] E. Etkina, Pedagogical content knowledge and preparation of high school physics teachers, Phys. Rev. ST Phys. Educ. Res. **6** (2), 020110 (2010). doi:10.1103/PhysRevSTPER.6.020110

[12] V. Otero, S. Pollock, and N. Finkelstein, A physics department's role in preparing physics teachers: The Colorado learning assistant model, Am. J. Phys. **78**, 1218 (2010). doi:10.1119/1.3471291

[13] D. E. Meltzer and P. S. Shaffer, eds., *Teacher Education in Physics: Research, Curriculum, and Practice* (American Physical Society, College Park, MD, 2011). http://www.phystec.org/webdocs/TeacherEducationBook.cfm

[14] See http://www.acs.org/content/acs/en/education/educators/chemistry-teacher-education-coalition.html

[15] See http://www.uteach-institute.org

The roots of physics teaching: The early history of physics teacher education in the United States

Amanda M. Gunning
School of Education, Mercy College, Dobbs Ferry, NY 10522

Keith Sheppard
Center for Science and Mathematics Education, Stony Brook University, Stony Brook, NY 11767

The present standard high school physics course and the system that prepares teachers to teach this course developed from the work of notable physicists and educators such as Edwin Hall, Charles Riborg Mann, Robert Millikan, and John Woodhull. During a period of extraordinary growth of high schools in the United States, these physicists and educators had experiences in education that positioned them to be champions of physics education reform. This chapter examines the development of physics teacher education from the post-revolutionary period through World War II, highlighting the importance that laboratory work played in shaping the course and in preparing physics teachers.

I. INTRODUCTION

Presently in the United States, high school teachers are purposefully educated and licensed to teach specific subjects and grade levels. However, this was not always the case. While much recent debate has focused on teacher accountability, there has also been a major push to increase the supply of appropriately qualified teachers, especially in science, technology, engineering, and mathematics fields. The Task Force on Teacher Education in Physics' 2012 report noted that the supply of qualified physics teachers was inadequate, and targeted physics teacher preparation as a major roadblock in resolving the issue: "[N]ationally, physics teacher preparation is inefficient, incoherent, and unprepared to deal with the current and future needs of the nation's students" [1].

Remarkably, when the task force authors wrote these words, the situation they described had already persisted for more than a century. This chapter outlines the compelling, though largely unknown, early history of high school physics teacher education in the United States, from the post-revolutionary period through World War II. We trace how the earliest physics courses became more mathematical and laboratory-based, and how this impacted the way that physics teachers were trained. This history highlights how, at the turn of the 20th century, the influence of physicists and educators such as Edwin Hall, Charles Riborg Mann, Robert Millikan, and John Woodhull helped establish the roots of present-day physics teacher education [1,2].

II. EARLY EDUCATION IN THE UNITED STATES, LATE 1700s TO MID 1800s

Schooling in the U.S. before 1800 was a vastly different enterprise from what it is today. Families, churches, and apprenticeships provided the backbone of education, and children were educated in an unsystematic way [3]. The type of education offered reflected the realities and needs of the time. As schools were created, they principally catered to younger students and covered basic subjects such as reading, writing, and arithmetic. At this time the nation was largely rural, and one-teacher-in-one-room schools predominated.

Teaching was a relatively low-status occupation, and teachers were often appointed (and dismissed) at the whim of local people. Who you knew was more important than what you knew, and teachers were selected based on factors such as their religious denomination, country of origin, and willingness to accept low wages and tolerate poor working conditions. There were no specialist teachers who exclusively taught science; instead, all were generalists who taught multiple subjects. In most cases, the ability to teach required that teachers have an education that slightly exceeded the level of the students whom they would be teaching. Teacher training was largely absent.

In the early 1800s, the few secondary schools that existed offered "natural philosophy," the forerunner of physics, as part of their courses of study. Natural philosophy was respected as a course that was useful to the public for explaining the increasingly technical world without including a rigorous mathematical treatment. Content addressed through these courses can be determined from texts of the era, which included physical science, the workings of machines and pumps, acoustics, and optics. Although there was great diversity in the length of natural philosophy courses, they were usually of short duration (less than a school year), and the subject matter often had a strong religious and moral undertone, with scientific phenomena presented as "outward expressions of the mystical" [4].

In these early schools, teaching methods were similar across subjects, with rote learning from texts as the norm. During class, students were assigned a lesson from a text; they were expected to learn the lesson by heart and recite it verbatim. The texts often contained the questions to be recited at the bottom of the page, making it possible for a teacher to conduct a lesson without personally understanding its content. Smith described the science classroom of the 1840s, which captured the nature of science teaching at the time:

> In science the instruction was wholly by catechism. There were illustrative diagrams in the text which might, or might not, be explained, and which might, or might not, be transferred to the blackboard...[there was] not a single article of apparatus, or a book of reference except the Bible [5].

However, the times were changing, as population growth and urbanization fueled the widespread development of academies and eventually of high schools [6]. The first high school established in the U.S. is generally considered to be the English Classical School, which opened in Boston in 1821. After its creation, high schools grew in number, slowly at first and then rapidly. High schools, especially in the Northeast, were compelled, either by law or by the conditions needed to obtain state funding, to teach natural philosophy, and in some cases to have documented science apparatus available in their schools. In New York, the Law of 1838 directed: "No academy shall participate in the annual distribution of the literature fund [public money set aside for educational resources] until the Regents shall be satisfied that such academy shall be provided with a suitable library *and apparatus* [emphasis added]" [7]. The academies, through fear of losing such funding, purchased equipment. An example of what the schools bought with state funds is shown in the equipment purchased for Fairfield Academy by the New York Regents [8].

Orrery	$20 00
Globes	12 00
Numerical frame and geometrical solids	2 50
Movable planisphere	1 50
Tide dial	3 00
Optical apparatus	10 00
Mechanical powers	12 00
Hydrostatic apparatus	10 00
Pneumatic apparatus	35 00
Chemical apparatus	25 00
One hundred specimens of mineralogy	10 00
Electrical machine	12 00
Instruments to teach surveying	80 00
Map of United States	8 00
Map of New York	8 00
Atlas	5 00
Telescope	40 00
Quadrant	15 00
Total	$309 00

FIG. 1. Equipment purchased by Fairfield Academy in 1835.

The list indicates potential changes in the way that the sciences were to be taught, though in reality, expensive physical equipment was usually kept in closets and not placed in the hands of students. The apparatuses were frequently used for demonstration and entertainment purposes. For example, Steele humorously described a typical natural philosophy lesson using such equipment:

> ...the apparatus was brought out of the case; the dust of the preceding year was brushed off, a withered apple was made plump, a frightened-half-to-death mouse was scientifically exterminated under the receiver of the old table air pump; a boy was put on the insulated stool and his hair caused to stand up like "quills upon the fretful porcupine"; next, the class, taking each other's hands, formed a ring and received the shock from a Leyden jar. So, the hour passed all too quickly with much fun and little science, and then the apparatus was carefully put away for the next yearly exhibition [8].

III. FROM NATURAL PHILOSOPHY TO PHYSICS WITH LABORATORY, MID 1800s TO LATE 1800s

By the 1850s, one-room schoolhouses were being replaced by organized school systems. Schools slowly moved from having one large class with students of many ages to grade-stratified, multiple-classroom schools, especially in urban areas [9]. During this time period, high schools proliferated, resulting in more than 6,000 high schools being established by the end of the century [10]. Teachers in these schools became increasingly responsible for specific subjects. This allowed for a more efficient division of labor and led to the familiar "chalk and talk" organization of instruction. But with the expanded availability of high school education, it became difficult to find teachers to fill the classrooms, as reported in 1850:

> The law requiring the establishment of high schools is rapidly creating a demand for a description of teachers which none of our institutions furnish...it is at this moment more difficult to procure suitable teachers for high schools than for any other class of school [11].

Normal schools, which were meant to provide "norms" for teaching, i.e., standards of acceptable practice, began to multiply to help relieve the shortage of teachers. However, while they helped increase the number of trained teachers, they did not focus specifically on the preparation of secondary teachers. It was widely believed that a teacher should be able to teach *all* subjects found in the school curriculum, which resulted in general programs being organized for teacher education. There were no distinctions between types of teachers, whether based on the age of their students or on the subjects that they would teach. Natural philosophy was part of the general education program given to all teachers. For example, in 1863, all prospective teachers attending the State Normal School of Ypsilanti in Michigan took a course in natural philosophy. During the same year, at the State

Normal School of Albany, New York, laboratory apparatus was provided for illustrating important principles in natural philosophy for teacher education.

Even though teachers studied physics, they did not have to be specialized to teach it. The concept of a *physics teacher* as someone with training in the subject and who predominantly taught physics did not exist. As natural philosophy texts of the time indicate, the courses were intended to be taught in a rote way [12]. The expectation was that anyone armed with a physics text, and with the ability to read and have students recite, would be able to teach physics. This was illustrated, a little later, by Robert Millikan's account of his own introduction to physics teaching in 1889: "My Greek professor...asked me to teach the course in elementary physics...To my reply that I did not know any physics at all, his answer was, 'Anyone who can do well in my Greek can teach physics'" [13].

Although Millikan was an unusual case in terms of his future physics prowess, the idea that any educated person could teach physics was firmly entrenched and enduring. It was not until the end of the century, when school populations had increased sufficiently to allow administrators to employ "single-subject" teachers, that specialized *physics teachers* were employed in high schools [14].

In the second half of the 19th century, teacher certification and licensing became more prevalent. Gradually the locus of control of teaching moved from small-town authorities to state authorities, though as late as 1900 only three states—New York, Rhode Island, and Arizona—actually controlled teacher certification. Before efforts were made to centralize certificating authority, individual schools or school districts had set their own rules and regulations dictating who could become a teacher and what kind of qualifications they needed for the position. The predominant way of obtaining a teaching certificate in the late 19th century was by examination. Examples of the type and nature of the questions asked of teachers regarding natural philosophy in the New York state teacher examinations held in 1878 include: "Illustrate the relation between the force of gravitation and the quantity of matter exerting it. State the Law..." and "How is light thought to be produced?" [15].

Examination questions were asked of all teacher candidates and were not aimed specifically at physics teachers. With increasing state control of licensing, there was a gradual increase in the physics content requirement for science teachers. Much like today, however, the regulations and requirements varied widely from state to state [16].

The organization of high school physics in the mid-to-late 19th century and the subsequent impact on how it was taught has been well documented, and many aspects of present-day physics teaching can be traced to this time period. A nationwide survey of 175 public high schools and 433 private secondary schools conducted in 1878 by the U.S. Bureau of Education (a precursor of today's Department of Education) found that "instruction in chemistry and physics is very generally given. Only a few cities report no teaching in these branches; while many state that natural philosophy, at least, is orally taught in grades lower than high school" [17]. Clarke reported that Steele's textbook, *Fourteen Weeks in Natural Philosophy,* was used in one quarter of the schools, and noted that the "short informational course" (i.e., a course lasting less than a school year) was well established at this time. Steele's textbook included topics such as properties of matter, motion, gravitation, machinery, "pressure of liquids and gasses," acoustics, optics, heat, and electricity [18].

In a follow-up national Bureau of Education study in 1884, results showed that "almost invariably physics has been one of those chosen for high school work" [19]. The study indicated that physics was commonly studied in the third year of high school, simultaneously with the study of geometry. Based on his findings, the author, Wead, recommended an inductive approach for teaching physics. He noted that many teachers had probably not been taught using inductive methods and were unfamiliar with how to teach in this manner. He called for major reforms to physics education and provided helpful suggestions, including a list of experiments to be performed.

Further insights into how high school physics was taught can be gleaned from the practices adopted in New York State. In the 1870s and 1880s, New York was the most populous state in the country and accounted for over one ninth of all U.S. students [20]. The widely known high school Regents examinations had been established in 1878, with the first science examination being given in natural philosophy, before taking the name of physics in 1879. The first physics syllabus was written in 1880 and was revised approximately every five years until 1905. The syllabus evolved from "encouraging" to "recommending" to "mandating" laboratory work as part of the course [21].

A. The impact of Harvard and importance of laboratory work

Many of the policies of the late 1800s that appear in the Clarke, Wead, and Regents annual reports stem directly from educational initiatives enacted at Harvard. In 1872, Harvard took the unprecedented step of accepting physics as an admission subject. By 1876 physics had become an entrance requirement, and in 1886 a laboratory-based physics course option for admission was allowed.

Edwin Hall, then a physics instructor at Harvard, was assigned the task of writing the laboratory requirement for high schools. Importantly, Hall had taught high school for two years before completing his Ph.D. in physics. He wrote, "I turned to science after two years of school teaching, because it was progressive and satisfied my standards of intellectual and moral integrity..." [22]. Hall wrote detailed procedural instructions for 40 physics experiments for use in high schools. The instructions became known as the "Harvard Descriptive List" [23]. The experiments were quantitative in nature, and focused on making precise measurements. The acceptance of laboratory work by Harvard was critical in promoting the widespread adoption of such work, and helped germinate the idea that laboratory work conducted by individual students was the preferred way to learn science. Importantly, it implied that physics teachers needed to have specialized knowledge to teach physics using laboratory methods [24]. Teachers who taught physics would have to be replaced by *physics teachers.*

The push for laboratory work combined with continued growth of the U.S. high school population in the 1890s led to newly built schools that included physics laboratories in their construction. Physics teachers were expected to have detailed knowledge of how to set up and conduct experiments, cementing the transformation of physics into a more demanding, laboratory-based, college-preparatory course. Recognizing the need to assist teachers to prepare for the Harvard laboratory requirement, Hall co-wrote a physics textbook especially for teachers [25]. He also was integral, in the 1890s, in establishing summer courses at Harvard in laboratory-based physics for school teachers [23]. These examples illustrate how, despite the general nature of teacher education in the second half of the 1800s, teacher education programs did provide specialized attention to physics instruction with laboratory work.

B. Colleges and universities enter teacher preparation

From the 1890s on, teacher training began to come under the control of colleges and universities. Columbia University's Teachers College (TC) in New York provides a well-documented early example of science teacher preparation that illustrates how the changes in the structure of high school physics were incorporated into teacher education. Even though TC is something of a special case, it is interesting to examine what was done there, in terms of practices of the time. From its formation in 1888, TC had a professor in charge of physical science: John Francis Woodhull. Woodhull had himself been a physics and chemistry teacher and principal from 1881 to 1885, before becoming a professor.

Woodhull had written textbooks on elementary physics; he was the major author of the syllabus for the New York State Regents physics course in 1891; and in 1899 he would earn what was probably the first ever Ph.D. in physics education, with a dissertation appropriately titled *The Teaching of Physics in Secondary Schools* [26]. In contrast to the practice of introducing and teaching physics to all prospective teachers, Woodhull designed and implemented a series of courses that were geared toward graduating more specialized science teachers and included courses that were specifically to prepare physics teachers. Candidates were warned that "being well read in [physics] was not sufficient preparation for the work...All candidates for admission will be given tests in the laboratory as well as in textbooks" [27].

The program includes some of the first courses designed specifically for the teaching of physics, as opposed to learning physics content. The courses included Physics for Secondary Schools, Physics for Elementary Schools, The Use of Tools for Making Homemade Apparatus, Homemade Apparatus for Teaching Chemistry, Physics and Physiology, and Laboratory Practice in Physics.

The progressive methods Woodhull introduced for science and physics teacher education are illustrated in Figure 2.

JUNIOR YEAR

	Monday	Tuesday	Wednesday	Thursday	Friday
9.00–9.10			CHAPEL SERVICE		
I { 9.15 9.55	Physics for Secondary Schools	Physics for Secondary Schools 2d half year	Physics for Secondary Schools	Physics for Secondary Schools 2d half year	Manual Arts — use of tools
II { 10.00 10.40	Zoölogy and Physiology for Secondary Schools		Zoölogy and Physiology for Secondary Schools		
III { 10.45 11.25	Psychology and General Method	History of Education	Psychology and General Method	History of Education	Psychology and General Method
IV { 11.30 12.15		Physical Training		Physical Training	
V { 12.25 1.10	Physics for Elementary Schools 1st half year	Physics for Elementary Schools 1st half year	Physics for Elementary Schools 1st half year	Physics for Elementary Schools 1st half year	Physics for Elementary Schools 1st half year
VI { 1.10 2.10					
VII { 2.10 2.55	Science Lecture		Laboratory Practice in Physics		
VIII { 3.00 3.45	Conference 1st hf.yr. Laborat'y Prac. in Zoöl. and Physiology 2d half year				

FIG. 2. 1895 junior year curriculum for prospective physical science teachers at Teachers College.

C. National committees influence physics instruction

As a result of two national committees meeting in the 1890s, proposals for how high schools could be restructured to make them more coherent were outlined. Both committees would promote the incorporation of science, especially physics, into the high school course of studies [28,29]. The committees' recommendations would cause considerable debate across the country and would have important ramifications for physics teacher education. The Committee of Ten met in 1892–1893 and was composed of leaders in education, most of whom were presidents of leading universities. Its physics, chemistry, and astronomy subcommittee made extensive recommendations about high school physics, including the proposal that it contain extensive laboratory work (50% of course time) and that it become a college admission requirement for all students.

The Committee on College Entrance Requirements (CCER) met from 1895–1899 and was charged with implementing the recommendations of the Committee of Ten. Its physics subcommittee was chaired by Edwin Hall, and the subcommittee, like that of the Committee of Ten, promoted laboratory work and the primacy of physics among the high school sciences. The recommendations of the CCER physics subcommittee would become the syllabus for the College Board physics exam. The first College Board physics exam, released in 1901, required students to hand in their laboratory books, which accounted for a substantial percentage of students' scores on the examination.

The content of the physics course was controversial. As chair, Hall had been unable to broker any agreement from the members of the physics subcommittee, and consequently he had submitted the Harvard Descriptive List as the high school physics course outline [2]. Not surprisingly, there was considerable debate among physicists and educators about virtually every aspect of the course, and the ensuing discourse was documented in journal articles and proceedings of national professional conferences. The new approaches to teaching physics soon came under fire for being overly technical and lacking in excitement and interest [29].

Prominent leaders in physics education called for reform. Charles Riborg Mann of the University of Chicago and John Woodhull of Columbia University were the most prominent of the critics of physics teaching during the first decade of the 1900s. Woodhull argued, "I cannot agree with those who would restrict physics to the select few who are mathematically inclined... Physics appears to me to be a subject that all pupils need" [30]. Woodhull was clearly an early advocate of physics for all. Despite the efforts of Woodhull and Mann, and of many physics educators afterward, the criticism that physics courses were too far removed from the interests of students and too technical to be accessed by all persisted.

IV. PHYSICS TEACHER TRAINING IN THE EARLY 1900s

In 1906, the article "A New Movement Among Physics Teachers" was published. It called for teachers of physics across the nation to discuss the problems with introductory physics courses [31]. The authors' aim was to make physics "more interesting, stimulating, and inspiring to the students and more useful." Responses from 275 teachers from 38 states were testament to the national interest in developing a more acceptable syllabus for physics. To meet the demand, a commission was formed, with Mann as the chairman. This commission collected and organized the responses to questions put forth in the circulars. The aim of the commission was to: "First, find out just what is wanted by the teachers as a whole for the improvement of physics teaching; and second, having found this out, to attempt to secure it for them" [32].

By 1908, the commission had devised a syllabus, which most of today's teachers would find strikingly similar to what is taught now (minus modern physics), and which defined what a "unit" in physics should be. This unit not only counted for student credit, but also was a way to measure a teacher's workload. The course was to be adaptable to local needs, yet maintain the level of uniformity needed to meet college entrance requirements. It required the study of a physics text and the completion of laboratory work. Additionally, the syllabus specified the amount of mathematics and the degree of precision in high school physics problem solving and laboratory work that the course should have.

The first 20 years of the 1900s was a time of reform; physics instruction became less focused on rote memorization and formal experimentation and more grounded in student interest and practical application of lab experiments. This was in line with the shift in thinking of the time toward having subject matter organized from a psychological rather than a disciplinary perspective. John Dewey notably contributed to the debate in a symposium entitled "The Purpose and Organization of Physics Teaching in Secondary Schools" [33]. Evidence of the shift toward more practical, hands-on applications can be seen through the purchases of science equipment required for the mandatory lab hours and also in the content of the textbooks of the time, especially those written by Millikan [34]. In 1918, the Commission on the Reorganization of Secondary Education published its report outlining how commission members thought secondary education should be reorganized. There were calls for increased relevance of science to students' lives and the importance of education's role as a vehicle for preparing effective citizens.

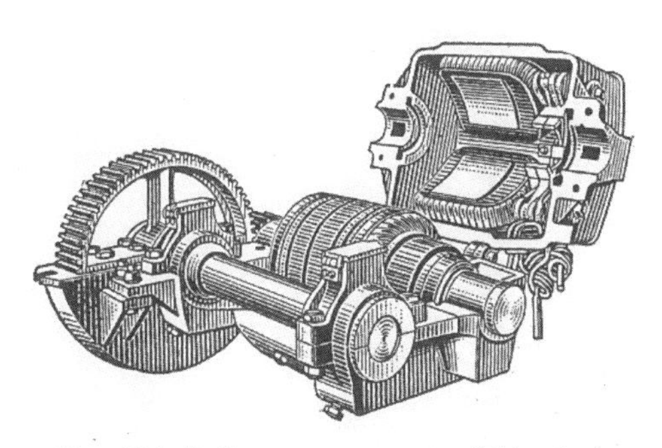

Fig. 345. Railway motor, upper field raised

FIG. 3. Technical drawings, such as this one from Gale and Millikan's textbook, were to add interest to the subject material for physics students.

During the same time period, discussion continued about whether teachers should have "professional" training or a more general education. Proposals were made for a generalized science program followed by specialized science courses. Although many teachers had general training in one-year normal schools, specialization in universities became increasingly popular. For example, examining the course offerings for 1910–1911 at TC, physics teacher candidates could choose a summer course taught by Mann that was "a review of the history of physics teaching, leading to a definition of present practices"; "practical work"; "laboratory work"; or "History of Physical Science." The practicum was described as "devoted to the investigation of problems connected with the teaching of physical science in the secondary and normal schools" [35].

Despite the attention provided to improving physics instruction, by the 1920s, the percentage of students enrolled in high school physics was not increasing [36]. The expansion of elective courses under the new push for more practical course offerings, the advent of a general science course, and the consolidation of the various biological courses into a one-year general biology course was making physics a less-elected elective.

V. LOW REQUIREMENTS FOR PHYSICS TEACHERS FROM THE 1920s TO THE 1940s

With relative high school physics enrollments declining, and with the majority of physics teachers possessing only normal school physics training, the future for physics education looked bleak. In a 1925 survey, Hughes investigated the experiences of teachers of physics in high schools with more than 550 students. He found:

> The "average" teacher of physics...has received training in college for his work as teacher of physics by taking 2.4 years of mathematics, 2 years of chemistry, 1.67 years of physics, and 1.5 years of education. The absence of such courses as special methods and practice-teaching is noticeable in the training received by most of the teachers [37].

Course catalogs from TC in the 1930s reveal a far more general approach to physics teaching than in previous years, and reflected the changing times, as well as, perhaps, Woodhull's absence; he had retired in 1922 [38]. Overall, standards of teacher training in the 1930s lacked rigor. There was widespread use of blanket or broad certification for science teachers. Some states (e.g., Maryland) required at least a minor in physics for a teacher to be licensed to teach physics, whereas others (e.g., Arizona and Colorado) maintained just general education requirements [39]. In any case, the majority of U.S. students were taught physics by teachers who were not physics specialists. Studies in California, Texas, and Minnesota showed that a significant proportion of high school science teachers had no college training in the subjects they taught and that half of the science teacher candidates had elected non-science majors in college [40–42]. Despite this obvious lack of preparation, many teachers actually met the low state requirements, which rarely exceeded 12 credits. Based on a survey of research between 1928 and 1943, it was recommended that teacher preparation colleges should

> upgrade the selection of teachers based on scholarship and personality as fast as possible, give them an improved broad, functional and somewhat professionalized general education, specialize them for teaching in broad fields rather than by subjects; increase the amount of well-supervised practice teaching or add a year of supervised internship; and lengthen the period of training [43].

The formation of national organizations for physics teachers, such as the American Association of Physics Teachers, formed in 1930, improved communication in the field, but did not result in a major overhaul of teacher education.

By the end of the 1930s, different certificates for elementary and high school teaching were issued by most states, based on college credentials. Eventually, high school subject-specialty certificates began to emerge. By 1937, the number of states issuing teacher certifications increased to 41 (of the then 48) states. By the end of World War II, 38 states required some form of subject certification, with two states requiring that the subjects be listed on the face of the certificate [44]. The increase of state certification was an improvement, but the prevalent standard of teacher education for that certification was a still a low requirement of coursework hours:

> States started mandating specific numbers of semester hours that teachers needed to obtain in college and university to teach specific subjects. While these mandated hours were generally low (some states only mandating 3 semester hours, or one class, to obtain a certificate to teach a subject) as time went by the number of hours mandated for certification slowly increased [45].

With "half the normal schools and teacher colleges still report[ing] two year curriculums," in today's terms, the equivalent of an associate's degree rather than a bachelor's degree was required to teach physics [43].

In 1945, 11 states still had blanket subject certification. Among the other states, there was a wide variation in requirements. Of the 22 states that mandated science work for science certification (few places had sufficient enrollments to allow for physics-only teaching, but it did exist; see the New York and Indiana example below), the lowest requirements were in Pennsylvania, with 12 semester hours in science (three semester hours each in physics, chemistry, zoology, and botany), whereas the highest standards were held in both New York and Indiana, with 36 semester hours of science. By mid-century, requirements for certification had changed greatly, and as a result, teacher education programs saw an increase in the number of semester hours prescribed for secondary teachers of science. However, as is true today, with varied circumstances within states and the prevalence of special circumstances and emergency certifications, the data on certification requirements alone do not reveal all the realities of teaching at that time. Physics teacher preparation nationally remained for the most part somewhat general, and in most cases lacked strong physics content requirements. Although the number of required credit hours for high school science certification increased, physics teachers practicing in the field were largely not physics specialists.

VI. CONCLUSION

For more than 200 years, physics has been an important part of public education in the United States. During this time, the course transformed from natural philosophy to physics; from conceptual, rote instruction to mathematical and laboratory-based instruction; and from a course in which a large percentage of students enrolled to one in which a low percentage of total students enrolled. To support these dramatic shifts in the course, multiple changes were made to the methods of physics teacher preparation. As laboratory work became valued for student learning, teachers were exposed to it during their training. As the course became more technical and math-based, and added student laboratory work, specialized training for physics teachers evolved.

Scientists and educators played key roles in the development of the physics course and of physics teacher education. Hall's creation of the Harvard Descriptive List of experiments was a pivotal moment in this history. Millikan used his understanding of physics to, during his time as an educator, create a student-centered textbook, bringing to life many of the wonders of the American Industrial Revolution through detailed drawings and explanations. Woodhull was also ahead of his time, utilizing innovative approaches for student-centered teaching that was grounded in real-world applications. Mann led the charge for reform efforts in the early 20th century and helped create a community of physics educators interested in reforming high school physics.

The work of this chapter illuminates these little-known aspects of the history of physics teacher preparation. Respected voices in both physics and education during the last two centuries have supported physics education reform. Leaders in physics education, such as Hall, Mann, Millikan, and Woodhull, all used their experiences with physics teaching to light the way. Through these men, there is a rich history connecting physicists and physics education. The effort continues today through professional organizations and publications such as this volume, to promote conversation, sharing, and reform. We hope that understanding the past will help illuminate our path for the future.

[1] D. E. Meltzer, M. Plisch, and S. Vokos, eds., *Transforming the Preparation of Physics Teachers: A Call to Action. A Report by the Task Force on Teacher Education in Physics* (T-TEP) (American Physical Society, College Park, MD, 2012). http://www.ptec.org/webdocs/2013TTEP.pdf

[2] K. Sheppard and A. M. Gunning, The other Hall Effect: College Board physics, Phys. Teach. **51** (6), 364 (2013). doi: 10.1119/1.4818378

[3] C. F. Kaestle, Common schools before the 'common school revival': New York schooling in the 1790's, Hist. Educ. Quart. **12** (4), 465 (1972). http://www.jstor.org/stable/367341

[4] D. A. Ward, *The History of Physics Instruction in the Secondary Schools of the United States*, M.A. thesis, University of Chicago (1911).

[5] E. Smith, Early history of the high school, in *Annual Report of the School Committee Prepared by the Superintendent of Schools* (City of Cambridge, Cambridge, MA, 1892), pp. 49-71. http://goo.gl/MxzdPC

[6] D. L. Angus, *Professionalism and the Public Good* (Thomas B. Fordham Foundation, Washington, DC, 2001). http://edex.s3-us-west-2.amazonaws.com/publication/pdfs/angus_7.pdf

[7] University of the State of New York, *Instructions from the Regents of the University to the Several Academies Subject to Their Report* (C. Van Benthuysen & Co, Albany, NY, 1845). http://hdl.handle.net/2027/loc.ark:/13960/t9p27mz95

[8] J. D. Steele, History of science teaching, in *Joel Dorman Steele: Teacher and Author,* edited by A. C. Palmer (A.S. Barnes, Cambridge, MA, 1900), pp. 190-215. https://archive.org/details/joeldoransteele00palm

[9] D. L Angus, J. E. Mirel, and M. A. Vinovskis, Historical development in schooling, Teach. Coll. Press **90**, 211 (1988).

[10] C. R. Mann, *The Teaching of Physics for Purposes of General Education* (MacMillan, Chicago, IL, 1912). https://archive.org/details/teachingphysics01manngoog

[11] Massachusetts Board of Education, *The Fourteenth Annual Report of the Board of Education* (Dutton and Wentworth, Boston, 1851), p. 59.

[12] J. E. Stout, *The Development of High-School Curricula in the North Central States from 1860 to 1918* (University of Chicago, Chicago, IL, 1921). https://archive.org/details/developmentofhig00stouiala

[13] R. A. Millikan, *The Autobiography of Robert A. Millikan* (Prentice-Hall, New York, NY, 1950).

[14] M. Cochran-Smith, S. Feiman-Nemser, D. J. McIntyre, and K. E. Demers, eds., *Handbook of Research on Teacher Education: Enduring Questions in Changing Contexts* (Routledge, New York, NY, 2008).

[15] New York Department of Public Instruction, Natural philosophy questions (1878), in *The New York Examination Questions* (Bardeen, Syracuse, NY, 1887).

[16] J. G. Hill and K. J. Gruber, *Education and Certification Qualifications of Departmentalized Public High School-Level Teachers of Core Subjects: Evidence from the 2007-08 Schools and Staffing Survey* (National Center For Education Statistics, Washington, DC, 2011). http://nces.ed.gov/pubs2011/2011317.pdf

[17] F. W. Clarke, *A Report on the Teaching of Chemistry and Physics in the United States* (Government Printing Office, Washington, DC, 1881).

[18] J. D. Steele, *Fourteen Weeks in Natural Philosophy* (A. S. Barnes, New York, NY, 1872). https://archive.org/details/fourteenweek sinn00steerich

[19] C. K. Wead, *Aims and Methods of the Teaching of Physics* (Government Printing Office, Washington, DC, 1884). http://bit.ly/1ylgyHt

[20] National Center for Education Statistics, *120 Years of American Education: A Statistical Portrait* (US Department of Education, Washington, DC, 1993). http://nces.ed.gov/pubs93/93442.pdf

[21] University of the State of New York, *Annual Report[s] of the Regents of the University of the State of New York* (University of the State of New York, Albany, NY, [1878-1906]).

[22] P. W. Bridgman, Biographical memoir of Edwin Herbert Hall, 1855-1938, Natl. Acad. Sci. Biographical Mem. **21,** 73 (1939).

[23] A. E. Moyer, Edwin Hall and the emergence of the laboratory in teaching physics, Phys. Teach. **14** (2), 96 (1976). doi:10.1119/1.2339318

[24] C. R. Mann, Book review of "The teaching of physics in the secondary school by Edwin H. Hall," School Rev. **11,** 157 (1903). https://archive.org/details/jstor-1075480

[25] E. H. Hall and J. Y. Bergen, *A Textbook of Physics: Largely Experimental* (Longman, Green & Co, New York, NY, 1891). https://archive.org/details/atextbookphysic06berggoog

[26] J. F. Woodhull, *The Teaching of Physics in Secondary Schools*, Ph.D. thesis, Columbia University (1899).

[27] Teachers College, Columbia University, *Circular of Information 1895-6* (Teachers College, Columbia University, New York, NY, 1896).

[28] K. Sheppard and D. M. Robbins, Lessons from the Committee of Ten, Phys. Teach. **40** (7), 426 (2002). doi:10.1119/1.1517887

[29] K. Sheppard and D. M. Robbins, The "First Physics First" movement, 1880-1920, Phys. Teach. **47,** 46 (2009). doi:10.1119/1.3049881

[30] J. F. Woodhull, *The Teaching of Science* (The Macmillan Company, New York, NY, 1918). https://archive.org/details/teaching science00woodgoog

[31] C. R. Mann, C. H. Smith, and C. F. Adams, A new movement among physics teachers, School Rev. **14** (3), 212 (1906). http://www.jstor.org/stable/1075618

[32] C. R. Mann, The new movement among physics teachers: Circular IV, School Sci. Math. **6** (9), 787 (1906). doi:10.1111/j.1949-8594.1906.tb02962.x

[33] J. Dewey, Symposium on the purpose and organization of physics teaching in secondary schools, School Sci. Math. **9** (3), 291 (1909). doi:10.1111/j.1949-8594.1909.tb03035.x

[34] R. A. Millikan and H. G. Gale, *A First Course in Physics* (Ginn & Company, Boston, MA, 1906). https://archive.org/details/afirstcourseinp01galegoog

[35] Teachers College, Columbia University, *Teachers College Announcement, 1910-11* (Teachers College, Columbia University, New York, NY, 1911). http://babel.hathitrust.org/cgi/pt?id=coo.31924070502004;view=1up;seq=3

[36] K. Sheppard and D. M. Robbins, Physics was once first and was once for all, Phys. Teach. **41** (7), 420 (2003). doi:10.1119/1.1616483

[37] J. M. Hughes, The history of physics instruction in the secondary schools of the United States, School Rev. **33,** 296 (1925).

[38] Teachers College, Columbia University, *Teachers College Announcements, 1930-40* (Teachers College, Columbia University, New York, NY, 1940). http://catalog.hathitrust.org/Record/000059910

[39] C. M. Pruitt, Academic requirements necessary to teach science, Sci. Educ. **17,** 48 (1933). doi:10.1002/sce.3730170108

[40] E. S. Boone and A. P. Jameson, The training of science teachers serving in California high schools, Calif. Quart. Sec. Educ. **9,** 350 (1934).

[41] C. M. Hicks, *The preparation of Texas science teachers for the scholastic year 1934-1935*, M.A. thesis, University of Texas (1935).

[42] H. R. Douglass and R. B. Stroud, The education and teaching load of science teachers in Minnesota high schools, Educ. Admin. Super. **22,** 419 (1936).

[43] W. E. Peik, Chapter IV: The preservice preparation of teachers, Rev. Educ. Res. **13** (3), 228 (1943). doi:10.3102/00346543013003228

[44] R. C. Woellner and M. A. Wood, *Requirements of Certification of Teachers and Administrators for Elementary Schools, Secondary Schools and Junior Colleges* (University of Chicago Press, Chicago, IL, 1945).

[45] American Association of Physics Teachers, Responsibilities of science departments in the preparation of teachers, A report of the committee on the teaching of physics in secondary schools, Am. J. Phys. **14** (2), 114 (1946). doi:10.1119/1.1990792

Case Studies of Successful Physics Teacher Education Programs

Seattle Pacific University: Nurturing preservice physics teachers at a small liberal arts school

Eleanor W. Close
Department of Physics, Texas State University, San Marcos, TX 78666

Lane Seeley, Amy D. Robertson, and Lezlie S. DeWater
Department of Physics, Seattle Pacific University, Seattle, WA 98119

Hunter G. Close
Department of Physics, Texas State University, San Marcos, TX 78666

The Department of Physics at Seattle Pacific University successfully recruits and prepares a large number of physics teachers for an institution of its size. Our goal in this chapter is to identify the main components that contribute to SPU's success in recruiting, preparing, and supporting new physics teachers. We have organized these components according to three themes: programmatic and structural supports, intellectual resources, and faculty and student dispositional commitments.

I. INTRODUCTION

The Department of Physics at Seattle Pacific University (SPU) successfully recruits and prepares a large number of physics teachers for an institution of its size. Most of the physics teachers who interact with the SPU physics department as undergraduates and are certified through SPU have a minor in physics and a major in another science, technology, engineering, and mathematics (STEM) field (rather than a physics major), and receive teacher certification through SPU after receiving their bachelor's degrees. Figure 1 shows physics certifications awarded by SPU to students who completed their undergraduate educations at SPU. Some of these students completed their certifications as undergraduates; the majority finished their bachelor's degrees at SPU without certification and then enrolled in one of SPU's master's degree programs offering certification. While these numbers do not capture students who choose to complete certification programs at other institutions, these data do represent students who take the most direct pathway from undergraduate physics classrooms at SPU to teaching K-12 physics students. SPU also grants physics teacher certification through our master's degree programs to students who completed their undergraduate degrees at other institutions; such students number between two and three per year over the past five years. Because our focus in this chapter is on students in our undergraduate program, we will not discuss students who completed their undergraduate degrees elsewhere.

As shown in Figure 1, the number of undergraduate SPU students choosing to become physics teachers has increased substantially over the past decade. Over this same time period, the physics department has transformed itself in several ways, including hiring new faculty with specialties in physics education research (PER), changing the curriculum to be more interactive, promoting a team attitude toward teaching in the department, and engaging in large research and teacher professional development projects. Our goal in this chapter is to identify the main components that we believe contribute to SPU's success in recruiting, preparing, and mentoring preservice physics teachers.

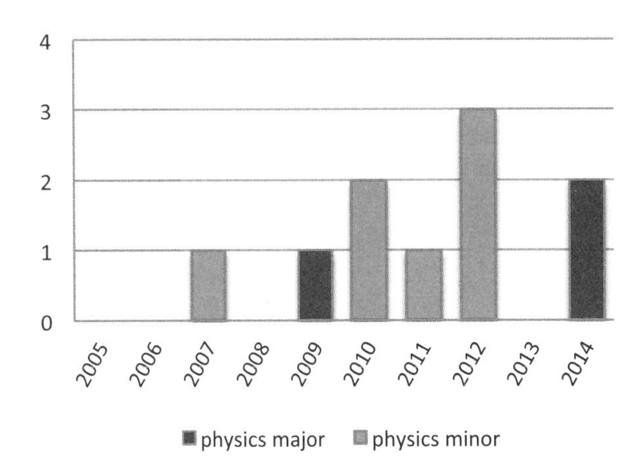

FIG. 1. Physics teacher certifications granted by SPU to students who completed their undergraduate degrees at SPU. (Certifications through both undergraduate and graduate programs.)

One important overarching theme is that we do not consider recruitment, preparation, and mentoring support to be a linear sequence through which to move students; rather, our model has been to create an inclusive community around the practice of physics education. The community has a low barrier for entry: Membership is defined by engagement in physics education; members of the community participate at whatever level is appropriate to their abilities and interests;

and members participate alongside other members of different levels of experience. When we refer to recruiting and preparing teachers, it is within this inclusive context.

We have organized these components according to three themes: programmatic and structural supports, intellectual resources, and faculty and student dispositional commitments. These categories are not orthogonal—for example, our choices in developing the intellectual resources of the department have been informed by our dispositional commitments—but we believe they are useful descriptors. In the sections below we will summarize the main features of what we believe are key program elements and how they support teacher recruitment and preparation, and we will provide a description of how those features are specifically instantiated at SPU.

II. STRUCTURES

In this section we describe key programmatic and structural features of the physics department, teacher preparation pathways, and physics teacher recruitment and preparation efforts at SPU. We describe structures on multiple scales, from university-wide structures to particular course and program structures within the physics department and teacher preparation programs. In each section, we will describe both the structures themselves and the impact these structures have on recruiting and retention of preservice physics teachers.

A. Programmatic change facilitated by small size of institution and department

SPU is a small, private, liberal arts university with approximately 4,000 students, most of whom are undergraduates. SPU's small size and administrative structure make it possible to implement programmatic changes swiftly and without much administrative burden. This has allowed the physics faculty to rapidly reshape the department's degree plans and course offerings in order to create and optimize pathways to teacher certification (including undergraduate, post-baccalaureate, and graduate programs), and to be responsive to the needs of students as they navigate these pathways. The small scale of the physics department, which currently has four tenure-track faculty positions and three non-tenure-track faculty, also facilitates consensus and coherence of mission.

Between 2003 and 2011, the Department of Physics introduced a number of new courses and developed two new majors. Four new upper-division courses were introduced as part of the Learning Assistant (LA) program, described in a later section; these were created as variable-credit, repeatable courses, allowing for ongoing flexible use as the Learning Assistant program evolved. In addition, two lab-based, inquiry-focused courses were created for the elementary teacher preparation program.

Prior to 2007, the department had only one degree path, leading to a B.S. in physics. In 2007, the department added two new B.A. degrees: the B.A. in physics, and the B.A. in physics with an education focus. The B.A. in physics with an education focus was designed both to incorporate the newly created LA pedagogy courses, and to allow students to simultaneously complete the undergraduate secondary certification program through the SPU School of Education. At the same time, the physics minor was modified to incorporate the new courses developed for the Learning Assistant program, making it simpler for students with experience in that program to complete the minor.

In summary, the administrative structures at SPU allow a level of programmatic responsiveness that might not be possible at larger institutions, and the Department of Physics has made use of this to reshape departmental degree and course offerings in order to encourage and support students in developing and pursuing an interest in teaching. At the same time, and similarly facilitated by the small scale of the department, the physics faculty have made significant changes to the instructional materials and methods used in the department's courses, beginning with the introductory sequences and expanding over time to affect all physics course offerings. This curricular renovation and the structures that facilitated it are described in the next section.

B. Curricular renovation facilitated by internal and external support

In 2002, the Department of Physics at SPU was composed of three tenure-track faculty, the most senior of whom had been at SPU for four years. Over the next few years, the department made major changes to the classroom setting, curricular materials, and instructional methods in most of its course offerings. In addition to improving the learning of all students in our courses, these changes created a need for undergraduate instructional assistants, leading to the creation of the Learning Assistant program (described in detail in Section I.D), which has been our primary source of physics teacher candidates. Two major structural supports facilitated this transformation: a renovation of the building in which the Department of Physics is housed, allowing complete redesign of departmental instructional spaces; and a curriculum implementation grant from the National Science Foundation (NSF), providing funds for instructional materials (including equipment) and faculty development.

The building renovation was funded by the SPU Science Initiative, a major fundraising campaign that entered its public phase in 2002. The three goals of this initiative were "to educate all SPU students to become scientifically literate citizens; to educate students for careers in science, medicine, and engineering; and to educate students to become influential science teachers" [1]. As part of the initiative, SPU constructed a new science building to house the departments of biology and chemistry, and extensively renovated the building housing the departments of physics, mathematics, computer science, and engineering. The physics department took advantage of this opportunity to create classrooms designed for integrated laboratory-lecture instruction, with students seated in groups of four to six around tables and with easy availability of computer-interfacing data collection devices. The department also strategically designed the spaces for no more than 32 students, effectively limiting class size and providing structural support for maintaining interactive and student-centered instructional methods.

In 2003, while the new physics instructional spaces were under construction, the department was awarded an NSF Course, Curriculum, and Laboratory Improvement grant, "Adapting and Implementing Research-Based Curricula in Introductory Physics Courses at Seattle Pacific University." This grant united the department in a complete restructuring of all introductory physics courses at SPU, both calculus- and algebra-based, in order to integrate elements from exemplary research-based curricula [2–4]. Since then, the average learning gain on standardized conceptual learning assessments (e.g., the Force Concept Inventory [5], Force and Motion Conceptual Evaluation [6], and Conceptual Survey on Electricity and Magnetism [7]) in SPU physics courses has ranged between 50% and 80% [8,9].

In addition to improving gains in conceptual understanding, our curricular renovation has dramatically impacted the learning environment in our introductory classes. Students are now expected to take active roles in every aspect of the learning process. A majority of class time is devoted to small-group activities in which the students work closely with peers and instructors to construct and test models and wrestle with new ideas. In this context, students practice articulating scientific ideas and listening critically to the ideas of their peers. This collaborative approach to learning now characterizes all physics courses at SPU, from introductory courses for non-science majors to upper-division courses for majors.

Curricular renovation has supported teacher recruitment and retention in two significant ways. First, the curricular reforms have included the implementation of an LA program, providing an opportunity for students to gain early experience with low-risk, highly supported teaching. Second, perhaps as a result of improved instructional experiences, the number of physics majors and minors at SPU has more than doubled, from an average of 1.8 majors (and 0.4 minors) graduating per year during the decade of 1993–2002 to an average of 5.0 majors (and 1.7 minors) graduating per year during the decade of 2003–2012.

The department's coordinated focus on the teaching and learning of physics led to collaborative efforts to support precollege physics teachers. These efforts are described in the next section.

C. Collaboration with School of Education supports teacher preparation efforts

The physics department's renewed and coordinated focus on the teaching and learning of physics, with an emphasis on making use of the results of physics education research, led to collaborative efforts with the SPU School of Education to support precollege physics teachers. The Department of Physics and the School of Education have worked together on grant proposals, revision of existing teacher preparation programs, and creation of new programs. The collaboration has expanded to include a joint faculty position (initially held by the first author), facilitating communication between the two units. These joint efforts, and the understanding created between physics and education faculty in the process of shaping and carrying out the projects, have greatly enhanced both the recruiting of future physics teachers and the resources and experiences available to them.

1. Building collaborations and recruiting a Resident Master Teacher

The first formal collaboration between the Department of Physics and the School of Education was a successful proposal to the Boeing Company in 2004 to develop a collaborative model of teacher preparation [10], which led to the hiring of a veteran elementary teacher (fourth author on this article) as Resident Master Teacher to work with faculty from the Department of Physics and the School of Education to co-teach and co-design teacher preparation courses. In 2005 the physics department began a five-year, $1.5-million NSF-funded Teacher Professional Continuum program, "Improving the Effectiveness of Teacher Diagnostic Skills and Tools" [11]. This collaborative project, combined with the continuing effort funded by the Boeing Company, had three significant impacts on the department, all of which have provided a context for subsequent physics teacher preparation efforts:

- Meaningful collaborative relationships with the School of Education, local school districts, and local teachers of physics;
- Additional faculty (including the Resident Master Teacher) with expertise in physics education research and extensive precollege classroom teaching experience; and
- Increased recognition on campus as a leader in STEM-focused educational research and innovative teaching.

Together these attributes led to the department's selection in 2006 as a Supported Site by the Physics Teacher Education Coalition (PhysTEC) project [12]. PhysTEC provided support for the SPU LA program (described in section I.D below) as well as two part-time Resident Master Teachers; in addition, PhysTEC membership provided access to scholarships for preservice teachers.

2. Securing scholarship opportunities for future secondary teachers

PhysTEC membership made it possible for SPU to become, in 2009, one of six institutions in the nation participating in the PhysTEC Noyce Scholarship program. Undergraduate students in their junior or senior years at any of these institutions are eligible to apply for PhysTEC Noyce Scholarships of up to $15,000, if the students are physics majors or minors and are pursuing physics teacher certification; students entering post-baccalaureate certification programs at PhysTEC Noyce institutions are also eligible to apply [13]. The revisions to the physics minor at SPU described above, as well as the upper-division physics credits available through participation in the LA program, make it easy for LAs to complete a physics minor and thus be eligible for this scholarship. SPU students have been very successful at winning these awards: Since the start of the program in 2009, over a third of the scholarships awarded (14 out of the total 40) have gone to students completing or having recently completed their undergraduate degrees at SPU [14]. This is particularly noteworthy given that several other PhysTEC Noyce institutions are large state universities with many more students than SPU.

In 2006, the Department of Physics led a successful interdepartmental effort to secure Noyce funding for scholarships for STEM majors seeking teacher certification at SPU at either the undergraduate or graduate levels [15]. Thus, since 2006, physics students at SPU have been able to apply for significant financial support from either the SPU Noyce or the PhysTEC Noyce. In addition, the collaboration with other STEM departments for this Noyce grant application laid the groundwork for the creation of a new STEM teacher certification program, described below.

3. Creating new programs for teacher certification

Between 2007 and 2010, the collaboration between the Department of Physics and the School of Education included the creation of two new programs: a new undergraduate major required of all elementary teacher candidates and a one-year master's degree and certification program for secondary STEM teacher candidates. Creation of the undergraduate major was a multidepartment effort led by a professor of mathematics with a specialty in elementary mathematics education; the joint position shared between physics and education facilitated involvement by the Department of Physics in this process. This led to a significant increase in the number of mathematics and science courses required of elementary teacher candidates. Physics faculty collaborated with the Resident Master Teacher to develop and co-teach the new physics courses included in the new major. These courses were designed to include LAs, and the major was designed to encourage teacher candidates to serve as LAs in these courses.

While physics faculty involvement in development of the new major for elementary teacher candidates did not directly provide structure for recruiting secondary physics teachers, it built relevant relationships and knowledge of institutional structures related to teacher certification, and contributed to the departmental momentum and focus on teacher preparation. The lab-based physics courses for elementary teachers have been continuously evaluated and revised for the past eight years, and the content has been realigned to address the current state standards as well as the Next Generation Science Standards, which are slated for adoption in Washington within the next few years. Designing and co-teaching these courses with the Resident Master Teacher has provided ongoing professional development experiences for physics faculty; these experiences have influenced teaching in courses for physics majors and minors and increased awareness of the professional needs and concerns of practicing K-12 teachers. This awareness was instrumental in the creation of a second new program.

Building on the expertise developed through previous grant and program development projects, and with financial support from a Noyce planning grant [16], the Department of Physics in 2009 took a lead role in creating the new Master in Teaching Mathematics and Science (MTMS) degree for secondary STEM teacher candidates. This one-year cohort-based full-time master's-plus-certification program offered through the School of Education was based on the structure of an existing program, with the addition of new STEM-specific courses. The MTMS program was approved in 2010 and graduated its first cohort of students in 2012.

In summary, the collaboration between the Department of Physics and the School of Education has supported recruitment and preparation of physics teachers in several significant ways: through PhysTEC membership, which provided support for a Resident Master Teacher and for the LA program as well as access to scholarships for future teachers; through grant funding facilitating faculty collaboration both between units on campus and with colleagues in K-12 education; and through the creation of new programs for teacher certification candidates. The LA program has been a particularly important mechanism for recruiting and preparing future physics teachers. This program is described in more detail in the next section.

D. Learning Assistant program provides transformative teaching experiences

One of the most successful PhysTEC-supported initiatives at SPU has been our LA program. This program has allowed us to combine the support of high-quality physics learning experiences on campus with our efforts to recruit and prepare excellent physics teachers. As described above, the department began recruiting undergraduate instructional assistants with the introduction of course reforms in 2003. Initially some pedagogical preparation was integrated into the weekly content preparation sessions. A separate pedagogy course was implemented in 2006, and the program has expanded to include LAs assisting with instruction in upper-division courses, courses for non-majors, and courses for preservice teachers. The number of LAs over time is shown in Figure 2; the current number of physics LAs per year is among the highest at any university in the nation [17]. Recently the physics department has been partnering with the SPU Center for Scholarship and Faculty Development to expand the LA program to a number of additional departments in the sciences, arts, and humanities.

The SPU LA program includes all three essential activities of the University of Colorado Boulder (CU Boulder) model: *content*, which consists of weekly meetings between LAs and faculty to work through curriculum and discuss instructional issues; *practice*, in which LAs facilitate dialogue in small groups over difficult physics concepts, staff physics help sessions, and assist faculty with grading; and *pedagogy*, consisting of participation in a seminar-style physics pedagogy course [18]. Some elements of our implementation of these three activities are unique to SPU and have facilitated recruiting and retention of physics teacher candidates. Here we will focus on these elements.

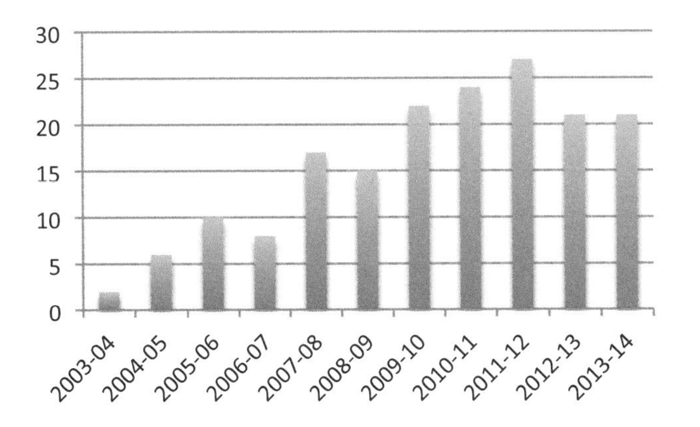

FIG. 2. The number of LAs hired by the physics department, 2003–2014; recent numbers are among the highest at any university in the nation.

Unlike the CU Boulder LA program, which serves many STEM departments, the LA program at SPU is limited to physics. We recruit LAs from all of our introductory course sequences, including the algebra- and calculus-based courses and the courses for future elementary teachers. Because all of these courses are LA-supported and use research-validated curriculum, any student who has completed an introductory sequence at SPU has significant experience with reformed instruction and is familiar with the role of LAs. In most cases, LAs assist in the sequence they took as students.

LAs receive both credit and pay for their work in the program: credit for weekly content preparation sessions and the pedagogy course, and pay for instructional practice. The Department of Physics created four upper-division, variable-credit, repeatable courses for the LA program: three corresponding to the three quarters of the algebra- and calculus-based introductory course sequences, and an additional course number for assisting in other introductory-level physics courses. Because introductory-sequence physics courses at SPU are offered only once each year (e.g., Mechanics is taught only during the fall quarter), most LAs are teaching similar topics concurrently, and the LA pedagogy course each quarter can focus on topics relevant to the LAs' current instructional experiences. Until 2011, LAs registered for either one or two credits each quarter they served: one credit for content preparation sessions with faculty each week, and the second credit for the LA pedagogy course, which only first-time LAs were required to take. Beginning in 2011, all LAs were required to register for the LA pedagogy course every quarter in the program; this decision was made in order to support the development of an LA community with a variety of levels of expertise.

In addition to the university structures that allowed the physics department to create appropriate courses for the LA program, the tuition structure at SPU facilitates student participation. Full-time students at SPU pay the same tuition for between 15 and 18 credits per quarter. Many students choose to register for 15 or 16 credits; these students can register for the LA course credits without paying additional tuition. For students who are already registered for 17 or 18 credits prior to enrolling in the LA courses, the university has created a special tuition exemption so that these students can participate in the LA courses without paying additional tuition. This removes a potential barrier to participation (the expense of paying for course credits), and demonstrates the support of the SPU administration for the department's teacher preparation efforts.

In summary, the SPU LA program provides students with the opportunity to earn hourly pay as well as up to six quarter-credits of upper-division physics each year, while participating in a tightly linked set of experiences enriching their content knowledge, instructional practice, and pedagogy. The program has a low barrier to participation and, like the CU Boulder program, has served both to recruit and to support and prepare future physics teachers. Once students decide to become K-12 teachers, SPU provides them with several options through which to become certified; these teacher certification pathways are described in the next section.

E. Multiple pathways to teacher certification provide flexibility

SPU has multiple pathways to teacher certification, including undergraduate, post-baccalaureate, and graduate programs. This provides flexibility for students to choose the program that best fits their needs; it also allows students to complete physics teacher certification at SPU even if they decide on teaching late in their undergraduate careers.

The undergraduate pathway consists of a secondary teacher certification program through the School of Education, which students take in parallel with courses for their content major. Students who choose this program can graduate in four years with a physics minor or major (via one of the new B.A. degree pathways) and a secondary teaching certificate with a content endorsement in physics. For students who enter SPU intending to become teachers, this is the fastest and least expensive pathway to certification. The majority of undergraduates who decide to become physics teachers do so later in their time at SPU, and so are more likely to choose one of the graduate degree pathways described below. It is also possible for students with a physics major or minor to complete secondary teacher certification coursework through a post-baccalaureate program without obtaining a master's degree; however, the time required for this program (four quarters) is approximately the same as that required for the graduate programs. The latter two pathways offer both certification and a master's degree, and students who seek certification after completing their bachelor's degrees typically choose one of these programs.

Students can choose between two formats of master's programs with teacher certification: a two-year, part-time program with a 10-week teaching internship; or a one-year, full-time, field-based program with a year-long internship. The part-time Master of Arts in Teaching (MAT) program allows students to maintain full-time employment while taking evening courses until their final quarter in the program, during which they complete a 10-week student-teaching internship. In the full-time program, students take intensive summer coursework followed by internship placement in K-12 classrooms every day for the full academic year; they take evening and online courses during the academic year and complete an additional summer of coursework to finish the master's degree. There are two versions of the full-time, cohort-based program: the alternative routes to certification/Master of Arts in Teaching (ARC/MAT), a general program with certification available in many content areas; and, beginning in 2011, the Masters in Teaching Mathematics and Science, which follows the same format as the ARC/MAT and also provides STEM-specific coursework and a cohort of colleagues preparing to teach STEM subjects.

As described in the introduction, many of the students who obtain their physics teaching certifications from SPU do not major in physics. These students often take introductory physics during sophomore or junior year as a requirement for another STEM major, then join the LA program the following year and discover an interest in teaching physics. For these students, the opportunity to complete both teacher certification and a master's degree in one year immediately following their undergraduate degree, at the same institution, creates a low-barrier pathway to following their new-found interest. The availability of Noyce Scholarships to fund teacher certification coursework for up to $15,000 makes the barrier even lower, and well-informed physics faculty ready to encourage applicants and write letters of recommendation further support entry into certification programs.

III. INTELLECTUAL RESOURCES

In this section we describe the intellectual resources of the SPU community, especially within the physics department, that we believe contribute to our success in recruiting and preparing physics teachers. These are intellectual in that they relate to the solving of problems or the answering of questions having to do with the learning and teaching of physics or with the processes of learning to teach physics. They are resources in that these skills, pieces of knowledge, concerns, problems, questions, practices, and social connections facilitate mutual engagement between physics faculty and novice teachers by serving as focuses of shared activity.

A. Extended community connects practitioners and researchers

One of the benefits of having several grant-funded physics education research projects at SPU has been an expansion of the faculty to incorporate a range of expertise. Eight researchers and faculty and one doctoral student currently call the SPU Department of Physics their professional home. In addition, a number of visiting researchers have spent extended periods of time collaborating with colleagues at SPU. During each of the past several summers we have also had approximately 10 visiting scholars spend two weeks at SPU as part of the Interdisciplinary Research Institute in STEM Education. As a consequence, an otherwise small, undergraduate-focused physics department has become a hub of active physics education research and collaboration.

One common thread among all of these teachers and researchers is a deep curiosity about student thinking and a positive regard for student thinking as sensible and potentially productive. Physics faculty enact their curiosity and regard by conducting research projects that emphasize the productivity of learners' natural conversation [19,20], by engaging in a responsive approach to instruction and to the professional development we offer [21,22], and by fostering a similar positive regard for student thinking among LAs and inservice teachers [22–26], as described in more detail in the next section. The strong culture of respect for learner ideas has supported a community in which everyone expects to learn from the contributions of all community members. Within this community, and with respect to understanding student thinking, LAs are treated as colleagues by faculty. We think that this environment is critical so that LAs can experience the teaching and learning of physics as a challenging intellectual endeavor that sometimes includes formal scholarship in collaboration with professional researchers (e.g., [21,23,27]).

B. Emphasis on responsive teaching develops novice teachers' repertoires

The LA Program is the central component of our preservice teacher recruitment and preparation efforts; therefore, key elements of our LA program are also key elements of our teacher recruitment and preparation efforts. This section describes one such effort: the LA pedagogy course emphasis on *responsive teaching*. Responsive teaching focuses attention on productive elements of student reasoning and discourse, as opposed to an exclusive focus on canonically correct results. This creates a low-stakes, high-curiosity environment that encourages development of both skill and confidence in teaching.

Teaching responsively involves attending to the *disciplinary substance* of student thinking—the "seeds of science" (e.g., mechanism and plausibility) that are inherent in many things students say and do [28–31]. When teaching responsively, teachers do not strictly evaluate student thinking for its canonical correctness; instead, they try to make sense of the meaning that the students are making of their physical experiences and to connect these meanings to the discipline. In addition to focusing on the content of ideas, responsive teaching also focuses on students' developing forms of participation, i.e., how they talk, draw, and act when engaging with physics, and seeks to mature those forms [32–34]. In such classrooms, teachers invite students to join in assessing ideas and in making decisions about which ideas to take up and pursue [33,35].

Much of the assigned reading for the SPU LA pedagogy course centers on attending and responding to the substance of student thinking (e.g., [29,36,37]). The discussion in the course often foregrounds the LA-generated questions, "Is it ever okay to leave students with a wrong answer?" and "How can we build on students' ideas in the classroom?" In the process of answering these questions, the LA community considers alternative focuses of assessment of student learning (e.g., mechanism, plausibility, productive sense-making, and argumentation), and reflects on the inherent tensions between different instructional goals [36].

Adopting responsive teaching as a stance toward learning and teaching provides novice teachers with distinctive opportunities for growth [38]. In such classrooms, the focus is much less on the correct answer or on the teacher as arbiter of disciplinary knowledge—a role that often intimidates novice teachers—and much more on listening and responding to the meanings that individual students are making. Thus, LAs see the classroom as a place where they can learn from students and experiment with their own teaching.

Teaching responsively, and searching for "kernels of correctness" or "seeds of science," can (and often does) foster a deep sense of curiosity about student thinking. For example, in a written reflection, one LA describes her search for these "kernels" as fostering a sense of appreciation for student thinking:

> This week, I was challenged to find a kernel of truth in every student response before I responded to the students that I was working with. This was a challenge because I think that I try to value student ideas, but often times, I dismiss them as incorrect...I also learned that when students are answering questions about how they think physics situations work, the students are giving me a small insight into their lives;...they are telling me what they have observed to be true previously, and are making sense of it in a new situation... There was one interaction where this particularly stands out to me. I was working with students on conservation of angular momentum...I asked them about whether the hoop or the disc won the race down the ramp the week before...They responded by telling me that the disc was going faster down the ramp, it was rotating faster, which meant that it had more $KE_{Rotational}$. Although this statement is not true, I decided to think about why they answered like this. I realized that they were saying that since the disc was covering more ground in the same amount of time, this meant that it must have had more rotations per second.

As demonstrated in this example, focusing on responsive teaching is challenging and rewarding for LAs and fosters a sense of teaching as an intellectually stimulating practice.

C. Video projects cultivate attention to student thinking

Since 2008, LAs have been asked to reflect on their implementation of responsive teaching and/or listening practices through pedagogy course assignments built around LAs watching video of themselves interacting with students. From 2008 to 2012, LAs completed the Physics Interview Project (PIP). From 2012 onward, LAs have video recorded themselves interacting with students in the natural flow of classroom activity. Each of these projects is described briefly below.

In the PIP, each LA interviewed a peer (typically a college student who had not taken physics) in order to study that person's thought process about a physical system (e.g., a ball rolling to a stop). The PIP gave LAs experience with the practice of listening for and describing the ideas students use to understand specific physical situations. Before conducting the interview, LAs read PER literature relevant to the interview topic and to the process of conducting a research interview. Based on this brief literature review, students constructed an interview protocol they believed would help them understand their interviewee's mental model of the physical system. The PIP was framed as an opportunity for LAs to practice the skill of *asking questions without teaching*—that is, the goal of the interviewer was to construct an understanding of the existing thinking of the interviewee, not to attempt to change that understanding in any way.

LAs performed and video recorded one or two interviews per academic term; transcribed part of an interview; and then reenacted the interview in class, projected part of the video recording, or analyzed the recording using specific theoretical tools. The LAs wrote a reflection paper that includes a characterization of the interviewee's thinking with supporting evidence from the video record and the LAs' thoughts on the difficulty and value of the experience.

In the more recent LA video project, cameras are placed on two separate tables in classrooms staffed by LAs, and interactions between LAs and students are recorded in the natural course of classroom activity. LAs select one five-to-10-minute clip (or several smaller clips) to transcribe. They reflect on their interactions in terms of their responsiveness to the substance of student thinking. In addition, they often connect their observations of themselves to their pedagogy class conversations, articulating which theories of knowledge they are enacting in the video they transcribe, what "seeds of science"—including specific ideas and practices—they see in the things students are saying and doing, and how they noticed and built on these "seeds" in the moment. The LA video project is framed as an opportunity for LAs to see what their interactions look like from the outside, and to reflect on the extent to which they are enacting their own values as teachers.

Both the PIP and the current video project put LAs—including the future physics teachers in the program—face to face (literally) with phenomena that are subjects of physics education research, and thus provide a particularly compelling entry point into that research. Importantly, LAs are exposed not only to the refined results of PER (as they might be in other LA programs) but also to the raw data. This gives LAs firsthand experience with both the complexity and intellectual value of students' existing thinking, and with the related intellectual challenge (and reward) of teaching with student ideas in mind [39,40].

D. Development of curricular knowledge empowers novice teachers

SPU's introductory physics courses draw largely from worksheets from *Tutorials in Introductory Physics* [2], a research-based and research-validated curriculum designed to address common student difficulties and to support conceptual understanding of physics [41-43]. To support LAs in implementing the *Tutorials,* our LA program explicitly focuses on the development of curricular knowledge [44], by which we mean an understanding of what the curriculum is "trying to do," not just the answers to the worksheet questions.

For example, in the tutorial "Motion in Two Dimensions," students are asked to respond to a sequence of questions prompting them to (1) draw position and displacement vectors for an object moving along a curve; (2) determine the direction of the average velocity of the object for a particular time interval; (3) consider what happens to this direction as the time interval is reduced; (4) use the limiting case to determine the direction of the instantaneous velocity of the object at the beginning of the time interval; and (5) generalize in order to characterize the direction of the instantaneous velocity at any point on the trajectory. As students answer these questions, the curriculum leads them through a logical chain of reasoning on the basis of which they can infer that the instantaneous velocity is tangent to the curve at any point. At the same time, the sequence includes a question addressing the common (incorrect) assumption that as the magnitude of the displacement goes to zero, so does the magnitude of the instantaneous velocity. These "decisions" to lead students toward a particular idea via particular steps, and to use a particular context to address a specific difficulty, are what we mean by "what the curriculum is trying to do" or the "curricular decisions."

The SPU LA program explicitly develops LAs' curricular knowledge in order to promote responsive teaching. The *Tutorials,* and many other research-based curricula, are designed to meet the needs of the generalized student; in the classroom, LAs partner with the *Tutorials* to implement or modify the decisions made by the curriculum on the basis of specific students' needs. In order to practice this kind of adaptive teaching, LAs need to understand the decisions the *Tutorials* make [45,46]. LAs spontaneously echo this in big-picture, video-recorded reflections on the course, saying, for example:

> Like, early in the beginning, when we weren't like quite prepared for the Tutorials,...we didn't know what was in the Tutorial, all these little strategies, like, all I was worried about was like, "do this, then this, then this,"...By the end of the quarter, I felt like we had much more freedom to be like, "oh, I know that the Tutorial did this or the previous one is going... something's going to happen that I'm aware of," so that I can change my teaching style to more address what's being talked about now because I know something else will be fixed later.

Logistically, we currently support the development of curricular knowledge by dedicating one hour each week to understanding the following week's *Tutorials.* LAs meet with course instructors for one hour to go over the content of the relevant *Tutorials,* and, immediately after, meet with the LA Program Coordinator (the third author of this article) to discuss the curriculum's "decisions" [21].

The development of curricular knowledge empowers novice teachers to teach responsively, rather than follow a teaching script. Several LAs (including the one quoted in the previous paragraph) have described this knowledge as freeing them up to really listen to students, whereas they were originally more immediately concerned with the right answer:

> Over the course of this quarter I have started to spend more and more time thinking about the curriculum and less time worrying if I have the right answer to every question a student can ask me…[W]ith the help of the tutorials, I am able to help out a student arrive at the right answer, even get past large obstacles just by paying attention to the questions they are asking and the questions the worksheet is asking, without having to be ready with the right answer before they even ask me a question.

Curricular knowledge is an often-ignored form of knowledge for teaching [44]. Explicit development of this kind of knowledge in teacher preparation programs offers teachers the tools to better understand and partner with curricula throughout their careers.

In addition, the development of curricular knowledge both makes visible *that* the content and structure of the curriculum are designed for specific purposes and *what* those purposes are, supporting teacher buy-in. One LA wrote:

> This process has been important to me because it has given me faith in the tutorials. Because I have this model on instructional strategies and can read the goals the strategy has from the literature I trust the tutorial. I also can see how my job as an LA works in with the tutorial to help the students. This faith is what gives me the freedom to let a student struggle through [an exercise].

This buy-in is important for the teacher's satisfaction, self-efficacy, and teaching practice, and for the classroom experience of his or her students [47].

In our program, the negotiation of a shared language to describe the *Tutorials'* curricular decisions has promoted the development of community among LAs. One LA reflected on the process, saying:

> This process has been important to me because it helps us develop our sense of community. It has taught us how to have adult conversations and how to disagree in a respectful manner. It has given the LA community a purpose to achieve together.

Active participation in a community develops participants' identities as members of that community [48]. Thus, the process of negotiating curricular knowledge promotes LAs' identification as physics teachers. Further supports for LA development of physics teacher identity, and the importance of identity for teacher recruitment and retention, are described in the next section.

E. Support for physics teacher identity encourages teaching career choice

The development of *physics* (or more general *science) identity* and (reform-oriented) *physics (science) teacher* identity is one of the explicit goals of the SPU LA program. We adopt a definition of identity as both seeing oneself and being recognized by others "as a certain 'kind of person' in a given context" [49], as well as participating in learned ways of being and engaging in a *community of practice* (broadly, a community organized around a shared set of goals and practices) [50].

Recent research has found that the physics identities of college students are closely related to their choices of majors in the physical sciences [51,52]; the literature suggests that identity is best developed through participation in a well-functioning community of practice. Active involvement in the work of the community supports novices in developing knowledge and skills [48,50]. In supporting students' development of physics teacher identity, we hope to increase both the number of students choosing K-12 science teaching as a career and their level of competence and confidence in the practices of reform-oriented science teaching.

Our program supports the development of *science* and *science teacher* identity among LAs in a number of ways. As described above, the practice of co-developing curricular content knowledge helps to create a shared sense of purpose, an essential element of a well-functioning community of practice [50]. Another essential element is the possibility for novices to participate in the work of the community (in this case, physics instruction) alongside more experienced and skilled members [48]. Not only do LAs in the SPU LA program benefit from opportunities to observe experienced instructors modeling reform-oriented teaching practices, their participation in the course presents multiple opportunities for mentorship from more expert members of the community (including faculty as well as more experienced LAs) who value the LAs as legitimate novice colleagues in the enterprise of physics teaching and learning; these relationships support LAs in development of science teacher identity [53].

LAs at SPU often attribute transformations in their perceptions of science teaching to their experiences with reform-oriented teaching practices. For example, one LA said, "I came to realize that students learn science much better through inquiry-based settings over lectures and reading, and it has transformed *who I am* as a teacher." (emphasis added)

Identification with science and with science teaching is central to our LAs' learning, empowerment, persistence, and self-efficacy. Intentional and explicit support for LAs' development of physics or science teacher identity is a central component of our efforts to recruit, retain, and support novice teachers.

IV. DISPOSITIONAL COMMITMENTS

In considering what we believe to be central elements supporting the success of physics teacher recruiting and preparation at SPU, we realized we could not describe the existing program without some discussion of the dispositional commitments that shape the community and inform our actions. Dispositionally, we place high value on teaching as a career, we value respect for teaching and teachers of all levels of expertise, and we enact concern for our students' development of teacher identity. The preceding values are held by every member of the physics faculty, albeit with varied expression from one to the next. This level of coherence of values among the faculty about teaching and teachers is unusual; it is partly due to the nature of SPU as an institution, and partly a matter of a coherent vision guiding the hiring of faculty and staff over the past 20 years. In this section, we describe some of the most important dispositional commitments of our community.

A. Institutional focus on service encourages choice of helping professions

SPU is a Christian university and identifies as "evangelical and ecumenical" [54], rooted in the Free Methodist tradition and with current faculty and students from over 50 different Christian denominations [55]. This identity shapes the campus conversation about students' choices of major and career: The university culture places a strong emphasis on vocational calling and service to the broader community. While it is difficult to quantify the impact of this element of institutional culture on students' choice of K-12 teaching as a career, we believe it does have some influence.

In its mission statement, SPU declares its institutional commitment to "engaging the culture and changing the world by graduating people of competence and character, becoming people of wisdom, and modeling grace-filled community" [55]. The career counseling office of the university is called the Center for Career and Calling [56], and the Office of University Ministries offers an annual competitive scholarship for students demonstrating service as a "servant leader" [57], a role that requires "a willingness to take on tasks regardless of recognition or challenge" [58]. The SPU Center for Scholarship and Faculty Development houses the grant-funded Spiritual and Education Resources for Vocational Exploration (SERVE) program, which has as its mission "to support theological thought and action around the concepts of vocation and calling for students, faculty, and staff" [59]; its description of vocational calling includes the statement that "a vocation-driven life strives to discover ways to serve the common good rather than to be driven by status and salary" [54].

We believe that this element of campus culture influences both students and STEM faculty to consider K-12 teaching as a legitimate vocational calling for STEM majors, in spite of the low pay and low prestige of teaching relative to other STEM-related careers such as medicine and scientific research.

B. Supportive, caring LA community fosters acceptance

The intellectual resources described in Section II are embedded in a general ethic of care between the course instructors (including the LA Program Coordinator) and the LAs, and among the LAs themselves [60–62]. This includes not only elements of care that are endemic to responsive teaching, such as seeking to understand another's ideas on *their* terms (i.e., empathy) [63]; it also means caring for the other as a whole person, including attention to affective experiences [64].

Caring and responsive teaching sustain and reinforce one another, since care for a person promotes care for the substance of their ideas, and understanding the substance of another's ideas implies understanding them more fully and thus reinforces caring. Research has demonstrated that caring relationships between teachers and students promote a sense of belonging [60], an affinity for learning [61,62,65,66], and autonomy or freedom in that learning [60]. Thus the dispositional commitment to establishing a caring community among faculty and LAs fosters learning and autonomy in the contexts of both physics content and physics teaching.

We make our caring visible by promoting respectful disagreement within our community and ownership of individual thoughts, feelings, and values, both about oneself as a teacher and about teaching and learning more abstractly. We accomplish this in several ways:

- Giving LAs significant agency over the content and direction of content preparation and pedagogy course discussions. LA questions, ideas, and experiences are taken up as the backbone of the classroom community's shared inquiry into teaching and learning. LAs are often involved in selecting topics or choosing assignments for themselves, and the pedagogy course instructor seeks to be transparent about the connections between course assignments and LAs' emerging ideas.
- Setting and enforcing respectful classroom norms; prioritizing caring and learning to care as outcomes of the course [61].
- Reflecting on successes and frustrations of practice in a safe space, and offering feedback that both encourages and challenges.

- Expressing enthusiasm to the LAs and among the faculty for the substance of LAs' thinking. We think of this approach as *disciplinary caring*, which blends interest in whole persons with interest in the ideas they generate.

That LAs' ideas drive the content and direction of the content preparation sessions and the pedagogy course communicates to the LAs that their ideas are seen as productive and sensible—i.e., that the LA course instructor (and eventually the community) expect these ideas to "get LAs somewhere," even if that somewhere is simply a better understanding of self and others. This acceptance and positive regard for LA ideas fosters a safe environment in which LAs can share their ideas and challenge the ideas of their peers; in such an environment, challenges are seen as opportunities to clarify and understand, rather than as potentially threatening [60,63]. This experience of having their ideas accepted fosters among LAs a sense of acceptance toward others' ideas, including the ideas of their students [22].

The experience of caring for others promotes LAs' self-actualization as novice teachers, and being cared for fosters acceptance of oneself and one's experiences during the affectively charged experience of teaching for the first time.

C. Inclusive science teaching community provides multiple role models

The community of physics teachers at SPU includes members with a wide variety of backgrounds, expertise, and levels of commitment to teaching. The physics faculty include former high school teachers and a Resident Master Teacher with over two decades of K-5 teaching experience, as well as Visiting Master Teachers with extensive secondary physics teaching experience. We welcome LAs into the community regardless of their professed interest in a career in K-12 teaching, and we bring in members of our extended community of practicing K-12 teachers, school district science coaches and administrators, and physics education researchers. We believe this commitment to broad inclusiveness supports the central practices of our LA program by reinforcing the value of each individual's ideas and skills.

One way in which this inclusiveness is demonstrated is by the commitment of the department to supporting preservice and inservice elementary teachers. The physics faculty at Seattle Pacific University believe that every K-5 teacher is a physics teacher. The Department of Physics therefore assumes the responsibility of preparing and supporting teachers at all grade levels (P-12) so that each aspiring teacher can realize her or his total potential as a physics teacher. We have deep respect both for the work of teaching at all levels and for the people doing this work. SPU has a strong undergraduate elementary teacher preparation program that for the past seven years has included physics courses designed for, and required of, most elementary certification candidates (physics faculty involvement in the development of the major for elementary certification candidates is described above in Section II.C). These courses now include LAs, most of whom are themselves elementary certification candidates choosing the natural sciences concentration in the required integrated studies major.

Elementary certification candidates who choose the natural sciences concentration are required to take several credits of "teaching practicum" in one of the science departments, and are strongly encouraged to meet this requirement by serving as physics LAs. Unlike the majority of students serving as LAs in the algebra- and calculus-based introductory physics sequences, these students enter the LA program identifying as future teachers; we have incorporated program elements designed to support their development of *science* teacher identity. Since these LAs have already had some exposure to ideas of classroom management and "best teaching practices," they are expected to take on more responsibility for the teaching and planning components of the class than are typically assumed by the LAs in the algebra- and calculus-based courses. For example, LAs in the physics courses for elementary teachers are often asked to lead whole-group discussions, introduce daily activities, and provide input on assignments, quizzes, and exams.

The impact of having an inclusive (P-20) science teaching community on recruiting and supporting secondary physics teachers is mostly indirect, but we believe it is important. The department's commitment to inclusiveness establishes a less hierarchical model of community than is often found at universities, with different and equally valued contributions from K-12 teachers, university faculty, and LAs of various backgrounds. The community is therefore also able to provide a variety of role models to future teachers and potential future teachers, allowing for development of professional relationships with experienced teachers as well as with science education researchers and experts in professional development. Through engagement in the community and nurturing of these relationships, novice K-12 teachers develop their physics teacher identities and are encouraged to see themselves and their peers as having the ability to make valuable contributions to the broader community of physics educators.

V. DISCUSSION AND CONCLUSIONS

Over the past decade, the Department of Physics and the community of physics learners and teachers at SPU has developed a strong commitment to teaching at all levels, and to K-12 teaching as a viable professional path. We believe that this transformation is the result of a combination of factors relating to structural supports provided by the university and the department, intellectual resources gathered into the community, and dispositional commitments of the people involved. While the exact structures and people that contributed to the changes in the SPU community may not be present at other institutions, we hope that this description gives a sense of the central components of our program. In this final section, we draw on our experiences to provide more general recommendations for faculty at other colleges and universities who are interested in expanding their involvement in physics teacher preparation and support.

To provide context for the recommendations below, we describe some central assumptions of our community, condensed from the more in-depth descriptions above. We adopt the following working assumptions:

- That students who can, or do, excel with interactive academic skills (e.g., argumentation, group collaboration) are equally, if not more, qualified to become teachers than those who excel at independent, individualistic academic skills (e.g., solo problem solving, traditional test taking).
- That students with a natural inclination toward or ability with interactive academic skills need formal learning environments that value and promote the development of these skills.
- That students who might become excellent physics teachers may not choose to major or minor in physics without encouragement to do so.

With these assumptions in mind, we provide the following recommendations for recruiting and preparing K-12 physics teachers, loosely ordered from general to more specific.

Provide opportunities for valuing and developing interactive academic skills in your courses. Ask yourself whether your courses value these skills enough that students can capitalize on their strengths to excel in the course overall. Traditional courses often privilege individualistic academic skills over interactive ones. Structure your courses (particularly at the introductory level) to provide opportunities for students to develop interactive skills. The promotion of this skill development may function as a crucial factor in students' choices to become physics majors or minors because of their affinity for interactive learning in physics.

Reflect critically on how you judge student ability in physics. In particular, recognize that it is common to think of students who do most of their work on their own, and who excel at individualistic tasks such as written exams, as the highest-ability students. Consider the value, both for learning physics and for teaching physics, of productive argumentation, creative thinking, intellectual empathy, and effective listening skills. We have found that many of our best LAs are not necessarily students who stand out on canonical (individualistic) measures of physics skill; rather, they stand out for their engagement in and support of productive interactions. The more your courses provide opportunities for interactive academic work, the more opportunities these students will have to demonstrate and build their skills.

Believe in the value, challenge, and reward of precollege teaching. Our students put great stock in our opinions and values. We communicate in both explicit and more subtle ways which vocational paths we value most. Convince yourself that precollege science teaching matters critically, and should be done by those with a deep understanding of STEM content and special training in STEM teaching, and you will be able to convince your students as well. Convince yourself also that for some students, the combination of intellectual and interpersonal dimensions of physics teaching can be highly motivating and rewarding. Personally telling a student that he or she could be an excellent physics teacher can be a pivotal influence on that student's thinking about career paths.

Encourage an ethic of service. At SPU, the ethos of service is closely tied to the university's Christian identity. We have therefore attempted to present the LA program as an act of Christian service. Whatever the service ethos at your university, you can contextualize your LA program as an act of service to the community of learners.

Raise awareness of the benefits of participation in the LA program for students who may not decide to be precollege teachers. College is a time of making difficult decisions, especially for students with diverse interests and skills. Many students might not apply to an LA program if they see it as being relevant only to future precollege teachers. We try to support our students in seeing every intellectually rigorous career as involving teaching and learning in some form. For example, students in pre-health professions tracks recognize the value of the LA program for improving communication between patient and caregiver, as well as for gaining physics knowledge and developing the ability to learn new ideas, both of which are valuable in preparing for the MCAT. Engineering students report that in job interviews, employers recognize the high value of the LA program as preparation for complex design work in interdisciplinary collaborative teams. Some students aim to be university STEM faculty, and see direct professional relevance in the program. Building a community of LAs with diverse interests creates opportunities for students to discover an interest in K-12 teaching and builds respect for teacher preparation in the STEM community at the university.

Nurture a collegial professional community. LAs at SPU appreciate the opportunity to work as partners with the physics faculty. Faculty intentionally encourage this spirit by sharing concerns and seeking LA input on strategic instructional decisions. We set high expectations for both collegiality and professionalism by having a rigorous application process and promoting community norms of intellectual engagement, demonstrating concern for others, and maintaining an attitude of resourcefulness and problem solving.

Seek ways to provide useful academic credit and remuneration to LAs. College students in general, and LAs in particular, have very full schedules. Our LAs are often at or above their course credit limit, and many have other jobs or volunteer responsibilities. While providing academic credit to LAs is easy (at least at our institution), making this credit meaningful is more challenging. Credit is only meaningful when it fulfills academic requirements in a student's program of study. At SPU, LAs receive upper-division physics credit for the LA pedagogy course that can count toward the physics major or minor. Another possibility is to make the pedagogy course count as writing intensive for institutions with a writing requirement for graduation.

Carefully consider the primary goals and structural constraints of your LA pedagogy course. Our course is small (10 to 15 students), enrolls only LAs who assist with physics instruction, and includes both new and experienced LAs. In contrast, the pedagogy course at CU Boulder is large (~60 students) and interdisciplinary. The scale of the SPU LA pedagogy course affords more dialogue and negotiation of activity between the instructor and the LAs, and makes it possible to give our LAs more input in the direction of the course. This allows us to enact our priority of supporting student agency by taking up LAs' ideas and experiences as the subjects of our inquiry. In addition, we have chosen to require all LAs, both novice and experienced, to enroll in the pedagogy course each semester (in contrast to CU Boulder, where only first-semester LAs enroll). This reflects our commitment to an inclusive physics teaching community and provides opportunities for ongoing mentoring at multiple levels.

Model the teaching culture you want LAs to adopt. This includes a general culture of respect for all learners, as well as more specific practices. We encourage our LAs to attend to students' ideas and to see those ideas (about physics) as sensible and potentially productive; in the same way, we are intentional about attending to our LAs' ideas about physics and about teaching and learning, and about seeing them as sensible and potentially productive. In addition, we adopt a non-didactic approach to theories and practices of physics teaching: We frame claims from articles and expert visitors as ideas to try on, not as authoritative voices about how to teach. This lowers the risk of "trying on" theories and practices we discuss, allowing LAs to explore challenging and sometimes counterintuitive ideas about teaching and learning without being forced to ignore possible tensions between the new ideas and their existing beliefs. This models the respect for students' existing ideas that we want LAs to enact in their own teaching.

Make space for uncertainty and provide support and appropriate affirmation. Frame the roles of LAs as co-learners in the physics classroom. Treating the classroom as a learning space for LAs (a space for learning about content, learning about students, learning about learning and teaching) allows LAs to become comfortable with "not knowing the right answer" and focusing on engaging in productive conversation with students. Find ways to recognize LAs for demonstrating skill and growth in their teaching. For example, we keep slips of paper that invite positive feedback for LAs in each classroom, and students and instructors are encouraged to spontaneously write notes to LAs when they notice something they want to affirm. These notes are read aloud in LA pedagogy course meetings. Recognize that introductory students often resist interactive instructional methods, and that LAs may feel unappreciated (or even disliked) by the students with whom they are working. Providing affective as well as intellectual support can help LAs maintain a positive outlook and continue to engage in challenging instructional environments.

While every institution has its own unique culture, priorities, and structures, we hope that these recommendations, drawn from our experiences at SPU, provide useful guidance for faculty at other institutions who wish to engage in the challenging and rewarding work of recruiting and preparing students to become excellent K-12 physics teachers.

ACKNOWLEDGMENTS

The authors gratefully acknowledge the support provided by NSF through grants DRL-0822342, DRL-122732, and DUE-1240036.

[1] See http://www.spu.edu/depts/uc/response/fall2k2/initiatives.html, retrieved Sept. 9, 2013.

[2] L. C. McDermott, P. S. Shaffer, and the Physics Education Group at the University of Washington, *Tutorials in Introductory Physics* (Pearson, Reading, MA, 2002).

[3] M. C. Wittmann, R. N. Steinberg, E. F. Redish, and the University of Maryland Physics Education Research Group, *Activity-Based Physics* (John Wiley & Sons, San Francisco, CA, 2004).

[4] D. R. Sokoloff, P. W. Laws, and R. K. Thornton, *RealTime Physics* (John Wiley & Sons, San Francisco, CA, 2004).

[5] D. Hestenes, M. Wells, and G. Swackhamer, Force concept inventory, Phys. Teach. **30**, 141 (1992). doi:10.1119/1.2343497

[6] R. K. Thornton and D. R. Sokoloff, Assessing student learning of Newton's laws: The Force and Motion Conceptual Evaluation and the Evaluation of Active Learning Laboratory and Lecture Curricula, Am. J. Phys. **66** (4), 338 (1998). doi:10.1119/1.18863

[7] D. P. Maloney, T. L. O'Kuma, C. J. Hieggelke, and A. Van Heuvelen, Surveying students' conceptual knowledge of electricity and magnetism, Am. J. Phys. **69**, S1 (2001). doi:10.1119/1.1371296

[8] J. Lindberg, L. Seeley, S. Vokos, and E. Close, *Adapting Existing Research-Based Curriculum for Use in our Local Environment*, presented at the meeting of the American Association of Physics Teachers (Sacramento, CA, August 2004).

[9] S. Vokos, L. DeWater, and L. Seeley, *Creating and Sustaining a Community of Learners and Teachers*, presented at the PhysTEC Conference (Fayetteville, AR, March 2006).

[10] E. Close, Leveraging corporate support for science education reform at Seattle Pacific University, American Physical Society Forum on Education Newsletter (Summer 2007). http://www.aps.org/units/fed/newsletters/summer2007/close.html

[11] S. Vokos, J. Lindberg, L. Seeley, P. Kraus, and J. Minstrell, NSF ESI-0455796, April 2005–March 2010.

[12] See http://www.phystec.org/institutions/seattle-pacific/index.php, retrieved Sept. 9, 2013.

[13] See http://www.phystec.org/noyce/about.php, retrieved Sept. 9, 2013.

[14] See http://www.phystec.org/noyce/scholars.php, retrieved Sept. 9, 2013.

[15] See http://www.spu.edu/acad/robert-noyce/about/index.asp, retrieved Sept. 9, 2013.

[16] J. Lindberg, E. Close, F. Kline, and R. O'Leary, *Development of Discipline Centered National Model for STEM Teacher Preparation*, NSF DUE-0934743, July 2009–June 2010.

[17] Dr. Laurie Langdon, CU-Boulder LA program (private communication).

[18] See http://laprogram.colorado.edu/, retrieved Sept. 9, 2013.

[19] A. R. Daane, S. Vokos, and R. E. Scherr, Learner understanding of energy degradation, in *2013 Physics Education Research Conference,* edited by P. V. Englehardt, A. D. Churukian, and D. L. Jones (American Institute of Physics, Portland, OR, 2013), pp. 109-112.

[20] S. B. McKagan, R. E. Scherr, E. W. Close, and H. G. Close, Criteria for creating and categorizing forms of energy, in *American Institute of Physics Conference Proceedings 1413*, edited by C. Singh, M. Sabella, and P. V. Engelhardt, (Physics Education Research Conference, Omaha, NE, 2011), pp. 279-282.

[21] A. D. Robertson, K. E. Gray, C. E. Lovegren, K. R. Rininger, and S. T. Wenzinger, The development of curricular knowledge among novice university teacher educators: Process and effects, Phys. Rev. ST Phys. Educ. Res. (to be published).

[22] A. D. Robertson, E. P. Eppard, L. M. Goodhew, E. L. Maaske, H. C. Sabo, F. C. Stewart, D. L. Tuell, and S. T. Wenzinger, Being a Seattle Pacific University Learning Assistant: A transformative experience of listening and being heard, American Physical Society Forum on Education Newsletter (Summer 2014). http://www.aps.org/units/fed/newsletters/summer2014/seattle.cfm

[23] C. E. Lovegren and A. D. Robertson, Development of novice teachers' views of student ideas as sensible and productive, in *2013 Physics Education Research Conference Proceedings*, edited by P. V. Englehardt, A. D. Churukian, and D. L. Jones (American Institute of Physics, Portland, OR, 2013), pp. 225-228.

[24] C. Alvarado, A. R. Daane, R. E. Scherr, and G. Zavala, Responsiveness among peers leads to productive disciplinary engagement, in *2013 Physics Education Research Conference*, edited by P. V. Englehardt, A. D. Churukian, and D. L. Jones (American Institute of Physics, Portland, OR, 2013), pp. 57-60.

[25] E. W. Close, R. E. Scherr, H. G. Close, and S. B. McKagan, Development of proximal formative assessment skills in video-based teacher professional development, in *American Institute of Physics Conference Proceedings 1413*, edited by C. Singh, M. Sabella, and Engelhardt (Physics Education Research Conference, Omaha, NE, 2011), pp. 19-22.

[26] J. Richards and A. D. Robertson, Identifying productive seeds for responsive teaching among university physics learning assistants (in preparation).

[27] L. M. Goodhew and A. D. Robertson, Investigating the proposed affordances and limitations of the substance metaphor for energy, in *2014 Physics Education Research Conference Proceedings,* edited by P. V. Englehardt, A. D. Churukian, and D. L. Jones (to be published).

[28] A. D. Robertson, L. J. Atkins, D. M. Levin, and J. Richards, What is responsive teaching?, in *Responsive Teaching in Science and Mathematics,* edited by A. D. Robertson, R. E. Scherr, and D. Hammer (Routledge, New York, NY, to be published).

[29] D. Hammer, Discovery learning and discovery teaching, Cognition. Instruct. **15** (4), 485 (1997). doi:10.1207/s1532690xci1504_2

[30] D. Hammer, F. Goldberg, and S. Fargason, Responsive teaching and the beginnings of energy in a third grade classroom, Rev. Sci. Math. ICT Educ. **6**, 51 (2012). http://spu.edu/depts/physics/documents/RTEnergyArticleRevSMICT.pdf

[31] D. Hammer and E. Van Zee, *Seeing the Science in Children's Thinking: Case Studies of Student Inquiry in Physical Science* (Heinemann, Portsmouth, NH, 2006).

[32] R. A. Engle and F. R. Conant, Guiding principles for fostering productive disciplinary engagement: Explaining an emergent argument in a community of learners classroom. Cognition Instruct. **20** (4), 399 (2002). http://vocserve.berkeley.edu/faculty/RAEngle/EngleConant2002.pdf

[33] J. L. Lemke, *Talking Science: Language, Learning, and Values* (Ablex Publishing Corporation, Norwood, NJ, 1990).

[34] E. Van Zee and J. Minstrell, Using questioning to guide student thinking, J. Learn. Sci. **6** (2), 227 (1997). doi:10.1207/s15327809jls0602_3

[35] K. Gallas, *Talking Their Way into Science: Hearing Children's Questions and Theories, Responding with Curricula* (Teachers College Press, New York, NY, 1995).

[36] D. L. Ball, With an eye on the mathematical horizon: Dilemmas of teaching elementary school mathematics, Elem. School J. **93** (4), 373 (1993). http://www.jstor.org/stable/1002018

[37] R. S. Russ, J. E. Coffey, D. Hammer, and P. Hutchison, Making classroom assessment more accountable to scientific reasoning: A case for attending to mechanistic thinking, Sci. Educ. **93** (5), 875 (2009). doi:10.1002/sce.20320

[38] M. L. Franke and E. Kazemi, Learning to teach mathematics: Focus on student thinking, Theor. Pract. **40** (2), 102 (2001). doi:10.1207/s15430421tip4002_4

[39] R. E. Scherr and H. G. Close, Transformative professional development: Cultivating concern with others' thinking as the root of teacher identity, in *International Conference of the Learning Sciences* (Chicago, IL, 2010).

[40] E. W. Close, H. G. Close, and D. Donnelly, *Learning how to listen: The Interview Project in LA pedagogy*, presented at the meeting of the American Association of Physics Teachers (Portland, OR, July 2013).

[41] P. R. L Heron, Empirical investigations of learning and teaching, part I: Examining and interpreting student thinking, in *Proceedings of the International School of Physics "Enrico Fermi," Course CLVI, Varenna, Italy, 2003,* edited by E. F. Redish and M. Vicentini (IOS Press, Amsterdam, 2004), pp. 341-350.

[42] P. R. L Heron, Empirical investigations of learning and teaching, part II: Developing research-based materials, in *Proceedings of the International School of Physics "Enrico Fermi," Course CLVI, Varenna, Italy, 2003,* edited by E. F. Redish and M. Vicentini (IOS Press, Amsterdam, 2004), pp. 351-365.

[43] L. C. McDermott, Millikan Lecture 1990: What we teach and what is learned - Closing the gap. Am. J. Phys. **59** (4), 301 (1991). doi:10.1119/1.16539

[44] L. S. Shulman, Knowledge and teaching: Foundations of the New Reform, Harvard Educ. Rev. **57**, 1 (1987). http://hepg.metapress.com/content/J463W79R56455411

[45] D. B. Harlow, Structures and improvisation for inquiry-based science instruction: A teacher's adaptation of a model of magnetism activity, Sci. Educ. **94**, 142 (2010). doi:10.1002/sce.20348

[46] D. B. Harlow, Uncovering the hidden decisions that shape curricula, AIP Conf. Proc. **1289**, 21 (2010). doi:10.1063/1.3515205

[47] R. M. Goertzen, R. E. Scherr, and A. Elby, Accounting for tutorial teaching assistants' buy-in to reform instruction, Phys. Rev. ST Phys. Educ. Res. **5** (2), 1 (2009). doi:10.1103/PhysRevSTPER.5.020109

[48] J. Lave and E. Wenger, *Situated Learning: Legitimate Peripheral Participation* (Cambridge University Press, New York, NY, 1991).

[49] J. P. Gee, Identity as an analytic lens for research in education, Rev. Res. Educ. **25**, 99 (2000). http://www.jstor.org/stable/1167322

[50] E. Wenger, *Communities of Practice: Learning, Meaning, and Identity* (Cambridge University Press, New York, NY, 1998).

[51] Z. Hazari, G. Sonnert, P. M. Sadler, and M.-C. Shanahan, Connecting high school physics experiences, outcome expectations, physics identity, and physics career choice: A gender study, J. Res. Sci. Teach. **47** (8), 978 (2010). doi:10.1002/tea.20363

[52] R. M. Lock, Z. Hazari, and G. Potvin, Physics career intentions: The effect of physics identity, math identity, and gender, AIP Conf. Proc. **1513**, 262 (2013). doi:10.1063/1.4789702

[53] A. L. Luehmann, Identity development as a lens to science teacher preparation, Sci. Educ. **91** (5), 822 (2007). doi:10.1002/sce.20209

[54] See http://www.spu.edu/depts/csfd/SERVEhistory.html, retrieved Sept. 9, 2013.

[55] See http://www.spu.edu/about-spu/spu-facts, retrieved Sept. 9, 2013.

[56] See http://spu.edu/depts/cdc/, retrieved Sept. 9, 2013.

[57] See http://www.spu.edu/depts/um/about/studentleaders/barnabas.asp, retrieved Sept. 9, 2013.

[58] See http://www.spu.edu/news/eNews/facstaff.asp?id=884#a14356, retrieved Sept. 9, 2013.

[59] See http://www.spu.edu/depts/csfd/serve_program.html, retrieved Sept. 9, 2013.

[60] M. Mayeroff, *On Caring* (HarperCollins Publishers, New York, NY, 1971).

[61] N. Noddings, An ethic of caring and its implications for instructional arrangements, Am. J. Educ. **96** (2), 215 (1988). http://www.jstor.org/stable/1085252

[62] N. Noddings, *The Challenge to Care in Schools: An Alternative Approach to Education*, 2nd ed. (Teachers College Press, New York, NY, 2005).

[63] C. R. Rogers, *On Becoming a Person: A Therapist's View of Psychotherapy* (Houghton Mifflin Company, New York, NY, 1961).

[64] L. Jaber, Attending to epistemic affect, in *Responsive Teaching in Science and Mathematics,* edited by A. D. Robertson, R. E. Scherr, and D. Hammer (Routledge, New York, NY, to be published).

[65] L. S. Goldstein, The relational zone: The role of caring relationships in the co-construction of mind, Am. Educ. Res. J. **36** (3), 647 (1999). doi:10.3102/00028312036003647

[66] A. J. Hackenberg, Mathematical caring relations in action, J. Res. Math. Educ. **41** (3), 236 (2010). http://profile.educ.indiana.edu/Portals/281/Hackenberg_MCRs_JRME10.pdf

Case studies of successful preservice physics education programs: Physics teacher preparation at the University of Arkansas

Gay Stewart and John Stewart

Department of Physics, University of Arkansas, Fayetteville, Arkansas 72701

Beginning in 1994, the University of Arkansas, Fayetteville implemented changes in its undergraduate physics program that dramatically increased the number of students graduating with a major in physics. Our physics major graduation rate grew from an average of one to two students per year for most of the years from 1990 to 1998, to 27 graduates in 2012. This change was accomplished by revising the introductory course sequence, improving in-department advising, and introducing flexible degree plans. With the selection of the University of Arkansas as a PhysTEC Supported Site in 2001, the number of physics students entering high school teaching also increased dramatically. The features that led to the increase in physics graduates were important to increasing the number of teachers graduated, but each feature required refinement to support future teachers. The refinements most important to increasing the number of highly qualified physics teachers graduated will be discussed. These refinements include ensuring that degree plans support a path to teaching; developing advising expertise in the requirements for licensing as a teacher and in the demands of teaching; and presenting teaching as a valuable, enjoyable profession in introductory physics classes. The complexity of helping students into the teaching profession is illustrated by case histories of four successful teaching candidates.

I. INTRODUCTION

The University of Arkansas, Fayetteville Department of Physics has seen dramatic growth in the number of physics bachelor's degrees awarded and in the number of those graduates choosing to enter the high school teaching profession. This growth began when a department-wide effort to revitalize the undergraduate physics major was undertaken in 1994. Figure 1 shows the number of physics majors graduating per year with a baccalaureate degree and the number of physics students, both graduate and undergraduate, entering the teaching profession. Until 2012, these students received a physical science teaching licensure; all taught physics, and many taught other science and mathematics topics.

As Figure 1 shows, the department was graduating very few physics majors and no teachers through most of the 1990s. The effort to revitalize the major began to have an effect in 1998, and the number of physics majors graduated has grown substantially since that time, and continues to grow. The department's success at revitalizing its undergraduate major has led to its reform initiative being featured in American Association of Physics Teachers (AAPT) and American Physical Society (APS) workshops on increasing physics graduation rates [1].

With the selection of the University of Arkansas, Fayetteville as a Physics Teacher Education Coalition (PhysTEC) Supported Site in 2001, the number of physics students going into high school physics teaching also began to increase dramatically. This chapter will discuss the reforms that led to the growth in the number of physics graduates. We will then discuss how these reforms were adapted to support students in the decision to become high school physics teachers.

The National Task Force on Undergraduate Physics published the Strategic Programs for Innovations in Undergraduate Physics (SPIN-UP): Project Report in 2003 [2]. The report offered a number of recommendations for the revitalization of undergraduate physics programs; these recommendations included strong departmental involvement in the undergraduate program, long-term leadership directed at undergraduate education, advising for undergraduates, institutional support at the dean and provost level, energetic recruitment of physics majors, flexible degree plans, career mentoring, an active Society of Physics Students chapter, and revised introductory classes. While the University of Arkansas physics department did not participate in the report, and while the revisions to our program pre-date the report, the modifications we made were very much in line with the suggestions of the SPIN-UP report.

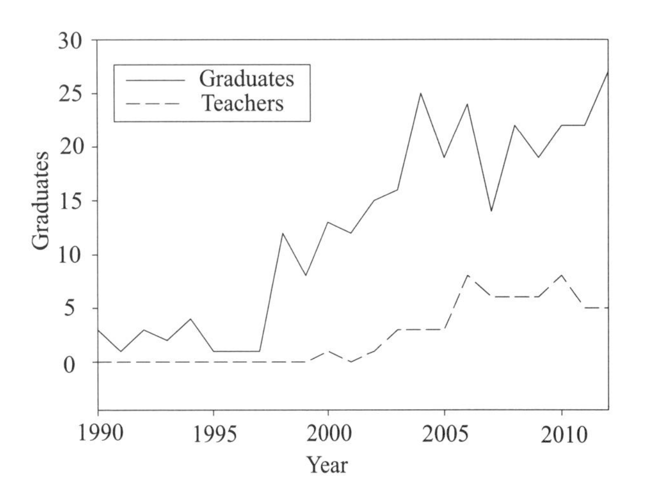

FIG. 1. Physics majors and physics teachers graduated per year at the University of Arkansas.

TABLE I. Program milestones. This table summarizes the major successes and outcomes of the University of Arkansas undergraduate physics program.

Milestone	Year
Hire of tenure-track physics education research (PER) faculty member	1994
Course and Curriculum Development grant to reform University Physics II course	1995
Selection as a PhysTEC Supported Site	2001
NSF Robert Noyce Teacher Scholarship Program grant	2007
$7-million NSF Mathematics and Science Partnership grant to provide professional development for math and science teachers and improve college attendance and matriculation rates	2009
Physics Today cites as "exemplary program" [5].	2009
NSF Scholarships in Science, Technology, Engineering, and Mathematics (S-STEM) grant to promote physics graduation and improve program diversity	2009
Implementation of UTeach Replication program	2012
Five-year average graduation rate of 22.4 physics majors and 6.2 physics teachers	2012

II. SUCCESSES AND ACHIEVEMENTS

Due to the changes described in this chapter, the University of Arkansas physics department went from graduating one physics major and zero physics teachers in 1995 to graduating 27 majors and five teachers in 2012. The average numbers from 1990 to 1995 were 2.2 majors and zero teachers; the average numbers from 2007 to 2012 were 22.4 majors and 6.2 teachers. Major milestones of the program are summarized in Table I.

Our graduates pursue a diverse set of career paths. Only 6% of our graduates with a terminal bachelor's degree in physics work in industry. The majority of our graduates pursue additional education, with 28% entering graduate school in physics, 26% entering graduate school in another discipline (usually in their second major), and 13% entering medical school or some other professional school such as dental school or physical therapy school. While teaching is an important outcome, only 16% of our graduates choose teaching as a profession. The balance of our graduates enter the military (3%) or work in industry in the field of a second major (8%).

Our physics teacher preparation program has been positively mentioned in the APS Congressional Quarterly [3] for its role in teacher preparation; it has been called a "thriving" program in the AAPT's *Interactions* publication [4]; and it has been cited as "exemplary" in *Physics Today* [5]. The program has excelled in producing award-winning students, with nine of the university's 28 Barry M. Goldwater Scholars coming from physics in the last decade, as well as one honorable mention. In that time, the department has produced 12 National Science Foundation (NSF) Graduate Research Fellowship winners, and we have seen the physics subject GRE score of our students consistently increase.

III. REVITALIZING THE PHYSICS MAJOR

The extremely low graduation rate of one to two physics majors per year experienced by the department in the early 1990s was unsustainable. Aggressive actions to improve the number of physics students graduated were taken, starting with the 1994 hire of a new faculty member (co-author Gay Stewart) with a specialty in physics education research (PER). This faculty member, working with the support of the department, instituted changes in the introductory course sequence, the department's degree requirements, and the department's advising, which ultimately produced the increased number of physics graduates shown in Figure 1. The increased number of majors and the presentation of physics teaching as a valuable career path produced a subset of graduates who became teachers. As a result, we view our increased graduation of teachers as a natural outcome of the growth in the physics major. This section will discuss the general features of the reformed program, and the next section will discuss how the general features were refined to support future teachers. For a more detailed discussion of the modified program, see the discussion in Stewart et al. [6].

A. Revised introductory courses

In 1994, immediately upon arriving at the university, the newly hired PER faculty member began revising the second-semester, calculus-based University Physics II (UPII) course to a reformed format focusing on hands-on learning using interactive engagement methods. The second-semester course was chosen as the first course to transform because its student population contained the students most likely to become physics majors, and because the learning outcomes of the course suggested it was the one most in need of immediate revision. With the support of PhysTEC, the first-semester, calculus-based introductory course, University Physics I (UPI) was also converted to the revised format in 2001. Since both courses were presented in the revised format and instructed by faculty members with a PER focus, the courses have been our primary venues for the recruitment of additional majors, and they have been the principal cause of the dramatic growth in graduation rate shown in Figure 1. The UPI/UPII course sequence is taken primarily by engineers, with a subset of students, around 20%, drawn from other disciplines, including chemistry, physics, mathematics, and biology.

The revised courses have a number of features that encourage students to consider physics as a career. The format of each course features two 50-minute lecture and two 2-hour laboratory periods each week. The UPII lecture has been presented by the same PER faculty member for over a decade, allowing substantial time for reflection on and refinement of the course. In UPI there have been multiple instructors, but the same PER faculty member has overseen the course. The laboratory period features a mixture of hands-on conceptual inquiry activities, small-group problem solving, TA-led problem solving, interactive demonstrations, and traditional experiments. The interactive engagement activities in the laboratory were inspired and informed by the Workshop Physics curriculum and the University of Washington's *Tutorials in Introductory Physics* materials; however, all materials were developed locally. Activity developers relied as much as possible on easily accessible equipment, so future teachers could replicate the learning experiences in their classrooms. The laboratory activities are closely timed with the lecture, which often introduces concepts used in an upcoming laboratory or makes use of the students' experiences in a previous laboratory to ground a concept in the real world.

One feature that makes the UPI/UPII course sequence a successful recruiting ground for physics majors and future teachers is the involvement of the lead lecturer in the laboratory component of the course. Two laboratory activities are presented in a normal week and, if possible, the lead lecturer teaches the laboratory section in which the activities are first presented during the week. This allows for effective mentoring of teaching assistants new to the course and of any undergraduate students helping out in the course, thus maintaining high-quality instruction with a constantly changing group of TAs. This instructional strategy has a very important additional benefit: It allows the lead lecturer to interact informally with relatively small groups of students, since there is a maximum of 24 students in a lab section. Because of the hands-on, inquiry-based nature of many of the laboratory activities, they generate substantial conversation within the lab groups and with the lab instructor. With students feeling a greater comfort level with the faculty member, they are more likely to attend office hours, often bringing students from other lab sections. The enhanced faculty-student interactions, which turn faculty members from distant, unknown quantities into people with whom the students have comfortable personal relationships, have many repercussions for recruiting. The faculty member is transformed into a humanized role model for the physics profession; the evident enjoyment that the faculty member derives from teaching highlights the personal benefits of an educational career; and when the faculty member tells the student that he or she has a talent for physics and should consider the physics major, the suggestion is much more authentic and carries much greater weight.

B. Flexible degree plans

Students taking the UPI/UPII sequence are often sophomores who have already completed many courses before taking their first physics class. The physics degree requirements were carefully revised to modify both physics and mathematics pre- and co-requisite requirements to allow students as much flexibility as possible in choosing the order in which to take mathematics and physics courses. The additional flexibility has been pivotal in allowing students to switch to the physics major or to double major; students have double majored in physics and a variety of second subjects, including mathematics, chemistry, engineering, history, and philosophy. Most paths to teacher licensure require additional courses that are not part of the physics requirement, so the additional flexibility was also vital in supporting the decision to teach. Complete eight-semester degree plans for both the Bachelor of Arts and the Bachelor of Science degrees can be found the at the University of Arkansas website.

C. Advising

The primary in-department physics advisers are also the lead instructors of the introductory course sequence, UPI/UPII/UPIII. This means that in many cases the adviser already has a personal relationship with the advisee before their first advising session. (Most freshman advising is done by the university's professional advisers.) The adviser also has a detailed impression of the student's abilities, strengths, and weakness. The in-department advisers have been working with students for many years, and have had time to develop the kind of detailed expertise often required to advise physics majors. The in-department advisers are also much better placed than the university advisers, and are much more likely to present physics teaching as a potential career choice. All students meet with a physics adviser at least once per semester after their freshman year. By using the familiarity developed in the introductory classes, physics advisers also conduct multiple short advising sessions, sometimes characterized as "drive-by advising," as they meet students in the halls of the physics building.

With physics graduates pursuing a diverse set of career goals, physics advisers must be knowledgeable about the career requirements for each potential goal. Many of these career paths have requirements beyond coursework, and students are often not aware of these additional requirements. To further complicate the advising process, students often change career goals multiple times during their undergraduate careers, or maintain multiple possible career goals for the majority of their undergraduate careers. This is particularly true of future teachers, who often make the decision to teach late in their undergraduate careers.

D. Summary

The revised introductory course sequence, flexible degree plans, and world-class in-department advising were vital to the revitalization of the undergraduate physics program at the University of Arkansas. The program's progress also depended on a number of additional features: a PER tenure-track hire in 1994 and a second clinical hire with a PER specialty in 2001, an active honors college, a strong tradition within the physics department of encouraging undergraduate research, support from the university's upper administration, and substantial external support. The two PER faculty members have served as "program champions," tirelessly promoting the undergraduate program and ensuring that undergraduate concerns were well represented. The National Task Force on Teacher Education in Physics identified a "champion" as a key feature of physics programs that have excelled in teacher preparation [7].

IV. SUPPORTING AND ENCOURAGING THE DECISION TO TEACH

Figure 1 shows that the increase in the graduation of teachers from our program did not coincide with the increase in the number of majors. The revisions that allowed the revitalization of the physics major also played important roles in increasing the number of students entering the teaching profession, but each revision required modification to support teachers. Increasing the number of students who plan on teaching also required two additional key features: a growing partnership with the university's College of Education and Health Professions (COEHP) and a growing partnership with local school districts. The modifications of existing features of the program and the development of new features were encouraged and partially supported by the PhysTEC program beginning in 2001.

A. Background

Very few of the majors in the physics department planned on becoming teachers when they entered college. The decision to teach, therefore, represents a career change, often from a career goal—research physicist—that the student has been pursuing for years. This change requires months or years of deliberation, discussion, and advising. The new career, teaching, has unexpected hurdles that present inflexible deadlines and requirements. While our physics department has committed, personable, in-department advisers, as described in the previous section, all advisers are traditionally trained physicists. Prior to the PhysTEC project, the department's advisers were substantially more knowledgeable about the requirements and hurdles of a graduate career in physics or a career in industry than those of a career in teaching. This did not mean that teaching was not respected as a career path, but rather that it was difficult for the advisers to discuss the career authentically and to answer detailed questions about deadlines, requirements for licensure, and the different career prospects of teachers with traditional versus alternate licensure.

With the growth of the major experienced in the late 1990s, high school teaching was identified as an attractive career path for a substantial subset of our majors. The career features the more flexible working conditions required by students who need to live in a particular geographic area or by students who have children and who therefore do not have the decade required to earn a bachelor's degree and a Ph.D. and find a permanent position in physics research. The career also offered the personal feedback, fulfillment, and personal interactions that are sometimes missing in a research or industrial career.

Arkansas is a poor state with a few major cities and many small rural school districts. To support the needs of these small districts, the licensure requirement for middle school and high school science teachers is a broad physical science licensure requiring mastery of physics, chemistry, and Earth science. The University of Arkansas, Fayetteville did not offer a four-year undergraduate program resulting in a high school science teaching license until 2012, with the introduction of UAteach, the University of Arkansas UTeach replication program. This chapter describes teacher preparation before the introduction of UAteach. Without a four-year program, the primary traditional path to licensure for our graduates has been the Master of Arts in Teaching (MAT) program offered by the College of Education and Health Professions. This one-year master's program provides extensive field experience along with directed class work, and results in a state license upon completion. For admission, the program requires a 3.0 GPA as well as the completion of 60 hours of classroom observations. Since the program graduates teachers with a broad license, the program also requires for admission the completion of a number of classes in physics, chemistry, Earth science, and atmospheric science. The state also offers an alternate licensure path that has its own very inflexible requirements. Fifty-two percent of our teacher graduates have pursued traditional licensure through the MAT; 48 percent have gone through an alternate licensure pathway.

B. Revised introductory courses: The effect on future teachers

The revised UPI/UPII course sequence is the primary recruiting ground for the physics major. It is also the point at which many students begin to harbor the thought that teaching is a possible career. Well after leaving the course, students often recall fondly the lively, highly personal, extremely interactive, hands-on nature of the laboratory experiences in these courses. The students' attachment to the introductory experience was evident both in informal conversations (they often spontaneously mentioned the course experience) and in advising sessions. As students move forward in their physics careers and begin to experience the requirements, demands, restrictions, and time scales of a research career, some look back on the very personal, human nature of the UPI/UPII learning experience and see roles for themselves more fitting with their personal tastes and goals. The classes are built around research-informed pedagogy, so the formative experience of these future teachers reflects the form of instruction that we would like to see continued in their high school classrooms.

For students who wish to seriously explore teaching as a career, we offer a Lab and Classroom Practices course, PHYS 400V. This course is elected for course credit and must be paid for like any other course, and can be used as an elective for the physics major. Students enrolled in the course assist in either UPI or UPII. They attend the weekly lab meeting, during which the instructional goals and pedagogical presentation of the following week's laboratories are discussed. The students act either as helpers to the teaching assistants of lab sections or as the lead teaching assistants of a lab section with graduate students in support. In this role, students participate in answering questions about the laboratory, help when activities go awry, and answer general questions about the physics content of the class or about physics in general. The students also hold office hours during which they work one-on-one with fellow students. PHYS 400V students help in the grading of examinations so that they can see the results of their efforts, and so they can begin to understand the importance of constructing a rubric and the impact that the types of questions asked have on the evaluation of student learning. These grading sessions are also very valuable for open discussions of learning and the practical limitations of education.

The Lab and Classroom Practices course has some relation to the University of Colorado Learning Assistant program [8], which our course predates. The programs differ in that our Lab and Classroom Practices course is not used to recruit physics teachers, but rather to help students already considering teaching to make an informed decision. Because it is elected for course credit, Lab and Classroom Practices requires no external support. The course was initially developed to support graduate assistants in teaching reformed courses; it was later extended to undergraduates who wanted better preparation for graduate school. The existence of the Lab and Classroom Practices course as a class that can be taken just like any other physics elective course is a powerful tool for advisers to help students with the often difficult decision to become a high school teacher.

C. Flexible degree plans supporting future teachers

Just as a student planning on a medical career has professional needs that differ from those of a student planning a research career in physics, teachers also have specific intellectual requirements that are not well supported by a one-size-fits-all physics degree. While many of our teachers eventually teach in large school districts where their primary teaching assignment is physics, such positions are often not open to beginning teachers. Many of our teachers start in smaller districts, and some teachers choose to remain in those districts even as other opportunities become available. These teachers have teaching assignments that often include, in addition to physics, chemistry, Earth science, and possibly mathematics, engineering principles, and biology. These teachers therefore need a broad knowledge of many topics in science. They also need a very deep conceptual grounding in introductory physics, including experience with the kinds of conceptual errors students make and how to help students overcome their conceptual difficulties. Such knowledge is often called pedagogical content knowledge.

The flexibility our program affords students to select physics elective classes and to use advanced classes in certain other sciences as physics electives allows physics students to tailor their undergraduate programs to develop the general knowledge of science required by effective teachers in small school districts. For example, a student may use chemistry classes to fill in some physics electives, or he or she may elect geology as an emphasis area to develop an understanding of Earth science.

To further provide a broad scientific background for teaching candidates, the physics department offers both a Bachelor of Arts (B.A.) in physics as well as a Bachelor of Science (B.S.) degree. Most high school teachers will make little use of the advanced mathematics skills that form a significant portion of the work in most junior and senior level physics classes required for a B.S. While not restricted to teachers, the B.A. degree option has been popular with future teachers; about 30% of our teachers graduate with a B.A. The B.A. degree does not require the most mathematically demanding advanced physics classes, and allows a broader sampling of physics electives. It fits well with the classes required by the MAT program. The B.A. degree provides a mechanism for a future teacher to develop the broad understanding of physics and astronomy required in the classroom and the flexibility to fit in the non-physics courses needed for the broad licensure.

D. Advising future teachers

By ensuring that teaching is one of the potential career options presented to students, physics advisers make the decision to teach possible. Most advising sessions include not only the selection of classes for the next semester but also a general discussion of career options. Advisers also monitor the general state of the student's internal career goals and try to match these goals with the student's academic performance. Advisers attempt to develop a clear picture of what the student wants to do with his or her life, and work to match this with what the student can do. If a student expresses a potential interest in teaching, the adviser presents the requirements, deadlines, and demands of the profession. This discussion is now informed and colored by long experience with graduates who are working teachers, and by the satisfaction these graduates report with their career choices. If a student wishes, a meeting with an inservice teacher can be scheduled to allow further discussion and opportunities for classroom observations. If the student wants additional experience with teaching to help him or her make the decision to teach, the adviser will suggest our Lab and Classroom Practices course, described above. The student may have already experienced the fulfillment of teaching in one-on-one sessions as part of Society of Physics Students outreach events. If the student decides that teaching will be his or her future career, the adviser will help the student rework his or her academic plan to make sure the entrance requirements of the MAT program are met.

Future teachers need substantial emotional support. The decision to teach is often a dramatic shift in a student's personal goal set and identity. Future teachers often need support though difficult points in their academic careers, and they require continuing support upon entering the classroom. The post-graduation support has taken many forms, such as long phone conversions after a difficult day in the classroom, placing a new teacher in contact with the correct mentor teacher, and including a new teacher in professional development activities at the university to build his or her professional confidence and network. Unlike our relationships with students pursuing other professions, who may send occasional emails updating their current status or drop by for visits during the holidays, our relationships with our teacher graduates continue to be close well after graduation. These close long-term relationships support teachers through the often difficult first years in the classroom.

The advising of teacher candidates is very personal and often very time consuming, and requires familiarity with the policies of the COEHP and with state licensure requirements. Such advising is a skill in which few physicists are trained, and must be actively cultivated by advisers seeking to help students enter the teaching profession. Most physics career paths have specific hurdles beyond simply earning the degree. Teachers seem to have more, and more inflexible, hurdles than other physics career options, and need more help from advisers to navigate the requirements of the career. For example, students wishing to enter the MAT program must first attend a required informational meeting in December, take the PRAXIS II test, maintain a 3.0 grade point average, and perform 60 hours of classroom observations.

E. Graduate students as teacher candidates

A significant number of our teaching candidates have been physics graduate students. The University of Arkansas physics program draws many of its domestic graduate students from small colleges in Arkansas, Oklahoma, Kansas, and Missouri. These students sometimes become disillusioned by the demands of a research career, which include the long time to degree, the need for multiple postdoctoral positions at multiple institutions, the geographic inflexibility of the field, and working conditions that often involve handling hazardous materials and little personal interaction. These graduate students at the same time have excellent teaching experiences in the reformed introductory classes. Some graduate students weigh the relative benefits of spending many more years in school to earn a Ph.D. against those of much more quickly entering the workforce either through alternate licensure or through the MAT program, often with Noyce Scholarship support. The personal fulfillment students experience in the introductory classes, combined with the professional flexibility that allows them to live near their families and often to support their own families, has caused a number of our graduate students to decide to become high school teachers. Like undergraduate students, graduate students require careful advising about licensure, MAT requirements, and possible financial aid. Like any teachers, they also need long-term mentoring and emotional support.

F. Teachers-in-Residence

The PhysTEC project allowed the hire of multiple Teachers-in-Residence (TIRs), who provided invaluable experience with the teaching profession and worked within the physics department. These TIRs informed the revision of introductory classes, fostered long-term partnerships with local school districts, and acted (and continue to act) as mentors to our teacher graduates. Our TIRs were also invaluable in helping physics advisers develop the expertise to correctly advise future teachers and to help physics faculty understand what was needed to properly support new teachers in the classroom.

G. Partnership with COEHP

The PhysTEC project also fostered a working partnership between the physics department and the College of Education and Health Professions. The partnership with the COEHP developed physics faculty's knowledge of and experience with the Master of Arts in Teaching program, the primary path to traditional licensure for our students. This expertise included knowledge of required deadlines, evolving entrance requirements, the expectations for classroom observations, classroom observation placements, and the teaching portfolio. Faculty from the COEHP talked about teaching and the MAT program at Society of Physics Students meetings. This relationship has continued to flourish after the end of PhysTEC support, and we continue to work together to draw students into the MAT program. The relationship has also resulted in a number of funded partnerships supporting scholarships for students and professional development for inservice teachers.

H. Partnerships with school districts

The introduction of our PhysTEC program in 2001 required working with local school districts to arrange for the hire of a Teacher-in-Residence in such a way that the TIR could return to his or her position after a one-year term. Later, as we began to graduate teachers and as they began to work in schools that were usually in the Northwest Arkansas region, we began to develop growing connections with our local schools. These connections were necessary to properly support inservice teachers, and have been invaluable in directing new teachers to promising positions. Connections with teachers and administrators have allowed us to identify strong mentor teachers when a new teacher is struggling, and to work through challenging in-classroom and interdepartment situations that sometimes arose. The relationship with local districts was also pivotal in creating the College Ready in Mathematics and Physics Partnership. This National Science Foundation Mathematics and Science Partnership (MSP) is a seven-million-dollar partnership involving 38 school districts and two University of Arkansas campuses, and provides high-quality professional development in physics and mathematics to the teachers of the districts, with the goal of increasing college attendance and college matriculation rates among the teachers' students. Former University of Arkansas physics majors have been active participants in the professional development programs, further strengthening our ties.

I. Ongoing relationships with inservice teachers

The relationships developed between the advisers in the physics program and students who become high school teachers continue well after graduation. These relationships are cemented by the often emotional process of deciding to teach and by the mentoring relationships that develop as teachers are supported through the difficult first years in the classroom. These relationships have been further strengthened by the retention of multiple TIRs through the PhysTEC program, by relationships developed while placing graduate students in classrooms through the GK-12 program (a program that places graduate students planning on a traditional research career as helpers in K-12 classrooms), and through involvement in the many professional development activities offered by the physics department.

Our teachers are very much our family. The close relationships we have with them help with placement of new teaching candidates, in the selection and management of field experiences for preservice teachers, and in building the partnerships with school districts, detailed above, that have enabled funding of major professional development initiatives.

J. Active Society of Physics Students

One subsidiary feature of our program that has a particular impact on the development of physics teachers is the on-campus Society of Physics Students (SPS) chapter. One of the PER faculty members acts as faculty mentor for the organization. Most future teachers are active members. The SPS chapter frequently invites working teachers to discuss the profession at meetings. Faculty members from the COEHP also speak at meetings and have discussed scholarship programs and the college's MAT program. The SPS chapter holds numerous outreach events each year. These events often give future teachers their first teaching experiences, and are sometimes brought up in advising sessions as one of the places where the idea of becoming a teacher germinated. SPS meetings often feature discussions of important career information and upcoming deadlines, and therefore act as another mechanism to reinforce information presented in advising sessions.

K. Financial support for future teachers

Financial considerations have often been important to the decision to teach, and in many cases, helping a student to obtain needed financial support has been critical to placing him or her in the classroom. Since the MAT program is a graduate program, many scholarships held by students will not support them in the program. The university did make the commitment that its Chancellor's Scholarship could be held for a fifth year to allow recipients to gain teaching licensure, but this affected few students. The MAT program is intensive, and working while progressing successfully through the program is very difficult. This makes the MAT year a significant barrier to entering the classroom. We have been fortunate to be able to lower this barrier for some students by providing Robert Noyce Scholarships both through a university-based grant and through the PhysTEC Noyce Program. The new four-year licensure program that became available in 2012 will remove this financial barrier to teaching.

V. CASE HISTORIES

Some common themes have arisen during external visits from PhysTEC program administrators, representatives of the National Task Force on Teacher Education in Physics, and the Science and Mathematics Teacher Imperative, each of which has observed our teacher preparation program. One theme was that having the introductory courses taught by physics faculty who respect teaching, who make a point that students' learning is important, and who show a commitment to ensure that this learning happens, communicates the message that teaching is important, and thus is a valid career choice. The second was that students comment on how fun and rewarding it was to work with their fellow students in the introductory classes, both as students and, in many cases, later as Learning Assistants in the Lab and Classroom Practices class. Yet, every teacher candidate from the University of Arkansas physics department has followed a unique path to his or her teaching career. In this section, we will discuss some of these paths, to show how various features of our program support the decision to teach. The names of the students have been changed to protect their identities.

Case 1: A B.S. student headed to industry

Paul's goal was a career in industry. He was pursuing a double degree in mechanical engineering and physics. At the beginning of his senior year, he expressed an interest in changing career paths. He said the "most fun he had ever had" had been in the peer-learning situations in University Physics, and asked if there was some way to find a career with the same sort of emotional experience. PhysTEC was relatively new at this time, so there was not a group of students pursuing teaching for Paul to join. But there was a network of teachers, our Teacher Advisory Group (TAG), with whom we had worked on various projects.

Paul's adviser discussed a possible career as a high school teacher, but Paul was unsure. We placed him in the classroom of one of our TAG members to observe a day in the life of a high school physics teacher. Paul came back ready to give it a try. He thought it would be hard, but if he failed, he could "always go make a lot of money as an engineer." Unfortunately, it was too late for Paul to meet the requirements for entry into the MAT or to take the Lab and Classroom Practices course. He was much better prepared for graduate school. So, upon graduation, he entered the Master of Arts (M.A.) in physics program and prepared for the non-traditional licensure (NTL) program. The M.A. is a flexible degree designed to help community college and inservice high school teachers deepen their content and pedagogical content knowledge. The NTL program is the state of Arkansas' alternative licensure program. The NTL program offers a minimal amount of training, with a two-week workshop in the summer and two weekend follow-up workshops during the academic year. A mentor is assigned at the school at which employment is found. This mentor does not generally have a specialty in science teaching, and the training is not science specific.

As an M.A. student, Paul was able to take the graduate Lab and Classroom Practices course and to target his graduate research on lesson development, so that his M.A. project would be useful upon entering the classroom. We carefully arranged classroom placements for him, as this was not part of the NTL program. When he first entered the classroom in the fall of 2005, we were able to aid him in finding a position in the same school as one of our TAG members, a future Teacher-in-Residence. She provided the school-based mentoring he needed.

Case 2: A non-traditional student

Mary decided to return to college after her daughter graduated high school. Mary had always been fascinated by science and was sure she wanted a career in research, but she found herself thoroughly enjoying building teaching demonstration equipment and working with her peers, and less interested in solitary lab time. Discussions with Mary revealed that she had been avidly involved in her daughter's education, enjoying her "classroom mom" experiences. She was excited to speak with our TIR about a career in high school teaching. With the receipt of our Noyce Scholarship grant, we were able to ease Mary's financial burden and begin preparing her for a career in teaching, where her loves of discovering how things worked and of working with people would be supported. Mary was able to complete the entrance requirements for the MAT with a B.A. in physics, and completed her MAT in 2009. She has returned almost every summer to complete a project with us, and has developed inquiry-based lessons that could be used in her classroom, thus furthering her preparation to teach physics and chemistry. Her school district is now in the process of implementing a more project-based science curriculum, and her administrators believe she has been a key to helping to make this happen.

Case 3: Two physics graduate students

Robert is from a small town in Arkansas. He never had to devote significant time to his studies as an undergraduate physics major at a small institution. He also married early in his graduate student career. He was willing to work hard, but found that much of his time was spent doing things he did not enjoy. He was very responsible about his position as a teaching assistant; he went well beyond the normal graduate student training to teach by sitting in with the TIR and the undergraduates in the Lab and Classroom Practices course for further discussion of teaching practices. Robert's intention was not to become a high school teacher, but to become a faculty member who could teach well, and possibly to return to an institution such as the one he had attended as an undergraduate. His wife Jessica, also a graduate student in physics, started interacting strongly with the undergraduate discussion group. Both started their research rotations. Robert soon approached us with the idea of doing a project in physics education research. As Robert and Jessica considered how much they enjoyed teaching, as well as their goals to have flexibility in location and to start a family in the near future, the decision to pursue high school teaching was a natural choice.

At first only Robert expressed an interest in teaching. We asked several of our TAG members to host him for a few hours and let him see what a public school classroom was like. Robert came back committed to the idea; Jessica then also expressed an interest. We worked with them to enter the NTL program. Jessica was able to find a position at a local private school. The pay was better in the big public school districts, so Robert held out for one of these positions. Because of the numerous physics teachers we have placed in the region, positions in the largest schools are rare. The positions at the largest local schools are sought by veteran teachers because of the higher levels of support, better pay, better facilities, and fewer preps. However, after a conversation between a physics faculty member and a teacher in a local school at a local school sporting event, the relationship with the teacher having been fostered through professional development activities, Robert's application was read at a large junior high school (grades seven through nine), and he was offered a position there.

Jessica's school has been amazingly accommodating, allowing her to change their instructional methods and to introduce a new physics class. Robert's experience was more normal. He was placed in an initial teaching assignment that was not optimal; his preparation as a physics master's student had not prepared him to navigate school politics. Several interventions, which connected Robert to effective in-school mentors, were required to help build productive working relations with his colleagues. Four years later, he was unwilling to consider leaving "his" school for a "prized" position at the high school. Each summer both Robert and Jessica return to participate in our inservice professional development activities.

A. Summary

These four students are only a few examples to illustrate the complicated process of placing well-qualified science teachers in the classroom. None of the students entered our program intending to teach. The reasons each turned to the teaching profession were different, with the decision to teach often taking months or years to complete. The students had to be helped through multiple different paths to licensure, but are now all productively providing exemplary education for the children of Arkansas.

All these students benefited from flexible, teacher-friendly degree plans; advisers who were supportive and aware of the challenges of teaching; and the example of the reformed instructional methods in the introductory sequence. Most also benefited from the Lab and Classroom Practices course. These examples were chosen to show the diversity of the paths to teaching of our students, but virtually any other student who completed our program and went into teaching would be an equally good example of this diversity. While most University of Arkansas students who go on to graduate school in science follow similar career trajectories, all of our teachers have been different, so flexibility in degree planning and advising have been critical for graduating teachers.

VI. TIPS AND LESSONS LEARNED

The teaching profession has challenges and rewards quite different from those of academic physics research or an industrial career in physics. Since the University of Arkansas began graduating physics teachers in 2000, we have had the pleasure of working with many fantastic teachers through their successes and through some very challenging times. Through these experiences, some lessons have stood out:

- Reformed introductory classes that provide an example of the power of exemplary instruction are an effective way to recruit new physics majors and future teachers.
- Teaching licensure is a complicated, shifting, political topic, and knowledge of the many paths to a teaching license is sometimes the key to placing excellent physics teachers in the classroom.
- Growing the number of majors in a department while keeping teaching as a possible career option for physics majors will naturally increase the number of teachers graduated.
- To properly support preservice and inservice teachers, connections beyond the physics department are required. These include relations with the College of Education and with local teachers and schools.
- Advising future teachers is a long-term relationship that should last far beyond graduation.

- There are sometimes institutional barriers to teaching. These can include degree requirements growing out of licensure requirements or funding barriers growing out of the need to enter a master's program to obtain traditional certification. The removal of these barriers can dramatically increase the number of students entering the classroom.

Many more lessons could have been added to this list; recruiting, advising, and supporting teachers has been a continuous learning experience.

VII. FUTURE

Our approach to physics teacher preparation at the University of Arkansas is currently being transformed by the implementation of UAteach, our UTeach replication program. This program, championed by the governor of Arkansas, has resulted in a change in state licensure requirements. The state has approved a physics/mathematics license and a physics/chemistry license that do not require Earth science. The UAteach program is a four-year program that allows a student to graduate with both a degree in physics and a license to teach, thereby removing the financial barrier of the fifth-year MAT program. This program began implementation in May 2012 and will transform our teacher preparation, hopefully further increasing the number of highly qualified science teachers graduating from the University of Arkansas.

VIII. CONCLUSION

Aggressive implementation of the suggestions of the SPIN-UP report can dramatically increase the number of physics majors graduated. The University of Arkansas physics teacher preparation program includes all the SPIN-UP report recommendations: strong departmental involvement in the undergraduate program, long-term leadership directed at undergraduate education, competent advising, institutional support at the dean and provost level, energetic recruitment, flexible degree plans, career mentoring, an active Society of Physics Students chapter, and revised introductory classes. The same changes, with some modification, can lead to an increased number of graduates going into teaching. Physics students are supported in the decision to teach and in entering the teaching profession by reformed introductory classes, consistent presentation of teaching as a career choice, teacher-informed advising, teacher-friendly degree plans, specific classes to help students make the decision to teach (in our program, Lab and Classroom Practices), and continued personal mentoring while students are in the licensure process and the classroom. Supporting students going into teaching requires development of additional expertise in issues of teacher licensure programs and the teaching profession; such expertise is not usually found in a physics department. The relationships that develop with graduates who are teachers can enrich physics departments in many ways, and are well worth the additional effort required to support teachers.

---

[1] American Association of Physics Teachers, *Spin-Up Workshop Archive* (2012). http://www.aapt.org/Programs/projects/spinup/spinuparchives.cfm, retrieved Sept. 9, 2013.

[2] R. C. Hilborn, R. H. Howes, and K. S. Krane, *Strategic Programs for Innovations in Undergraduate Physics: Project Report* (2003). http://www.aapt.org/Programs/projects/ntfup/index.cfm, retrieved Sept. 9, 2013.

[3] American Physical Society, PhysTEC prepares future physics teachers, APS Capitol Hill Quarterly **3** (3), (2008). http://www.aps.org/publications/capitolhillquarterly/200809/upload/200809.pdf

[4] K. Krane, Help Wanted: What physics departments have done, can do, and should do to increase student enrollment and better prepare physics majors for the workforce, Interact. Phys. Educ. **37**, 48 (2007). http://www.aapt.org/publications/upload/ia_marapr_2007_web.pdf

[5] T. Hodapp, J. Hehn, and W. Hein, Preparing high-school physics teachers, Phys. Today **62** (2), 40 (2009). doi:10.1063/1.3086101

[6] J. Stewart, W. Oliver III, and G. Stewart, Revitalizing an undergraduate physics program: A case study of the University of Arkansas, Am. J. Phys. **81** (12), 943 (2013). doi:10.1119/1.4825039

[7] D. E. Meltzer, M. Plisch, and S. Vokos, eds., *Transforming the Preparation of Physics Teachers: A Call to Action. A Report by the Task Force on Teacher Education in Physics* (T-TEP) (American Physical Society, College Park, MD, 2012). http://www.ptec.org/webdocs/2013TTEP.pdf

[8] V. Otero, S. Pollock, and N. Finkelstein, A physics department's role in preparing physics teachers: The Colorado learning assistant model, Am. J. Phys. **78**, 1218 (2010). doi:10.1119/1.3471291

A case study in preservice teacher education: Middle Tennessee State University

Ronald H. Henderson and Brian Frank
Department of Physics and Astronomy, Middle Tennessee State University, Murfreesboro, TN 37132

Physics teacher education became an emphasis at Middle Tennessee State University (MTSU) when the Department of Physics and Astronomy began offering career-specific pathways in an effort to increase the number of physics graduates. Prior to this effort, no physics majors had graduated as teachers in at least 20 years. An external evaluator highlighted this fact and helped motivate the department to transform the physics teacher education program. The department secured a grant from the Physics Teacher Education Coalition, and used those resources to design a new concentration in physics teaching while the university concurrently became a UTeach replication site. A physics education expert was hired to develop new physics-specific pedagogy courses and to lead redesigns of introductory and intermediate physics content courses. As of 2015, the department is projected to have graduated 12 physics teachers and have another 14 in the pipeline, and the five-year average of all physics graduates will have more than doubled, from four to 11. The process to achieve these successes included educating both faculty and students, and has resulted in new attitudes toward careers in physics teaching. These attitudes have been an important influence contributing to our initial success.

I. INTRODUCTION

Prior to the fall of 2008, the attitude of the Department of Physics and Astronomy toward physics teaching as a profession was well-described as indifferent. While faculty were certainly aware that physics teachers existed in local high schools, we had operated with the assumption that high school teachers were a product of the education college. In the span of about a year, department faculty developed a sense of obligation to prepare qualified physics teachers, and revised our department vision to make this a priority. A number of events helped to build momentum for these efforts, and led to changes in our department faculty profile and physics curricular offerings in a way that introduced new opportunities for MTSU physics majors interested in physics teaching. The purpose of this chapter is to describe how a combination of departmental vision, programming changes, and grant successes transformed the physics teacher education program at MTSU and helped grow the overall number of physics majors.

II. BACKGROUND

Middle Tennessee State University is a comprehensive university serving 23,000 students in the geographic center of the state. The university offers Ph.D. programs in a number of disciplines, but a Bachelor of Science is the highest degree available in physics. The physics program was small in 2008, with a five-year average graduation rate of 4.2 students, when we began considering ways to increase the number of graduates by targeting specific careers. An average graduation rate below 10 degrees per year is designated as "low producing" by the state of Tennessee, and this was a cause for some concern during each annual review. In addition, the department had graduated zero physics teaching majors in the prior 20 years. Physics majors had the option of following a standard graduate school curriculum or pursuing a concentration in astronomy. An external program evaluation took place that year, and the evaluator (Theodore Hodapp of the American Physical Society) highlighted both our low graduate counts and an inattention to careers in physics teaching. Hodapp suggested that developing a program to graduate physics teachers could both address a national need for physics teachers and provide some assistance in growing the number of physics graduates. The need for qualified high school physics teachers both locally and nationally was revealed through a report by the American Association of Employment in Education (AAEE) [1]; the combination of this information coupled with the external evaluator's feedback began to change departmental faculty attitudes toward physics teaching as a viable career option for our physics graduates.

An outside marketing consultant, with experience working with physics departments, was hired in 2010 to conduct focus groups to help direct efforts in marketing the physics degree in general, and also to specifically support the recruitment of future physics teachers. The primary truth revealed in the report was that no one in our sphere of influence had significant knowledge of potential career trajectories for physics graduates; this included high school students enrolled in physics, their high school physics teachers, general MTSU undergraduates taking physics, MTSU physics majors and, to a large degree, physics faculty. Each group had difficulty identifying careers for physics graduates beyond attending graduate school. Based on this experience, our department decided to create a collection of concentrations under the physics bachelor's degree to help guide students along potential career paths. Along with avenues for students interested in graduate school and astronomy, concentrations in applied physics and physics teaching were planned.

III. CHANGES TO THE TEACHER EDUCATION PROGRAM

The first step in planning the physics teaching concentration was to gather information through both literature searches and interviews with experts. A number of resources were utilized to guide this process, including discussions with faculty at universities graduating non-zero numbers of physics teachers, advice from leaders in the Physics Teacher Education Coalition (PhysTEC), and the online project PER-Central [2]. However, the volume of available material was almost overwhelming. Ultimately, the most useful exercise for our department involved identifying physics programs at other universities that had experienced sustained growth in the number of high school teacher graduates, and inviting a key innovator to visit our campus for an intense day of one-on-one interviews. The department invited Gay Stewart, previously at the University of Arkansas, to serve as a mentor in setting goals and identifying programming and curricular changes that would be required. The mission was to build an excellent physics teaching concentration starting with a blank slate, as opposed to trying to tweak or modify the existing teacher education pathway.

A. Initial efforts

Early in the program design process, one task was the creation of a pathway for MTSU physics majors to achieve a degree in physics and licensure to teach high school physics in a traditional four-year degree plan. A number of largely philosophical discussions occurred during the 15 months of program design that eventually led to a concentration in physics teaching under the physics Bachelor of Science degree. Conversations generally followed the following line of reasoning: The best high school physics teachers will pursue our standard graduate school preparation program, add physics pedagogy courses, and complete the secondary education minor. However, this program, without pedagogy courses, had been available for 20 years with no physics teacher graduates. By contrast, most physics classes in high schools are taught by instructors who have limited physics content knowledge, with many teachers having been trained in chemistry, mathematics, or biology. The principal question centered on the value of offering a curriculum that occupied a space in between these extremes, and the department was comfortable with designing a new program option.

The beginning point in building the required classes for the physics teaching concentration was an examination of the Tennessee Licensure Standards and Induction Guidelines [3]. A study of this document revealed a list of competencies that encompassed the full academic requirements for becoming licensed to teach physics in grades 7–12. Faculty then matched courses currently available in the department with the requirements to achieve licensure, and it was determined that four lecture courses and a modern physics laboratory would satisfy all physics content competencies. Further mathematics and general science requirements motivated the addition of classes in mathematics (four), biology (one), chemistry (two), and astronomy (one). This represented the minimum content, and helped identify a degree of flexibility for achieving a pathway to physics licensure within the 124-credit-hour maximum allowed by the university for teacher education programs (see Appendix). Curricular modifications in the physics department, which are described in the next section, focused on three areas: working with the education college to learn how to best prepare our graduates for careers in teaching, strengthening physics content in the new concentration beyond the baseline required to achieve licensure, and introducing courses in physics pedagogy based on recommendations from our expert consultant and PhysTEC.

Another initial effort was submitting and receiving (in 2010) a grant from PhysTEC in support of efforts to transform our physics teacher education program. PhysTEC is a partnership between the American Physical Society and the American Association of Physics Teachers, with the goal of improving and promoting the education of future physics teachers. PhysTEC has identified 10 “key components” that are shared by successful physics teacher education programs [4]. While over 300 universities are members of PhysTEC, funds were available through a competitive grant process for a more limited number of universities to become supported sites. After our grant proposal was selected for funding, PhysTEC provided support to implement or improve each of the key components. The areas of greatest need at MTSU were recruitment, curriculum reform that addressed preservice teacher pedagogical content knowledge, and improving collaboration with the College of Education. Our involvement with PhysTEC has yielded many benefits beyond the funded activities, but the most important has been interactions with the community of experts in the broader coalition of sites at the annual PhysTEC conferences.

A final task involved collaborating with the education college on the secondary education curriculum leading to licensure to teach high school. Most universities with successful physics teacher education programs have close ties with the education college, and it is no coincidence that these collaborations are integral to one of the key components identified by PhysTEC. While our department was designing the physics teaching concentration, the dean of the science college at MTSU concurrently began pursuing funding to replicate the UTeach secondary teacher education program developed at the University of Texas at Austin [5]. Through this application process, representatives from each science, technology, engineering, and mathematics discipline met often with colleagues in the education college. The physics faculty gained an appreciation for the need to involve future teachers in early teaching experiences to help these students make informed decisions regarding their career trajectories, and also gained input regarding skills and competencies to include in the pedagogy courses. The MTSU UTeach replication effort was successful, and MTeach was launched in 2010. The process for becoming licensed to teach high school physics involves obtaining a bachelor's degree in physics and satisfying the secondary education minor that is managed through MTeach.

Table I shows the major activities that are now regularly implemented as part of MTeach, the details of which are described in the sections below.

TABLE I. Major activities involved in physics teacher education program reform at MTSU.

Category	Major activities and events
Curriculum modification	• Course reform in introductory sequence • Physics pedagogy courses developed • Intermediate-level course reform
Marketing and recruitment	• Career-focused marketing • Noyce Scholarship targeting physics majors
Capacity and sustainability	• Physics education researcher hire • Monthly meetings with area physics teachers • Learning Assistant program grant awarded

B. Curriculum modification

Our department's efforts in physics teacher preparation were significantly structured around the development of a physics teaching concentration. After design of the concentration, efforts were directed at curriculum modification to support the new concentration; these efforts included introductory course reform, modified intermediate physics course offerings, and the development of a sequence of physics teaching courses that would be taught in the department and count as upper-level physics electives.

1. Introductory physics course reform

Physics majors, including those in the physics teaching concentration, can complete either the algebra-based or calculus-based introductory physics sequence. Because these sequences typically represent students' initial exposure to physics teaching and constitute a vital recruiting ground for physics teaching majors, both the algebra- and calculus-based introductory sequences make extensive use of interactive engagement methods and curricular materials based on physics education research. The introductory algebra-based physics sequence had previously been reformed in a manner similar to the SCALE-UP model, with two 2.5-hour weekly laboratory meetings [6]. Students in these courses are involved in guided discovery experiments, practice small-group problem-solving exercises with whiteboard presentations, have meaningful Socratic conversations with instructors, and wrestle with conceptual problems. Even though most physics majors complete the calculus-based course sequence, almost all of our majors end up helping in the algebra-based sequence as undergraduate TAs in a required course (Physics Practicum), where they have responsibilities to facilitate collaborative problem solving. Our introductory calculus-based physics sequence underwent additional reforms in 2011 as a result of PhysTEC funding. This sequence now utilizes a similar SCALE-UP model, and the course content is structured around the *Matter and Interactions* text [7], with video analysis being used extensively throughout the course. Currently, while the majority of our majors take the calculus-based sequence, the majority of students in the physics teaching concentration have completed the algebra-based sequence. Allowances for majors to complete either sequence have been critical to the successful recruitment of future physics teachers.

2. Courses specific to physics teaching

Developing physics pedagogy courses was more of a challenge. There were no physics education research (PER)-trained faculty at our university, and expert advice was critical. The PhysTEC grant provided release time for the development of physics pedagogy courses, and also provided access to off-campus PER experts through technical conferences and monthly video meetings with other PhysTEC Supported Sites. Before the department was able to hire a faculty member with expertise in PER, we developed two courses specifically designed for physics teachers: a one-credit seminar focused on preparing students for the PRAXIS II physics content exam, and a three-credit pedagogy course primarily consisting of readings and discussion on physics teaching methods.

After our PER faculty member was hired, the physics teaching courses were revised and a third class was added. The former PRAXIS course was removed in favor of a pair of one-credit courses (Physics Licensure I and II) focused primarily on further developing students' conceptual understanding and reasoning with introductory physics material. The first of these two courses focuses on topics in mechanics, while the second focuses on electricity and magnetism.

Students take these courses either in their sophomore or junior years, and they are provided with opportunities to revisit, deepen, and broaden their understanding of the physics content they will most likely go on to teach. In these two courses, students collaboratively work through *Tutorials in Introductory Physics* [8] to develop deeper conceptual understanding and reasoning. In parallel, they begin to learn about specific student difficulties in these content areas by examining and discussing examples of students' written work on tutorial pretests, and they strengthen their problem-solving skills by working through Advanced Placement (AP) physics exam problems. Deepening content knowledge, familiarizing students with research-based curriculum materials, and learning about student thinking provide a foundation for the third course in our sequence. The three-credit Teaching of Physics focuses more specifically on developing knowledge and practices for teaching physics through inquiry and interactive engagement. In this course, students further develop knowledge of student thinking, learn about physics-specific pedagogies, and engage in repeated cycles of preparation, enactment, and evaluation of their own teaching. This course continues to undergo significant revisions as we learn more about the diverse needs of our preservice teachers and how our courses interface with those in secondary education.

3. Modified intermediate physics course offerings

For our intermediate- and upper-level physics content, three new courses were developed that provide an additional focus on conceptual understanding of upper-division topics. Since low-enrollment courses are vulnerable to cancellation and new physics content courses designed for future teachers are virtually guaranteed to have small numbers of students, these new upper-division courses have lectures that are held concurrently with the legacy content courses, but with the emphasis on conceptual understanding highlighted in out-of-class assignments and exams. The three courses include modern physics, introductory quantum mechanics, and thermodynamics. Students enrolled in these three courses have more opportunities to consider conceptual problems in homework and exams designed to further develop students' understanding and reasoning in a manner that is similar to the approach that Physics Licensure I and II provides for introductory topics. A fourth new course was developed with no association to any legacy courses. Intermediate Physics emphasizes topics in electricity and magnetism and waves, specifically for students who complete the algebra-based introductory sequence.

C. Marketing and recruitment

1. Career-focused marketing campaign

During the on site visit from the marketing consultant, focus group interviews were conducted with a variety of groups. These included high school students enrolled in a physics course, high school teachers, general undergraduates at MTSU who were not majoring in physics, physics majors, and physics faculty. As mentioned previously, the primary truth revealed was that high school students, general undergraduates, and physics majors were all unaware of careers that are available to physics graduates. As a result, the department decided that efforts to recruit physics teachers would best be combined in a framework that addresses multiple physics careers. A second lesson learned from the marketing consultant was that our methods of advertising were sorely outdated. Brochures, letters, posters, and T-shirts need to provoke interest, and not try to answer all conceivable questions about the department or about particular careers. To reach high school students and general undergraduates, new marketing vehicles were designed. The old brochure and letters that had copious amounts of text were abandoned for new colorful and picture-rich documents, with the primary purpose of informing the reader of the variety of careers available to physics majors, one of which is physics teaching. The new brochure is used when potential students tour the campus, at departmental fairs with current MTSU students, and in direct mail to high school juniors and seniors. For students who are in the science college, especially those enrolled in introductory physics and calculus, a short presentation was developed that showcases MTSU physics graduates, including those who have become physics teachers. Photos of graduates are shown, along with their year of graduation and current occupation. This technique of displaying testimonials has proven to be very effective in generating interest in the physics major in general, and physics teaching in specific, as indicated by students seeking further information.

2. Noyce Scholarships

A primary incentive for students to consider careers in teaching has been scholarships from the National Science Foundation (NSF) Robert Noyce Teacher Scholarship program [9]. The department secured a Noyce grant in 2010 that specifically targeted teacher licensure in physics and/or mathematics. Students seeking a single endorsement area (physics or mathematics) receive a stipend amount that is below what is offered to students seeking endorsement in both disciplines. The monetary incentive for dual endorsement in math and physics, combined with the understanding that dual endorsement would improve a student's future employability, resulted in a number of physics majors choosing to take additional classes to add the mathematics endorsement. While there is a great national need for physics teachers, most high school physics teachers will need to teach other courses in related disciplines to fill their workload, since few schools offer sufficient sections of physics in a given year.

D. Capacity building and sustainability

Several key events have helped in building capacity and strengthening sustainability, including hiring a tenure-track faculty member in PER, establishing a professional learning community with local physics teachers, and using funds from a state grant to implement a Learning Assistant (LA) program [10].

1. Tenure-track hire in physics education research

A consequence of the department's involvement with PhysTEC was that university administrators viewed the grant as an outside endorsement of our efforts in physics teaching by an organization with national credentials. It is a near certainty that this recognition helped influence the on-campus decision to approve a new tenure-track faculty position in physics education research, which was filled by one of the authors (Brian Frank). It would be difficult to overstate the benefits of having a PER expert in the department to help with the details involved in both introductory sequence redesign and development of the courses specific to physics teaching. Beyond traditional coursework, all graduates of our physics program are required to complete a year-long research project and write a formal thesis. With the hire of a physics education researcher, almost all of our preservice teachers have chosen to be involved in projects related to physics education research. More generally, the sense of community among our physics teaching students has improved as well. These majors see that physics teaching as a career is valued highly within the department, and appreciate that providing a PER mentor was a departmental priority.

2. Professional learning community

Another component of our program that was supported by the PhysTEC grant was development of a professional learning community to help with mentoring and induction of our physics teacher graduates. Our department had previously experienced very positive, yet infrequent, interactions with local physics teachers. Faculty and students involved in the local chapter of the Society of Physics Students had visited high school physics classrooms to present physics demonstration shows; we also interviewed high school students and teachers during the marketing focus groups, and we even participated with several teachers on a Math and Science Partnership grant. However, there was no coherent structure to facilitate meaningful interactions among teachers from different schools, and our future teachers were not involved at any level.

To facilitate a strengthening of this community and to provide mentoring opportunities for our students, the department assumed responsibility for engaging local physics and physical science teachers in continuing education while also introducing our future teachers to these local experts. Our Teacher Advisory Group meetings now occur monthly, and often involve a time for sharing favorite classroom demonstrations, a small catered dinner, and some type of professional development that helps satisfy the continuing education obligations required by teachers' school districts. These meetings were initially funded through the PhysTEC grant, but the department has now assumed the financial responsibility.

Establishing a renewed relationship with local high school physics teachers has had immediate rewards in terms of recruiting majors interested in physics teaching and induction of physics teaching graduates. The inservice teachers are in a position to influence high school student choices of both career trajectory and university, and teachers having a close relationship with members of the MTSU physics program has resulted in more high school students choosing to matriculate at MTSU with a major in physics. A second benefit has been that job openings at local schools are made known to our department by the local high school teachers, and the department is often asked to help recruit our physics graduates to apply for these positions.

3. Learning Assistant program

The Learning Assistant program was developed at the University of Colorado Boulder, and involves the placement of undergraduate students as classroom assistants. Students who have been successful in a particular course are recruited to the program, where they are trained in both content and pedagogy in preparation for their Learning Assistant roles. The responsibilities of student LAs vary, but usually involve having the LAs act as a mentor and resource to students in a lecture or laboratory setting. Many of the activities performed by LAs parallel the efforts of middle and high school teachers, and, as a result, some LAs choose to pursue a career in teaching. Because the Noyce Scholarship grant funds were available for only five years, the department chose to pilot a Learning Assistant program in select sections of our introductory algebra-based physics sequence. During our pilot semester, we applied for and received funds from a state grant to expand to all algebra-based sections. It is our hope that this program will become a robust and sustainable

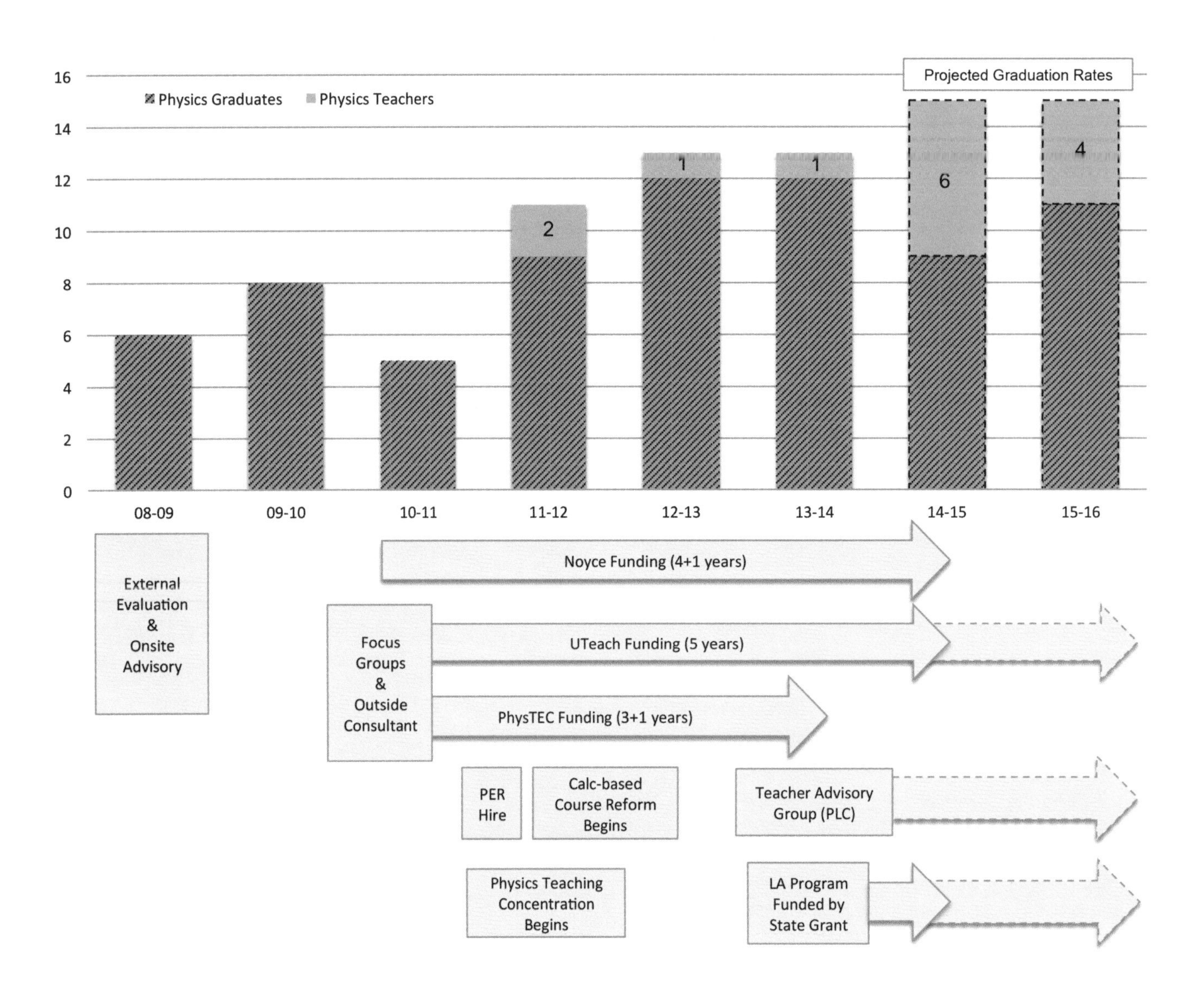

FIG. 1. Graph depicting numbers of physics graduates and physics teachers with accompanying timeline of major events.

recruiting avenue for preservice physics teachers, and also a mechanism for continued course revision.

IV. INDICATORS OF SUCCESS

The Department of Physics and Astronomy has pursued the goals of graduating physics teachers through a transformation of the physics teacher education program at MTSU, and of helping grow the overall number of physics majors. During this process, courses were reformed, new courses were created, a concentration in physics teaching was developed, and many activities were initiated. The following section examines some metrics that are being used to gauge the degree to which the department has achieved success towards realization of these goals.

A. Physics teachers and physics graduates

By far the most significant indicators of success are a growing physics teacher pipeline and increases in the numbers of physics teachers prepared, as well as in the number of physics majors graduated in general. Indicators of the growth of physics teachers and physics majors since 2008 are provided below:

- The physics teacher pipeline grew from zero to a current size of 14.
- Six new physics teachers are projected to graduate during 2014–15, equaling the total number of physics majors graduated in 2008–09.
- The five-year average for physics degrees awarded has increased from four in 2008 to a projected 11 in 2015.

B. Curricular reform

Efforts to increase the numbers of physics majors and the number of physics teaching graduates involved reforming existing courses, introducing new courses, and developing a new concentration in physics teaching. Student performance in the reformed introductory sequence, as measured by the Force Concept Inventory (FCI) exam, marginally increased; however, the DFW rate (the percent of students who earned a D or F or withdrew from the course) showed significant improvements. Changes were made to intermediate courses to incorporate an emphasis on conceptual understanding, and three new courses were introduced specific to the physics teaching concentration, as described in the curriculum modification section above. Enrollments in new classes have not yet met the minimum to avoid the "low enrollment" designation, but numbers have grown to the six-to-eight student range. Outcomes from curricular reforms since 2008 are highlighted below:

- Conceptual emphasis in intermediate courses;
- Increasing enrollment in three courses specific to preservice physics majors;
- Twelve graduates expected by 2015 in new physics teaching concentration; and
- DFW percentage for algebra- and calculus-based physics dropped from 50% to 20% with no decrease in FCI gains.

C. Reaching future and current teachers

Potential physics teaching majors are recruited through direct and indirect marketing strategies. High school students in Tennessee who complete the ACT and indicate an interest in physics, astronomy, or science teaching receive our physics career-oriented brochure and are pointed toward an online interest survey whose completion triggers receipt of a physics T-shirt. This offer has led to a 10% rate of survey completion. The department is also reaching out to science and mathematics majors once they reach campus. The pilot of the Learning Assistant program in the algebra-based physics sequence is an indirect method to market physics teaching as a career, and presentations made in introductory science and mathematics courses are a more direct means of advertising physics teaching as a career option. In addition to recruiting students to become physics teachers, the department's Teachers Advisory Group provides an open channel of communication with high school teachers, with the added benefit of introducing future teachers to current inservice teachers. The inservice teachers become strong advocates for the physics department in their interactions with high school students, and the teachers make the department aware of job opportunities for new physics teachers. Successes related to reaching future teachers and maintaining contact with inservice teachers are provided below:

- State support received for Learning Assistant program;
- Participation in the Teacher Advisory Group among local high school teachers is strong and growing; and
- Students targeted with information from ACT respond to direct mail.

V. TIPS AND LESSONS LEARNED

The path to building a physics teachers program was neither easy nor direct, and others interested in similar efforts will no doubt have an experience that is different from ours. Working with the College of Education to develop early teaching experiences, for instance, was made straightforward due to the university's replication of UTeach. Recruiting students, on the other hand, continues to require significant time and effort. The national Task Force on Teacher Education in Physics [11] addresses issues on a national scale, and our site-specific conclusions mirror many of their findings.

A. Recruit continuously

To recruit students into the physics major, both high school students and undergraduates need to hear about potential physics careers on a regular basis. There are too many high schools for our faculty to visit regularly, so we utilize periodic letters sent to students who have indicated an interest in physics on ACT's Educational Opportunity Service survey. We also rely on members of the Teachers Advisory Group to help promote the physics major, and physics teaching as a possible career, in their high schools. Information sessions about physics teaching at the university may interest a few dozen students, but very few follow up on their curiosity after a single presentation. The best recruiting efforts occur often and with regular frequency. A version of the presentation for students enrolled in introductory courses is shown during Physics I, Physics II, and Modern Physics. This information is provided to physics majors in the freshman seminar course, and we have piloted reaching out to students who are completing Calculus II. The general theme seems to be that large efforts lead to small results, and small efforts achieve almost nothing.

B. Encourage multiple endorsements

Reports from the AAEE [1] clearly show a nationwide need for individuals with licensure to teach high school physics. Schools in rural areas and those that offer few physics sections, however, are unlikely to hire a teacher who is licensed only to teach physics. For this reason, our department strongly encourages all physics teachers to add additional endorsements to their license. Many will choose to add mathematics, due to the large number of courses they have already taken in that discipline, but chemistry is another good choice. Both mathematics and chemistry are perpetually listed with physics at the highest level of need, or "considerable shortage," in the AAEE report. The process for adding endorsements varies by state, but can be as simple as passing an additional content exam.

C. Provide support as students transition

The transition to the workforce begins before graduation. One lesson learned is that students can use guidance regarding interviewing for positions and in setting expectations for their first assignments. For instance, seniority is respected in school districts, and new teachers should not expect to have a single preparation of AP Physics. Rather, they should understand that a new teacher is likely to teach less desirable classes and more than one preparation during the first year.

It is widely believed that the most difficult period for a teacher is the first year. As a result, both PhysTEC and UTeach require participating institutions to develop a plan for the induction process that often involves periodic meetings with a mentor at the university, faculty visits to the high school classroom, and communication through electronic means. For most new teachers, these efforts combine to help ease the transition year. There are occasions, however, when new teachers need additional time and attention. MTSU had a December graduate choose to take a position teaching physics at a school where the prior person left in the middle of the year. Faculty should have counseled this new teacher to investigate the situation before accepting the position. MTeach program faculty spent many hours helping this new teacher survive that initial semester, and the department learned a valuable lesson.

D. Hire individuals with expertise in PER

It is natural to think of hiring a physics education researcher when building a physics teaching program. That said, the benefits of having someone trained in PER extended well beyond our initial expectations. It was anticipated that a PER expert would contribute to the design of physics pedagogy courses, help mentor future teachers, and forge relationships with colleagues in the College of Education. What was not foreseen was the degree to which a PER expert would benefit the department in course improvements outside of formal redesign, and how the attitudes of faculty would move toward continuous course improvement. Faculty in the department began to critically analyze their own teaching methods because there was a resource in the department who could vet ideas and plans. Some faculty have added clicker questions to their lectures, others have sought opportunities to involve Learning Assistants, and many more have become interested in how their physics sections performed on pre- and post-tests compared to sections taught by other faculty members. The department was previously content with the course reforms in algebra-based physics, but with the addition of a PER faculty member, the department instigated a new cycle of improvement—not out of obligation, but because we want to provide the best possible education for all physics and astronomy students.

VI. CONCLUSIONS

Development of a physics teacher education program at MTSU has enabled the department to graduate a number of high school physics teachers, and has also facilitated many more benefits. Physics major graduation rates have increased, continuous course improvement has become the department's mantra, and faculty attitudes have shifted beyond preparing great candidates for graduate school. The department is becoming an important resource for local physics teachers, and is preparing our physics teaching graduates for a career that can be both challenging and fulfilling.

ACKNOWLEDGMENTS

The authors would like to thank Theodore Hodapp of the American Physical Society and Gay Stewart, previously of University of Arkansas, for advice in designing the physics teacher program at MTSU, and Vic Montemayor for the time spent on course reform and curriculum design. Financial support was provided by PhysTEC and the NSF Robert Noyce Teacher Scholarship program.

---

[1] American Association of Employment in Education, *Educator Supply and Demand in the United States* (2008).

[2] See http://www.per-central.org

[3] Tennessee Department of Education, *Tennessee Licensure Standards and Induction Guidelines* (2009). http://www.state.tn.us/education/licensing/docs/lic_educator_licensure_standards.pdf

[4] See http://www.phystec.org/keycomponents, retrieved Sep. 8, 2013.

[5] See http://uteach-institute.org/replicating-uteach

[6] R. Beichner, North Carolina State University: SCALE-UP, in *Learning Spaces,* edited by D. G. Oblinger (Educause, Boulder, CO, 2006), pp 29.1-29.6. http://net.educause.edu/ir/library/pdf/PUB7102.pdf

[7] R. W. Chabay and B. A. Sherwood, *Matter and Interactions*, 3rd ed. (John Wiley and Sons, Hoboken, NJ, 2010).

[8] L. C. McDermott, P. S. Shaffer, and the Physics Education Group at the University of Washington, *Tutorials in Introductory Physics* (Prentice Hall, Upper Saddle River, NJ, 2002).

[9] See http://nsfnoyce.org

[10] See http://laprogram.colorado.edu

[11] D. E. Meltzer, M. Plisch, and S. Vokos, eds., *Transforming the Preparation of Physics Teachers: A Call to Action. A Report by the Task Force on Teacher Education in Physics* (T-TEP) (American Physical Society, College Park, MD, 2012).

APPENDIX

MTSU Physics Teaching Concentration Curriculum

Course	**Credit hours**
Physics Freshman Seminar	(1)
Physics I, II	(8)
E&M/Thermodynamics	(3)
Modern Physics	(3)
Introduction to Quantum	(3)
Advanced Laboratory	(1)
Physics Licensure I, II	(2)
Teaching of Physics	(3)
Physics Research	(2)
Physics Thesis	(2)
Physics Practicum	(1)
Physics Senior Seminar	(1)
Astronomy	(3)
Calculus I, II	(8)
Differential Equations	(3)
Statistics	(3)
Biology I	(4)
Chemistry I/II	(8)
General Education (remaining)	(30)
Secondary Education	(33)
TOTAL	**(122)**

Recruiting and Retaining Future Physics Teachers

Recruitment and retention of physics teacher education majors

Carl J. Wenning
Department of Physics, Illinois State University, Normal, Illinois, 61790

There is a large and growing problem associated with high school physics teaching. There simply is not an adequate supply of secondary-level physics teaching graduates to fill the void created by the retirement and attrition of experienced high school teachers. In order to fill that void, physics teacher education programs around the U.S. must both increase and improve their efforts to recruit, prepare, and retain the next generation of high school physics teachers. This chapter focuses on the successful efforts of one post-secondary institution, Illinois State University, to help fill the void. Effective recruitment and retention practices at the university have resulted in a teacher education sequence within the Department of Physics that has attracted teacher candidates in growing numbers. This chapter uses a historical approach in describing program growth and provides a summary of lessons learned. It attempts to make clear that effective recruitment and retention practices cannot be separated from the need for a high-quality high school physics teacher education program.

I. INTRODUCTION

Something needs to be done to address the growing problem that America's high schools have an insufficient supply of qualified physics teachers. Fortunately, there is a large supply of interested and altruistic individuals among today's science students who can and will join the science teaching profession provided someone encourages and promotes this career choice to them. Without support from inservice teachers, and from community college and university science faculty, it is unlikely that the physics teacher supply problem will be solved.

In Illinois, crossover science teachers (e.g., biology teachers with little or no physics background providing physics instruction) fill a significant number of physics teaching positions. According to the Illinois State Board of Education, nearly 2,500 high school *science* teaching positions will need to be filled by qualified teachers during the next five years [1]. The number of *science* teachers graduating from Illinois teacher preparation programs is far less than the necessary 500 per year, and the number of authentically qualified physics teachers is fewer still. The implications of these data are disturbing.

Illinois State University (ISU), working with the Illinois Section of the American Association of Physics Teachers (ISAAPT), has been on the forefront of developing the Illinois physics teacher education "Pipeline Project" to help solve the problem of too few authentically qualified high school physics teachers [2]. Now, after many years of effort, ISU is seeing a growing number of undergraduate physics teacher education (PTE) majors with an increasing number of program graduates. We offer here our experiences for improving the recruitment and retention of PTE majors, in the hope that faculty at other institutions will find this example useful.

II. PHYSICS TEACHER EDUCATION AT ILLINOIS STATE UNIVERSITY

ISU was a "normal school" from its founding in 1857, as its original name—Illinois State *Normal* University—indicated. As the oldest public university in Illinois, ISU has a strong tradition of graduating schoolteachers; that tradition continues to this day. Teacher preparation is a university-wide effort that now includes five colleges. About one fourth of the 20,000 students who attend the University are enrolled in a teacher preparation program. ISU ranks as one of the top 10 largest teacher preparation institutions among American Association of Colleges for Teacher Education (AACTE) institutions. Approximately one in eight current Illinois public school teachers graduated from ISU. This tradition as a "normal school" has some relevance in students choosing to become high school physics teachers by enrolling at ISU, but it is by no means a primary influence.

The PTE major at ISU is one of four sequences offered by its physics department. The other sequences are physics, engineering physics, and computer physics. PTE majors share a common core of content courses with the other sequences, and are generally required to take the same content courses as other physics sequences through their sophomore year. Starting with the junior year, PTE majors begin taking more education-related courses and fewer physics content courses than do students in the other sequences. There are six required physics teaching methods courses, the first of which is taken during the autumn of the sophomore year by "native" students. Junior-level transfer students take their first physics teaching methods course during their first autumn at the university. The remaining five methods courses are taken during the junior and senior years.

It has long been the tradition at ISU that preparation of high school teachers is a joint effort of the various subject matter departments and the College of Education. The

College of Education provides generic teaching and social context courses as part of the overall teacher preparation process; the subject matter departments provide the specialized content and methods related to their respective majors, conduct advising, manage clinical experiences, and supervise student teaching. This arrangement allows a large degree of freedom in the preparation of teachers, and one that the ISU Department of Physics capitalized on to create its large and successful program.

As of Spring 2014, the ISU Department of Physics had 11 faculty members, including the chairperson and five academic support staff. Among these are two full-time physics teacher educators and one experienced PTE advisor (semi-retired). There were 41 undergraduate PTE majors in a department of 126 majors. PTE majors make up nearly one third of all declared majors within the department.

Until 1994, when the ISU PTE program came under new leadership, it was much like many other PTE programs around the nation. There was one PTE graduate in 1973 and another in 1977. There were no PTE graduates from 1978 through 1986. From 1987 to 1994, there were 10 graduates. This gave an average graduation rate of about 0.8 majors per year from 1973 through 1994. Over the past decade, the average number of PTE majors graduating per year was 6.6. Figure 1 shows that, over the past three decades, the number of new high school physics teachers graduating from ISU has trended upward, having grown to total more than 100 by 2014.

The growth in the number of PTE majors since 1995 has been roughly linear, but began to level off in 2006. Figure 2 shows a steadily increasing number of majors over the years, and includes several drops in the number of majors following years with larger graduating classes.

III. DEVELOPMENT OF THE ISU PHYSICS TEACHER EDUCATION PROGRAM

There appears to have been a "threshold event" that led to the increasing numbers of PTE majors and program graduates at Illinois State University. That event was a broad and sustained commitment from the physics department to improve the quality of its physics teacher education program. There was among department faculty a critically important belief upon which all efforts to improve recruitment and retention were predicated—that *significant improvements in recruitment and retention would occur if and only if the physics teacher education program increased in quality.*

Starting with the 1986–87 academic year, the Department of Physics took increased responsibility for preparing PTE majors from the College of Education. That year a faculty member who previously had taught high school physics took charge of the department's PTE program as a small part of his regular duties. A new three-semester-hour course—PHY 301, Teaching High School Physics—was offered during autumn semester of the senior year. As a low-enrollment course, it was taught informally using a Personalized System of Instruction format [3]. Students completed independent readings using a variety of resources. These resources generally consisted of a number of short guidebooks from the National Science Teachers Association: *Safety in the Classroom*, *Legal Concerns of Teachers*, and so forth. The course instructor then discussed these readings with teacher candidates in periodic office meetings. Eleven major topics were addressed during the semester, and student knowledge was assessed through a series of examinations. Students also completed clinical experiences in area middle schools and high schools, as well as student teaching, under College of Education supervision.

At the start of the 1993–94 academic year there were six undergraduate PTE majors in a department of some 120 majors. A new, one-semester-hour physics teaching methods course dealing with computer technology—PHY 302, Computer Applications in High School Physics—was introduced. This course consisted solely of students building a photogate for use with an Apple IIe or IIGS computer. This second course was taken during the same semester as the physics teaching methods course PHY 301.

In July 1994, a new PTE program director replaced the retiring director. The new director was a certified high school physics teacher who had been a member of the Department of Physics since 1978. Revisions to the existing PHY 301 and 302 courses were rapidly made, with the intention of graduating better-prepared physics teaching candidates. Supervision of clinical experiences and student teaching became the purview of the PTE program at this time. These activities would now be directed, supervised, and evaluated in ways more consistent with expectations of the department.

During the Department of Physics faculty retreat in the summer of 1998, the PTE program director was given the opportunity to explain why replacing one physics content course *elective* with a *required* physics teaching methods course would be desirable. The goal was to supplement PHY 301 and 302 with a readings course. After receiving a detailed clarification of needs, the faculty agreed to allow the creation of a new required course, PHY 310, Readings for Teaching High School Physics.

The general education program at ISU changed with the 1998–1999 academic year, and PHY 101, Exploring the Universe, was replaced with PHY 205, Origin of the Universe. This new general education course was much more physics intense and served both general education and major requirements. PHY 301 was completely redeveloped and was replaced by PHY 311, Teaching High School Physics, during the 1999–2000 academic year.

Beginning with the 2001–2002 academic year, student teachers were required by the university to be full-time. Prior to the 2001–2002 academic year physics student teachers took only eight semester-hours of credit during student

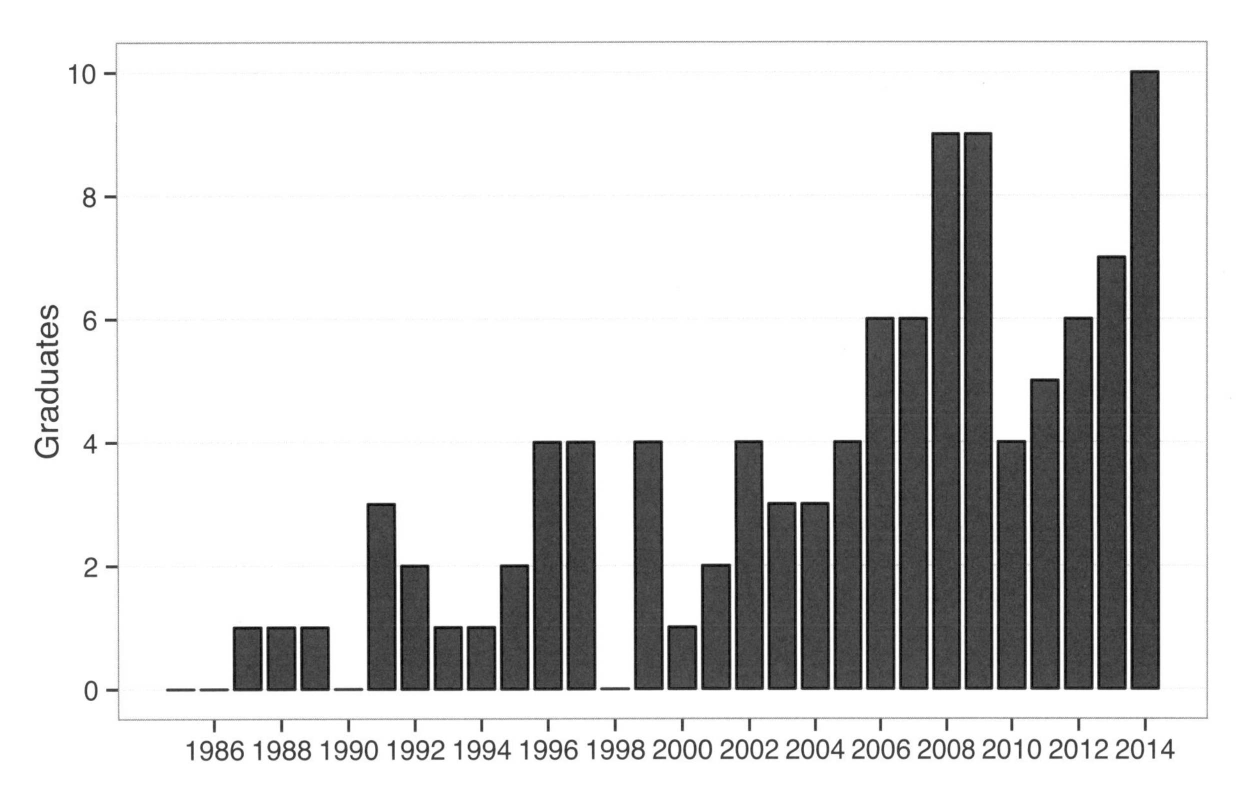

FIG. 1. Number of PTE graduates per year since 1985.

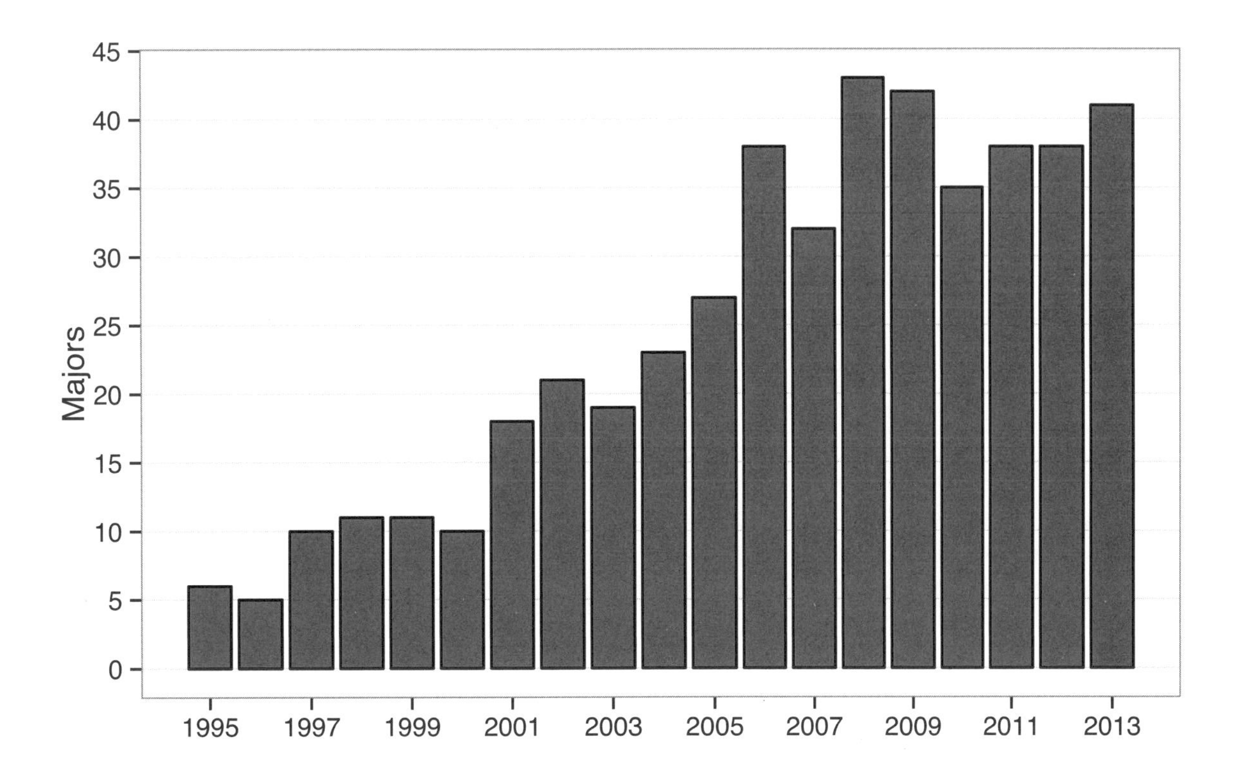

FIG. 2. Autumn enrollments of PTE majors by year since 1995.

TABLE I. Comparison of changing science content course requirements.

39 Semester-Hours of Required Science Courses during Academic Year 1994–1995	43 Semester-Hours of Required Science Courses during Academic Year 2013–2014
PHY 101 (3) *Exploring the Universe*	PHY 107 (2) *Frontiers in Physics*
PHY 110 (4) *Physics for Science/Engineering I*	PHY 110 (4) *Physics for Science/Engineering I*
PHY 111 (4) *Physics for Science/Engineering II*	PHY 111 (4) *Physics for Science/Engineering II*
PHY 112 (4) *Physics for Science/Engineering III*	PHY 112 (4) *Physics for Science/Engineering III*
PHY 217 (3) *Methods of Theoretical Physics*	PHY 205 (3) *Origin of the Universe*
PHY 220 (3) *Mechanics I*	PHY 217 (3) *Methods of Theoretical Physics*
PHY 240 (3) *Electricity & Magnetism I* or	PHY 220 (3) *Mechanics I*
PHY 284 (3) *Quantum Mechanics I*	PHY 240 (3) *Electricity & Magnetism I*
PHY 270 (2) *Experimental Physics*	PHY 270 (2) *Experimental Physics*
PHY 2XX (3) Any 200-level physics course	
Also required: Chemistry 140 (5) *General Chemistry I* and Chemistry 141 (5) *General Chemistry II*	Also required: Chemistry 140 (4) *General Chemistry I,* Chemistry 141 (4) *General Chemistry II,* Biological Science 101 (3) *Fundamental Concepts in Biology,* and Geology 100 (4) *Earth Science Systems*
Adding teaching methods courses PHY 301 (3) and 302 (1), the total number of science-related courses was 43 semester-hours.	Adding teaching methods courses PHY 209 (1), 302 (1), 310 (3), 311 (3), 312 (3) and 353 (1), the total number of science-related courses is 55 semester-hours.

teaching (making them part-time students). The new mandate meant that they subsequently had to enroll for 12 semester-hours of credit during the student teaching semester. This four-semester-hour deficit allowed for the creation of two new courses to be taken during the five weeks immediately prior to student teaching but during the same semester. These courses were PHY 312, Teaching Physics from the Historical Perspective, and PHY 353, Student Teaching Seminar. The start of student teaching was delayed until early February, but it still provides adequate time for completion prior to ISU graduation. The PTE sequence now requires a total of 12 semester-hours of physics pedagogy distributed over six courses over four to five semesters. A comparison between the catalog requirements of 1994–95 and today can be found in Table I.

While today's PTE program at ISU is geared toward preparing teaching physics for the high school level, it also employs a broad-field science preparation model. In 2006 the Illinois State Board of Education added content course requirements to all secondary science teaching majors. All high school science teacher candidates must now complete a core of science content courses consisting of introductory-level biology, chemistry, geology, and physics. Astronomy and chemistry were existing requirements of the PTE program, so only biology and geology had to be added. Now, after graduating from the ISU PTE program, our candidates are legally permitted to teach not only physics but also biology, chemistry, earth and space science, and environmental science at introductory levels. Outside of chemistry, which many of our graduates will teach, our physics graduates are rarely called upon to teach other disciplines, due to the comparatively large numbers of graduates in these teacher preparation fields.

Room for the additional biology and geology courses, as well as for new physics teaching methods courses, came from reduced requirements in the university's general education program, from replacing a 200-level physics content course elective with a required physics teaching methods course, from reduction in credit hours associated with Chemistry 140 and 141 (which both went from five semester-hours to four), and from optional courses required of students to meet the 120-semester-hour graduation requirement. Over the course of many years, every opportunity to make change in plans of study was utilized to the benefit of the PTE program.

IS THE ISU SITUATION UNIQUE?

This is one of the first questions those seeking to reform their physics teacher education programs might ask. It relates strongly to the question of whether the results achieved at ISU can be replicated elsewhere. Clearly the ISU reform was site-specific in many ways. PTE at ISU is under the control of the Department of Physics and not the College of Education. This afforded excellent opportunities for change. PTE leaders took advantage of rare—but not unique—changes foisted upon it by new requirements in teacher education, university and departmental accreditation, certification law, general education requirements, and changes in credit hours associated with certain courses. Rather than seen as something to be resisted, the required changes were embraced as opportunities for reform. PTE also took advantage of existing physics options within the established sequence.

Were all these changes foreseen as part of an original plan of PTE reformers to come up with a 12-semester-hour physics methods course sequence? Assuredly not; it just turned out that way. And as changes took place, the program gradually improved.

Will the situation be the same at the reader's home institution? The answer is almost certainly in the negative. Nonetheless, those who take the long view will likely encounter opportunities to make changes, with each change leading to an improvement in the quality of the teacher preparation process.

No matter what changes take place at other institutions, what counts is that the quality of the program increases as a result. Effective recruitment and retention will likely occur only in the presence of a quality PTE program.

In 2008, a master high school science teacher was hired to direct ISU's PTE program; the former program director (now semi-retired) continued to advise and assist him for the next four years by teaching some of the physics methods courses during this transition period. The new program director has 25 years of high school science teaching experience and received a 1993 Presidential Award for Excellence in Mathematics and Science Teaching. In 2012, a newly hired faculty member joined the current director in teaching within and managing the growing PTE program. Today the program is formally associated with 2.0 FTE positions, with other faculty members providing content area preparation.

It has been a long-held belief of the ISU program directors that only a high-quality teacher preparation program would serve as sufficient enticement to successfully recruit prospective PTE majors, and retain them once in the major. Still, we have found that providing a quality physics teacher education program is not the only thing required to achieve growth in the number of PTE majors and program graduates. Conscious efforts also must be taken to recruit and retain PTE majors.

ONE SIZE FITS ALL?

Do all PTE programs have to look the same as that at ISU? Assuredly not! Different models of teacher education likely will lead to different results—some better and some worse. Different things work for different programs and different institutions. For instance, Teachers-in-Residence and Learning Assistants might work well at some institutions and not so well at others. Each PTE program will be unique, and this provides an opportunity for experimentation, with results at multiple institutions constitute independent laboratories to study change. To the author's knowledge, there is no proven best way to prepare teachers.

IV. TEACHER CANDIDATE RECRUITMENT

Considerable attention has been paid to high school physics teacher candidate recruitment and retention at ISU since the mid-1990s. Special concern continues to be shown for the broader physics teacher "pipeline" that conducts graduating high school students through teacher preparation and channels them back to the high school classroom as teachers.

A. Indirect recruitment

Much of the recruitment for the ISU PTE program is of a long-term, indirect variety. The goodwill of our program graduates and that generated through summer teacher workshops for inservice physics teachers appear to have had significant positive impacts on enrollment in the PTE program. Based on their experiences with our program, inservice physics teachers will often advise students seeking to become high school physics teachers to enroll at ISU.

Professional development opportunities for inservice teachers have been numerous. From 2001 to 2005 ISU was an American Association of Physics Teachers (AAPT) Physics Teaching Resource Agent rural center, offering summer professional development activities for inservice physics teachers within about 100 miles of campus. During 2002, 2003, and from 2005 to 2007, two- and three-week Modeling Method of Physics Instruction workshops were presented for inservice teachers. Professional development workshops such as Physical Science with Math Modeling, Problem-Based Learning, Physical Science Alliance, and Mini Modeling were offered in 2001, 2002, and yearly from 2009 through 2014. All these activities were supported with grants.

The leadership of the ISU PTE program has been actively engaged with the Illinois Section of the American Association of Physics Teachers since 1994. During the section's semi-annual meetings, numerous workshops and presentations have been given about physics teacher preparation at ISU. These activities engage community college teachers who compose much of the association's membership. They too become informal recruiters for our PTE program and are helpful in directing transfer student to the ISU campus.

During the past several years, the ISU physics department has worked with the ISAAPT to establish local physical science alliances within Illinois. ISU maintains such an alliance, through which some recruitment activities are channeled. Other long-term, indirect recruitment efforts addressing young children include the outreach program Physics on the Road, with hands-on activities; an active physics department planetarium; an annual "Expanding Your Horizons through Math and Science" program for middle school girls; and involvement with the local Children's Discovery Museum and Challenger Learning Center. Most recently, the semi-retired PTE coordinator has started teaching an inquiry-oriented introductory high school physics course for home-schooled students, several of whom are especially gifted.

Other indirect recruitment makes use of the World Wide Web. Today's high school students are Internet savvy. They know how to use search engines to learn about careers and about preparation for those careers they choose to follow. ISU maintains an extensive PTE web presence that we believe is an effective recruitment instrument. It shows with clarity that ISU has dedicated time, treasure, and talent to the task of preparing the best high school physics teachers possible [4].

B. Direct recruitment

The ISU physics teaching program has been actively involved with the ISAAPT in direct teacher candidate recruitment for the past few years as well. In 2004 the ISAAPT established a standing committee for coordinating statewide efforts in the recruitment, preparation, and retention of high school physics teachers. Starting with its "Pipeline Project," the ISAAPT along with three other Illinois professional organizations (the Chicago Section, American Association of Physics Teachers; the Illinois Science Teachers Association; and the Illinois Association of Chemistry Teachers) collaborated to create, publish, and distribute a six-panel recruitment brochure for students and an eight-page recruitment booklet for teachers [5,6]. These were disseminated to nearly every high school science teacher in Illinois over the next two years. The group also created (and continues to maintain) a recruitment web page for prospective science teachers [7].

Complementing all these efforts are direct efforts aimed at graduating high school seniors. These include campus visits, personal phone calls from student majors, and a departmental scholarship program. Numerous ISU-specific brochures for all four sequences within the physics department are also widely distributed.

V. TEACHING MAJOR RETENTION

Each year PTE programs "lose" students as a result of graduation. At other times—and this is the real problem of retention—students drop out of the teaching major or the university prior to graduation. The reasons for doing so are as varied as the students themselves. Some students switch to another physics-related sequence within the department due to changing interests after learning more about physics. Others leave the major due to loss of subject matter interest, poor academic performance and/or emotional distress, and lack of financial resources.

If recruitment efforts are to pay dividends, adequate attention must be paid to retention in the major by addressing as many of these situations as possible. At ISU we have instituted a number of retention practices that we believe help reduce the loss of PTE majors. These practices include quality advising, motivational support, early PTE coursework, and scholarships.

A. Quality advising

Two types of students enroll at ISU with the intention of becoming PTE majors: native students (those who arrive as freshmen directly after high school graduation) and transfer students (typically enrolled with junior status and coming mostly from community colleges). In both cases we make every effort to seek out, meet, and advise these teacher candidates as early as possible. At the first formal meeting with PTE advisees, advisors give a hearty welcome, introduce the PTE program in detail, and work out a plan of study that will most cost-effectively allow students to graduate in a timely fashion—typically four years for native students and two years for transfer students. Advisors point out that the PTE plan of study can be a minefield for the unwary. Because so many courses in the major are in a lock-step sequence, failing to take a particular course at a critical point in time—or a grade of less than C in any required course—can lead to a delay in graduation. We stress this point during advising sessions, noting that mistakes can result in, perhaps, another year in school and the concurrent loss of income from the first year of teaching. Advising sessions are fully documented so that students walk away with a written record of what they should or might do.

Another focus of the advising meeting is academic and personal success. Advisors introduce PTE majors to SAAMEE (Success = ability • ability • motivation • effort •

environment), which is a hypothetical model that we believe can lead to increases in academic success [8]. This model states that a student's academic success is a function of innate ability, learned ability, motivation, effort, and environment. Advisors point out that there are significant differences between the behaviors of successful students and those who are less successful.

Advising sessions are holistic. Advisors not only check with each student about academic progress, but also try to address any problems—academic or social—that can lead to the loss of teacher candidates. Students are frequently directed to a web page titled "Dr. Wenning's Advice for PTE Majors" that was assembled following more than three decades of post-secondary teaching [9]. The web page includes content on (1) achievement and goals; (2) studying, learning, and time management; (3) professional development; (4) personal development; and (5) personal and professional integrity.

B. Early PTE coursework

Our experience has shown that majors sometimes will drop out of the PTE program early if associated physics methods coursework is delayed until later in the college career. One of the strengths of the current PTE program is that we offer the first physics teaching methods course (PHY 209, Introduction to Teaching High School Physics) during the autumn semester of sophomore year for native students and the first autumn semester for transfer students. The course is designed to allow current and prospective PTE majors to test their actual or possible decisions to become teachers.

During PHY 209 students learn about teaching, experience model inquiry lessons, review and discuss professional teaching standards, review indicators for assessing teaching performance, and relate findings from clinical experiences. Clinical experiences (consisting of both observations and interactions) are assigned shortly after the first week of class. Clinical students turn in written reflections for each clinical observation, following a structured format that includes three types of written information: descriptive, analytical, and reflective. This helps clinical students think systematically about teaching and increases learning associated with each clinical experience. There are two frameworks for each student's clinical experiences: a 25-hour service-learning project and an eight-hour urban studies field trip to Chicago. The high school service-learning project often solidifies a teacher candidate's desire to become a high school physics teacher. The urban studies field trip inspires many students and shows them the need for making a difference in the lives of students that propels so many of our majors to become teachers in the first place. Each year we lose one or two PTE majors from among those who take this course. Some decide based on these experiences that they do not want to become high school teachers. Still, of the 92 PTE majors who took the course from 1999 to 2010, 79.3% went on to graduate with a physics teaching degree.

Seniors, juniors, and a select few sophomores enrolled in the PTE major are also encouraged to work as teaching assistants and lab supervisors within the physics department. Nearly all PTE majors will serve as volunteer tutors through the ISU Physics Club. These activities also inspire and strengthen our majors for the hard work ahead.

C. Motivational support

Each of our PTE courses has activities that we believe serve to positively motivate our majors. These activities include an urban studies field trip (PHY 209); a 25-hour service-learning project in area high schools (PHY 209); a capstone research project using appropriate technology (PHY 302); a Learning Assistant experience and/or study group mentoring (PHY 310); model inquiry lessons (PHY 310, 311, 312); whiteboarding and Socratic dialogues (PHY 310, 311); a Japanese lesson study project (PHY 311); a theory-into-practice project (PHY 311); nature-of-science case studies (PHY 311, 312); and a social context project (PHY 353). All of these learning activities—many of which are conducted cooperatively with other majors—include scoring rubrics and authentic assessments that help students understand the science teaching profession and build camaraderie, and encourage them to remain in the program.

Another significant aspect of all physics teaching methods courses at ISU is our assessment-as-learning policy. This policy, adapted from Alverno College in Milwaukee, communicates high academic standards and provides strong motivational support to PTE majors. Assessments of students' performances on essays and other projects are used not only to assign grades but also to improve those performances. Unsatisfactory work is returned to students for improvement. Students' scores can be improved following appropriate revision and resubmission, as long as deadlines are met. Students who have submitted work for review two to three times before submitting the final copy typically produce the best work in a course. This policy was established consistent with the belief that all teacher candidates must demonstrate the ability to meet expectations in all areas of teaching.

Our students almost always remark very positively on their learning experiences in area high schools, where they work with students, teachers, administrators, and support staff. While the minimum number of hours of clinical experience required prior to student teaching is set at only 100 by state law, our PTE majors often accumulate as many as 200 hours in the classroom prior to beginning student teaching. Many of these activities are conducted in classrooms where they will student teach, thereby providing a natural transition process.

D. Scholarships

Many PTE programs have students with financial difficulties. A lack of adequate financial support causes some students to drop out of school for a time or even altogether. Recently, at the prompting of and with the support of advisors, several PTE majors have received Robert Noyce Teacher Scholarships. These scholarships, available to ISU mathematics and science teaching majors since 2009, provide junior- and senior-level students with the financial support necessary to stay in school.

In recent years again, following faculty nominations, three of our majors have received AAPT's Barbara Lotze Scholarship. In addition, three more PTE majors have receive ISU's highest academic accolade, the Robert G. Bone Scholarship. One of our PTE majors recently received the largest cash scholarship ever given by the university's College of Arts and Sciences to an enrolled undergraduate—the inaugural Laurine Reiske Scholarship—valued at some $9,000 at that time.

Can we say that any one of the above efforts is more important than the others when it comes to successfully retaining PTE majors in the program? Not really. All of these activities exhibit to teacher candidates the dedication that the faculty and staff of the ISU physics department have for our physics teacher education majors. These activities work synergistically to give this impression.

VI. POST-GRADUATION RETENTION IN TEACHING

Retention in the physics teaching major is meaningless if retention in the field of high school teaching post-graduation does not occur. The ISU Department of Physics has remained in contact with most of its PTE graduates over the past two decades, since major revisions in the PTE program began. From Spring 1995 through Spring 2013 there have been 90 program graduates. Today some 72% to 74% of these graduates are working in schools. This includes those who have moved up administratively to become principals or other types of administrators but who are still involved in education. These numbers are high in relation to the well-recognized fact that almost half of all new teachers leave the teaching profession within about five years of starting [10]. Research shows that regardless of the school poverty level, science teachers have especially high departure rates and are more likely to leave the teaching profession compared to other subject matter teachers [11]. This might be due to significant opportunities outside of teaching for people with science education backgrounds.

A. Transitional support

Most of ISU's PTE program graduates come from large metropolitan centers within Illinois, especially the Chicago area. It is therefore only natural for graduates to return to the same areas, where they will teach only physics in large high schools with adequate teaching materials, supportive physics teaching colleagues, and physics teaching networks or alliances. These graduates rarely need transitional support from ISU. This is often not the case for our graduates who choose to work in small towns and rural areas, where they sometimes turn out to be the only science teacher in the region's high school. PTE faculty at ISU not infrequently receive emails and phone calls from these teachers seeking advice and sometimes asking to borrow equipment. We do the best we can to meet their needs.

It has been reported by many recent graduates—no matter where they teach—that one of the best forms of transitional support comes from the resources found on the PTE program's website. Here students are able to access massive amounts of information about a wide array of topics that they previously encountered during the physics teacher preparation process. Information about these resources, including syllabi hyperlinked with extensive teacher education resources including scoring rubrics and assessments, can be found online [12].

B. Professional development

Offering high-quality professional development courses helps cement the bond between ISU and our "frontline" recruiters—inservice high school physics teachers—no matter if they are ISU graduates or not. To this end we regularly offer the following summer courses, which are sometimes associated with a master's degree offered through the College of Education:

- PHY 400, Independent Study
- PHY 413, Teaching High School Physics II
- PHY 429.02, Introduction to Modeling Method of Instruction
- PHY 429.03, Second Semester Topics in Modeling
- PHY 498, Professional Practice

VII. INDICATORS OF PROGRAM QUALITY

Just how successful is the ISU physics teacher education program? There are multiple indicators that might be reported that go beyond enrollment and graduation numbers. For instance, the program is clearly innovative with its six required PTE courses and wide array of innovative approaches to student learning. Other indicators that might be considered include:

Program accreditation. Illinois State University's teacher education program is fully accredited by the National Council for the Accreditation of Teacher Education (NCATE) and its physics teacher education program is fully accredited by both the National Science Teachers Association (NSTA) and the Illinois State Board of Education. NCATE and NSTA standards for teacher preparation are demanding.

Retention in teaching. Data from graduates starting in 1987 and extending through 2013 (the last year for which we have a complete set of data) show that we had 93 PTE graduates. *Of those who began teaching following graduation* (83, or 89%), 80 are still teaching or otherwise working in education. That represents better than a 90% retention rate in the area of education. We attribute this exceptional rate in part to the belief that our graduates are well prepared for the vocation that they have chosen.

Other indicators. While the above data provide interesting insights, the question remains about just how good our graduates really are. Again, lacking detailed data, we refer to the following recent indicators. During the past three years alone, one or more of our PTE grads have distinguished themselves in a number of prominent ways: earning National Board Certification from the National Board for Professional Teaching Standards, being recognized as an Outstanding High School Science Teacher by the Illinois Science Teachers Association, becoming an AAPT Physics Teaching Resource Agent, being named an Albert Einstein Distinguished Educator by the U.S. Department of Energy in cooperation with Triangle Coalition, and being nominated as an Illinois finalist for Presidential Award for Excellence in Mathematics and Science Teaching by the National Science Foundation. One recent graduate has served as an invited speaker to lecture and conduct workshops on an international level. Another recent graduate was hired by ISU's University High School to help continue preparing PTE majors.

Admittedly, all this information constitutes anecdotal evidence for the successful nature of our program. While there are other metrics for measuring program success, the most important of them—quality teaching—is very difficult to assess. It is a well-recognized fact that nothing is more important to student learning than highly qualified teachers. Admittedly, we do not have the hard, empirical evidence to show convincingly just how good our graduates actually are. Nonetheless, we do have a program of an increasing size with a steadily growing number of graduates who stay in teaching to suggest that we are on the right track.

VIII. PROGRAM CHANGE PRINCIPLES

Thomas Kuhn, writing about scientific revolutions, mentioned that new ideas rarely triumph over old ideas; it's just that new ideas come to dominate when adherents of old ideas pass from the scene [13]. While that might be the case with revolutionary ideas, the evolutionary changes that occurred within the ISU PTE program allowed for gradual modifications of the program without requiring program directors to wait for generational change.

Program change at Illinois State University was made possible by a set of conditions prevailing within the Department of Physics and the university during this time. These changes, which involved adding new physics teaching methods courses (sometimes at the expense of a physics content course), were predicated on the acceptance by the department's faculty that more physics content knowledge—beyond a certain minimum—does not necessarily make for a better high school physics teacher.

HOW MUCH PHYSICS IS ENOUGH?

Does having more physics knowledge beyond a certain minimum make for a better physics teacher? This question from skeptics of PTE program revision often has to be addressed. In the case of ISU, the following two responses were given: (1) Typical high school physics teachers who spend most of their school year teaching introductory topics rarely need advanced knowledge provided by courses such as quantum mechanics, solid state physics, thermodynamics, electronics for physicists, mathematical methods of physics, or more advanced topics in mechanics, electricity, and magnetism. Such sophisticated physics subject matter is rarely taught at the high school level. (2) If more physics content knowledge makes for better physics teachers, then those with Ph.D.'s necessarily should be better physics teachers than those with B.S. degrees. Remember who recruits physics majors for universities; it is high school physics teachers that inspire the most students to consider physics-related careers. In ISU's experience, few physics majors of any stripe are recruited from introductory physics courses at the university level.

Teacher educators at other institutions wishing to improve the both size and quality of departmentalized programs at their own institutions will benefit from five change principles that characterize conditions for growth experienced at Illinois State University:

- Change Principle 1: *A team of people is needed who are personally committed to improving the teacher preparation process.* Experience has shown that one or more "champions" are often needed to lead transformation of the teacher preparation process [14].
- Change Principle 2: *A team of people is needed who deeply understand the teacher preparation process.* It is important to have qualified physics teacher educators who possess the appropriate content, pedagogical, and pedagogical content knowledge, as well as experiences and skills described by the Center for Mathematics, Science, and Technology Commission on the National Institute for Physics Teacher

Educators in "Professional Knowledge Standards for Physics Teacher Educators." [15]

- Change Principle 3: *A team of people is needed with adequate release time for the process of properly educating teacher candidates, for incorporating external standards, and for participating in and providing professional development activities.* University, college, and departmental faculty status committees need to give appropriate credit toward tenure for those responsible for improving PTE programs.
- Change Principle 4: *A team of people is needed who are dedicated to and capable of quality teaching and who effectively model such teaching.* Physics teacher educators should keep in mind that many teachers teach the way they are taught [16]. If PTE programs are to improve the quality of physics teacher preparation, they must be prepared to demonstrate the performance expectations that they have for their students.
- Change Principle 5: *A departmental faculty is needed who understands the procedures and worth of physics teacher education, and who supports the efforts of the physics teacher education leadership.* Change is often dependent upon a departmental faculty member who is willing to make programmatic changes required to implement a quality PTE program.

All of these change principles and the history of the changes within the ISU PTE program were fully explained and documented elsewhere [2,17,18]. It is unlikely that significant changes in existing PTE programs will occur unless a majority of these conditions are met.

IX. ACHIEVING SUCCESS

The success of any PTE program will depend not only on the above principles, but also on a variety of other factors. It is the synergistic effect of many of these factors that will result in the growth of such programs. At least this has been the experience at Illinois State University.

It is important to restate that program revisions at ISU have been evolutionary, and not revolutionary. If PTE sequence changes had been revolutionary rather than evolutionary, it is not likely that they would have occurred, as the faculty would most likely have resisted them in an effort to ensure that teachers were "highly qualified" to teach high school physics. Besides, it would have been difficult to convince faculty that dramatic program changes would make a difference. This is quite understandable, because no one wants to invest lots of time, effort, and resources in a program if no one is going to enter it. The growth of our program occurred along with its development; one did not precede the other. Evolutionary change is more likely to succeed than revolutionary change. Taking advantage of every conceivable opportunity for growth in association with dedicated efforts and reasoned explanation will be helpful in gaining faculty support.

Those who are intent on growing their university's PTE program should keep in mind that others have gone this way before. In doing so, they have gained many valuable experiences and can offer useful advice. Consult with the champions of successful PTE programs around the nation—there are several [14]. Keep in mind, too, that there is no need to reinvent the wheel. Check with these individuals about instructional resources, rubrics, assessments, clinical experiences, course syllabi, educational philosophies, knowledge bases, and other such things, which all high-quality PTE programs should have. Contact the Physics Teacher Education Coalition for assistance as well [19]. Feel free to contact the author of this article, who would be more than delighted to help by providing advice and resources for program development [20].

TAKE-AWAY LESSONS

What lessons should the reader take away from this chapter? The author believes that the following three points are the most crucial: (1) Effective recruitment and retention cannot be separated from program quality; (2) Program growth will mirror improvements in the quality of the program and efforts to promote it; and (3) Successful program change will be evolutionary rather than revolutionary. Evolution helps to solve the "chicken and egg" paradox concerning which comes first. That is, how can a physics teacher education program have lots of majors without the existence of a quality program, and how can a quality program be built unless there are lots of majors? Only a long-term, dedicated commitment to developing a quality program is likely to bring about the desired changes in recruitment, preparation, and graduation of PTE majors, and in the retention of inservice teachers.

ACKNOWLEDGMENTS

The author wishes to acknowledge the assistance of Dr. Richard F. Martin, chair of the ISU physics department (through June 30, 2014), in the preparation of this chapter. He provided numerous valuable insights in describing the PTE curriculum reform process. The author also wishes to thank anonymous peer reviewers for advice on how to improve this chapter. Any errors or omissions within this chapter are the responsibility of the author.

[1] Illinois State Board of Education, *Educator Supply and Demand in Illinois: 2011 Annual Report* (2011). http://www.isbe.state.il.us/research/pdfs/ed_supply_demand_11.pdf

[2] C. J. Wenning, Repairing the Illinois high school physics teacher pipeline: Recruitment, preparation, and retention of high school physics teachers, J. Phys. Teach. Educ. Online **2** (2), 24 (2004). http://www.phy.ilstu.edu/ptefiles/publications/illinois_pipeline.pdf

[3] C. P. Frahm and R. D. Young, PSI for low-enrollment junior-senior physics courses. Am. J. Phys., **44** (6), 524 (1976). doi: 10.1119/1.10390

[4] See http://www.phy.ilstu.edu/pte/

[5] Illinois Section of the American Association of Physics Teachers, *A Career in Science Teaching? Think About It!* http://helios.augustana.edu/isaapt/teach/brochure3.pdf

[6] Illinois Section of the American Association of Physics Teachers, *Recruiting the Next Generation of Middle and High School Science Teachers.* http://helios.augustana.edu/isaapt/teach/booklet4.pdf

[7] See http://www.illinois-science-teaching.info

[8] C. J. Wenning, SAAMEE: A model for academic success. J. Phys. Teach. Educ. Online, **3** (2), 29 (2005). http://www.phy.ilstu.edu/~wenning/jpteo/issues/jpteo3%282%29dec05.pdf

[9] See http://www.phy.ilstu.edu/pte/advice.html

[10] National Commission on Teaching and America's Future, *The High Cost of Teacher Turnover* (2007). http://eric.ed.gov/?id=ED498001

[11] R. M. Ingersoll, *Teacher Turnover, Teacher Shortages, and the Organization of Schools* (2001). http://depts.washington.edu/ctpmail/PDFs/Turnover-Ing-01-2001.pdf

[12] See http://www.phy.ilstu.edu/pte/

[13] T. S. Kuhn, *The Structure of Scientific Revolutions* (University of Chicago Press, Chicago, IL, 1962).

[14] D. E. Meltzer, M. Plisch, and S. Vokos, eds., *Transforming the Preparation of Physics Teachers: A Call to Action. A Report by the Task Force on Teacher Education in Physics* (T-TEP) (American Physical Society, College Park, MD, 2012). http://www.ptec.org/webdocs/2013TTEP.pdf

[15] C. J. Wenning, K. Wester, N. Donaldson, S. Henning, T. Holbrook, M. Jabot, D. MacIsaac, D. Merrell, D. Riendeau, J. Truedson, Professional knowledge standards for physics teacher educators: Recommendations from the CeMaST Commission on NIPTE, J. Phys. Teach. Educ. Online, **6** (1), 2 (2011). http://www.phy.ilstu.edu/ptefiles/publications/knowledge_standards.pdf

[16] L. Darling-Hammond, Are our teachers ready to teach?, Qual. Teach. **1**, 6 (1991).

[17] C. J. Wenning, Change principles for physics teacher education programs. J. Phys. Teach. Educ. Online, **2** (1) 7 (2003). http://www.phy.ilstu.edu/ptefiles/publications/change_principles.pdf

[18] C. J. Wenning and R. F. Martin, Development of the physics teacher education program at Illinois State University, American Physical Society Forum on Education Newsletter (Fall 2005). http://www.aps.org/units/fed/newsletters/fall2005/development.html

[19] See http://www.phystec.org

[20] cjwennin@ilstu.edu

Where do physics teachers come from? Recruitment and retention of preservice physics teachers

Ronald S. Hermann, Jim Selway, and Cody Sandifer
Department of Physics, Astronomy and Geosciences, Towson University, Towson, Maryland 21252

There has been an ongoing shortage of highly qualified physics teachers across the United States. In an effort to reduce this shortage, universities must attract students to physics teacher preparation programs and retain those students once they are enrolled. However, there is a dearth of information on the best practices for recruiting and retaining preservice physics teachers. In recent years we have more than tripled the number of students in physics teacher education programs at Towson University near Baltimore, Maryland. In this chapter we describe our efforts to recruit students into physics teacher education programs. We offer suggestions based on the lessons we learned, in hopes that other universities may also have greater success with teacher recruitment and retention.

I. INTRODUCTION

The need for highly qualified physics teachers throughout the United States has been well documented in the 2012 Task Force on Teacher Education in Physics (T-TEP) report [1]. However, scholarship related to the effective practices for recruiting and retaining physics teachers has been documented to a much lesser extent. To address these issues at a national scale, colleges and universities must recruit students into teacher preparation programs that lead to a physics teaching certification, and document their efforts in doing so. Moreover, colleges and universities must retain students they have recruited into these programs.

In this chapter, we describe our efforts, some successful and others not as successful, to recruit and retain preservice physics teachers. During our three years as a Physics Teacher Education Coalition (PhysTEC) Supported Site, we have made tremendous progress in increasing both the number and quality of our future physics teachers. While our efforts have occurred at one institution—Towson University near Baltimore, Maryland—we feel the lessons we have learned are pertinent to other institutions seeking to increase the number of highly qualified physics teachers graduating from their programs.

A. Background on the problem

There is an ongoing and well-documented shortage of qualified science and mathematics teachers both nationally and locally [2–5]. According to national rankings of the need for teachers in academic disciplines (not including special education), physics suffers from the greatest shortage of qualified teachers [6]. Only approximately 35% of first-year physics teachers have a degree in physics or physics education [7]. Over half of all first-year physics teachers who are new to teaching physics are career changers who are also new to high school teaching [1]. The other half are inservice teachers who taught other subjects prior to teaching physics [1]. School systems across the country have experienced great difficulty locating qualified science teachers to fill vacancies. The National Science Teachers Association (NSTA) conducted a nationwide survey in 2000 and reported that 61% of high schools and 48% of middle schools experienced difficulty filling science vacancies [8]. Nationally, five of 30 reporting states had 30% of their students taking physics prior to graduation, while 13 states had less than 20% of student take physics. Nationally only 74% of mathematics and science teachers are certified to teach physics [9]. Thus, although more high school students are taking physics, there is a greater chance that they are being taught by a non-certified physics teacher. This is especially disconcerting since researchers have found that students learn more from teachers who majored and studied teaching methods in the subject they teach than from those who did not [10,11].

While the causes of the current shortage of science teachers are many and complex, the recruitment and retention of highly qualified science, technology, engineering, and mathematics (STEM) teachers is paramount among them. This is evidenced by a recent report stating that the new teacher supply pipeline was more than sufficient to replace losses due to retirement, but was barely sufficient to replace the larger portion of teacher attrition due to dissatisfaction [4].

There is a dearth of research on early recruitment experiences in science education [12]. Active recruiting by faculty has been suggested to significantly increase the number of students who consider teaching as a career [13]. There are also reported recruiting benefits of letters from the dean to students emphasizing the importance of teaching, initial field experiences, financial incentives, effective advising and mentoring, and coordinating with other programs [14]. The Internet, career fairs, institutional reputation, and advisors are reported to be the best mechanisms for disseminating information about teacher education programs [15]. Unintentional and intentional gatekeepers such as attitudes of advisors, paperwork deadlines, time to complete the program, financial cost, life stage, and different levels of Internet savvy reportedly keep students from completing applications to teacher education programs [15]. With these findings in mind, we developed recruiting methods to limit gatekeepers and took the advice of Artzt and Curcio and provide a "personal touch" to our recruitment [16].

B. Background on Towson University

Towson University (TU) began as a teachers college, and the tradition of developing highly qualified teachers continues, as TU graduates the highest number of education majors among all universities in Maryland. Currently 2,382 undergraduate and 1,698 graduate students are enrolled in teacher education programs at TU. Over 1,000 of those undergraduate students are working toward an elementary education or early childhood education degree. By comparison, there are 98 undergraduate physics majors, and only 13 of those students are working toward a physics teaching certification. These numbers are important, as they highlight a difference in the way various departments may think about recruitment and retention of students. As we will describe in this chapter, the physics department has a significant need to recruit students, whereas other departments with an abundance of majors may not emphasize recruitment as much. Likewise, we are very deliberate in our efforts to retain students, whereas other departments may not have as strong an impetus to ensure students are provided adequate support to complete their programs of study. To better understand the efforts undergone at TU, we first provide an overview of the programs leading to certification, in order to highlight both the support mechanisms and the possible barriers to becoming a physics teacher.

1. Departmental structure and college structure

There are several different ways that universities structure programs by which students become certified to teach physics, so it is important to describe the attributes of our programs to provide a context for comparison. Towson University offers majors in biology, chemistry, Earth-space science, and physics, and each of these majors also has a secondary education concentration. For example, there are four concentrations for physics majors: applied physics, astrophysics, general physics, and secondary education. Each student enrolled in one of the concentrations is a student in the Fisher College of Science and Mathematics (FCSM) within the Department of Physics, Astronomy and Geosciences (PAGS). After completing a minimum of 45 credits and maintaining a minimum GPA of 2.75, students are eligible to apply for acceptance into the teaching certification program through the College of Education (COE), specifically within the Department of Secondary and Middle School Education. The department requires several general education and methods courses be taken by all majors in a teacher certification program.

After introductory secondary education courses are completed, the prospective middle or secondary teacher enters the student teaching term. Methods, philosophy, techniques, and practice are combined to provide a thorough preparation for teaching, and such preparation must be demonstrated prior to student teaching through a minimum GPA requirement and a passing score on the Praxis I exam.

Undergraduate physics secondary education students are physics majors enrolled in a secondary education concentration. In addition to their general education requirements, physics secondary education students complete 39 credits of physics courses, 26 to 27 credits of non-physics science courses, 29 credits of education courses, and 15 credits of science education courses. The science education courses comprise 12 credits of student teaching and three credits of secondary science methods coursework. Secondary education students receive around 20 fewer credits of physics coursework than physics majors in other tracks, though the exact credit difference depends upon which physics concentration is compared. This is mainly due to education requirements required to adhere to the Maryland State Department of Education requirements for certification. Considerable effort has been made to meet this requirement and still remain not too far over the 120 credits required to graduate, so that students can graduate in four years. The teacher candidates are jointly advised by PAGS science educators and faculty in the Department of Secondary and Middle School Education to ensure that they fulfill their content and education coursework requirements as well as other educational requirements such as Praxis testing.

TU also offers a Masters of Arts in Teaching (MAT) program. Students in this program have already earned an undergraduate degree in physics or in a closely related field, and complete education-specific coursework and student teaching in order to apply for a teaching certificate. MAT students are majors in the College of Education and, though we interact with them, we do so to a much lesser extent than with undergraduate secondary education students. As a result, most of our recruitment efforts are targeted at undergraduates, though our retention efforts target both undergraduate and MAT students.

2. TU support mechanisms, past and present

Like many universities, TU has historically recruited physics majors into the secondary education concentration. In 2000, only 34 students were enrolled as physics majors, and it was abundantly clear that an increase in the number of physics majors in all concentrations was needed. In recruiting physics majors broadly, the expectation was that a subset of those students would enter the secondary education concentration. Exit interviews with the department chair indicated that approximately 90% of the students who changed their majors from physics to other subjects did so in their first year. This was due to improper freshman advising leading to students not taking physics classes during their first year, a lack of a peer support group, and a lack of connection between faculty and students. In response to these findings, several key initiatives have been instituted over the past 10 years to retain physics majors at Towson University and increase physics teacher enrollment.

(a) *Pre-admission interactions.* Prior to their acceptance at TU, potential physics majors are contacted by a physics faculty member, and many of these students attend pre-enrollment meetings to discuss the program. After students arrive at campus, this faculty-student contact is reinforced throughout the year through other orientations, meetings, and social events.

(b) *Housing.* Once accepted into this program, all freshman students seeking on-campus housing participate in the Science Scholars Housing thematic community. This is a program that houses all physics, mathematics, and chemistry freshman on a common floor of Prettyman Hall. This residence hall is strategically selected because it has been demonstrated in the past that students living in it form strong connections with each other, and because it is located very close to the physics building, the library, and food facilities on campus. The purpose of this housing arrangement is to develop a large cohort of students sharing common interests, courses, and needs. A specially selected resident assistant works with physics faculty, student clubs, and housing officials to design a series of science-based activities and field trips, and to apprise the students of opportunities available to them.

(c) *First-year courses.* All incoming physics freshmen, as well as new transfer students, are required to take PHYS 185, Freshman Seminar. This is a one-credit course intended to provide students with a better understanding of what opportunities are available in our department and what career options are available with a physics degree, including a teaching career. It provides a way for us to assemble all new students in our program and for students to get to know our faculty, our department, and each other. Additionally, special sections of the first two calculus-based introductory physics courses are offered just for physics majors. These sections have smaller class sizes (25 or fewer students), allowing inquiry-based pedagogies like Workshop Physics to be implemented.

(d) *Financial support.* PAGS and the FCSM have secured funding for scholarships for students. Most relevant for physics teaching are National Science Foundation-funded Noyce Scholarship and S-STEM grants, which help students who are declared physics secondary education majors pay their tuition. Towson University's financial aid office prides itself on the extensive list of grants and scholarships compiled and made available to students; many of these support STEM teacher preparation.

(e) *Mentoring.* All freshmen at TU are assigned a First Year Experience (FYE) advisor, who is not necessarily a faculty or staff member in the student's proposed major. PAGS has worked with the admissions office and student advising to assign all freshman physics students to an FYE advisor in our department. Additionally, all declared majors are assigned a physics faculty advisor who works closely with the FYE advisor. This ensures that students receive the correct advice for classes during the critical first few semesters. Our FYE advisor also works in conjunction with Dr. Ron Hermann, a PAGS science education faculty member, who officially begins advising students during their sophomore year. Through advising, students make personal contacts with department faculty at a time when they may need advice the most.

(f) *Connections with students.* A "Welcome, majors" pizza party is hosted by the department near the beginning of the semester. Majors are also encouraged to join our active Society of Physics Students chapter, and a room has been aside in the department for sole use as an SPS meeting space. The SPS email distribution list disseminates information about internships, scholarships, and summer opportunities, and students are invited to attend department seminars and maintain an active social calendar. Furthermore, majors are encouraged to join the department Facebook page.

As a result of these efforts, the number of physics majors increased steadily from 34 in 2000 to 77 in 2009. However, the number of students in the secondary education concentration remained low. Only two undergraduate students graduated from TU with a physics secondary education degree between 2002 and 2009. In response, we applied for PhysTEC funding, and became a PhysTEC Supported Site in 2010. Through this effort we further increased the number of physics students in the secondary education concentration.

II. RECRUITMENT EFFORTS

Over three years of grant-funded support, the number of students in the physics teacher pathways increased from three pre-PhysTEC to 13 post-PhysTEC, a trend consistent with that of other PhysTEC Supported Sites [17]. During that time span we maintained the recruiting activities we had developed prior to PhysTEC, initiated new activities, and learned of additional recruiting ideas based upon our observations of how some students entered the teacher education program regardless of our activities. Here, we describe several effective practices for recruiting students into physics teacher education programs, and present them as suggestions that may be useful for other institutions.

A. Be champions of the cause

At TU, Dr. Ron Hermann has advised all second-year-and-beyond undergraduate physics majors in the secondary education concentration since 2009. Having a designated point of contact for all prospective physics teachers has been invaluable in helping students better understand program requirements and available opportunities, such as early teaching courses, scholarships, Learning Assistant (LA) positions, and other resources. Prior to having a dedicated physics secondary education advisor, interested students were not directed to any one person who was knowledgeable about all facets of the program and about teaching in general. In fact, many students didn't know if they were majors in the Department of Physics, Astronomy and Geosciences or in the Department of Secondary and Middle School Education, and some students were taking courses that were not needed for graduation, extending their time at TU. Dr. Hermann has 11 years of high school teaching experience, so initial meetings with students cover not only information related to program coursework, but also a candid discussion about what high school teaching entails. Introductory meetings are intended to inform prospective students about all the support mechanisms and opportunities that are in place, while also helping students decide if teaching physics will be a viable and enjoyable career option. Initial discussions with prospective students are critical in establishing a professional relationship.

Deciding upon a major is one of the most important decisions a college student can make, and advisors who can inform students about all aspects of the program and about secondary science teaching can help students make well-informed career decisions. Sadly, transfer students and those changing majors often tell us that previous advisors were unwilling or unable to spend as much time with them or to answer their questions in meaningful ways. An academic advisor who is also mindful of the need for more physics teachers can be the hub of several activities that may attract students to the physics teacher education programs. Many of the activities the advisor should support are listed in the remainder of this section on recruitment and in the following section on retention of preservice physics teachers.

B. Make faculty and staff across campus aware of your efforts

Begin and maintain a conversation with physics faculty and education faculty about your efforts to increase the number of highly qualified physics teachers (i.e., those with a physics major or minor and physics teaching certification). While this may seem self-evident, the need to prepare more highly qualified physics teachers has long been implicit and, therefore, rarely discussed at our university. It wasn't until our PhysTEC project began that we explicitly communicated our efforts to attract more students to the physics secondary education concentration. Through the initiation of conversations about the nationwide need for physics teachers with physics faculty, education faculty, and administrators, we helped faculty realize that in addition to teaching courses that included students working toward becoming physics teachers, they should also help recruit students into the physics secondary education concentration and better support students already in the concentration. One physics professor in particular has embraced the cause and now consistently discusses physics teaching with students in his class. From our discussions with physics majors, we have found that when a physics professor tells a physics major that he or she would make a fantastic physics teacher, it can have a profound effect on the way that student thinks about teaching as a career.

We now have regular conversations with the Department of Secondary and Middle School Education, and students who ask education faculty about teaching physics are quickly directed to Dr. Hermann. As simple as this sounds, it was not happening even a few years ago. Additionally, our Department of Secondary and Middle School Education has well-established networks with local K-12 school personnel that physics faculty often lack. Because of our dialogue with secondary education faculty, when a local school administrator contacts them inquiring about physics teacher candidates, the message is forwarded to us, so we can put current or former students in contact with the school. Partly as a result, all of our graduates have received teaching positions after graduation, a fact that helps recruit students to the major at a time when students in other majors may be unsure about their career prospects. Maintaining clear and sustained lines of communication with various entities on campus is vital to recruiting and retaining future physics teachers.

C. Utilize diverse recruiting strategies and embrace flexibility

Along with Jim Selway and Lisa Rainey, our Teachers-in-Residence (TIRs), we have recruited majors through one-on-one and group discussions about physics teaching as a career with current physics majors, and by talking to high school students; non-major, introductory-level physics students; and interested community college students. We also developed a poster to initiate a dialogue with students about teaching physics. The poster included a picture of a TU student and a middle school student completing an activity related to the scale of the solar system. We placed several posters around the FCSM and the COE. The poster was designed to announce meetings for anyone interested in learning about science teaching. In addition, the meetings were announced in an email that students receive daily that lists campus activities. These meetings were held several times per semester and often included light refreshments for students attending. While a few students attended each meeting, the response was not as high as we had hoped, and many of those who attended had already decided to become science teachers, though most did not consider teaching physics. From this response we realized we needed to diversify our recruiting strategies.

During discussions we had with students in the physics secondary education concentration, many students stated they entered the program due to other motivating factors. Some students entered the program because they had been motivated by their high school physics teachers. This aspect really came to our attention when we learned that two of our physics students were motivated to become physics majors after taking a high school physics course taught by the same teacher. We then questioned other students and often found that they too attributed their decisions to become physics majors to particular high school physics teachers. This motivating factor was the impetus for a second recruiting poster designed for high school students. The poster provides a problem-solving strategy branded with the Towson University logo. We made these posters available to local physics teachers at a state-wide science education conference, and personally delivered some to physics teachers with whom we have worked closely. The purpose of the poster was to remind high school physics teachers to occasionally spend some time discussing physics teaching as a career option, and to make students aware that TU offers a physics secondary education concentration.

We also determined that other students were informed of the program by faculty in other departments or by physics professors who encouraged and motivated them to consider teaching. In one instance, a physics professor discussed physics teaching as a career option in class and subsequently introduced a student to Dr. Hermann for further discussion about the physics teacher education program. In another instance, a professor in an elementary education methods course realized a student had an interest in teaching physics and initiated a discussion about teaching physics. The professor then put the student in contact with Dr. Hermann for a follow-up discussion. Based on these cases, we made a concerted effort to encourage faculty to discuss physics teaching careers with physics majors and education majors.

Over time we diversified our recruiting strategies by maintaining the traditional activities we had in place prior to PhysTEC, while also remaining flexible and exploring other activities based on what we learned about how several students actually became motivated to teach physics. Whereas in the past we did not look to high school physics teachers to identify students, or to elementary education students for possible recruits, recruiting from these areas is now commonplace.

D. Partner with local high school teachers

We learned that an effective practice is to establish relationships with local physics teachers in an effort to help them see the need to motivate their own students to consider teaching physics. Our experiences suggest that physics teachers spend more time encouraging students to consider physics careers other than teaching than they do encouraging their students to consider teaching. We have initiated a dialogue with local physics teachers about the physics teacher shortage and the prominent role they can play in developing the next generation of physics teachers. This was done in a variety of venues, including one-on-one and group meetings. We also presented a physics lesson at the state National Science Teaching Association meeting and concluded the session by handing out the aforementioned posters and discussing the physics teacher shortage.

We found that some of the teachers were encouraging students to consider teaching careers in implicit ways, like stating that teaching is a career option for physics majors or stating how much they enjoy teaching physics. Through our discussions we discovered ways teachers could be more explicit. For example, one high school physics teacher calls upon each student over the course of a marking period to explain to the class how he or she solved a problem. The student explains the solution and is responsible for answering questions from other students. As a result of our dialogue, the teacher felt this was the perfect situation to informally observe students' teaching potential and explicitly encourage those who seem to enjoy teaching and have a natural talent for it.

E. Personalize your attention to individual student needs

Personalize your interactions with students and tailor your recruitment and retention to the individual student: One size does not fit all. Each student comes to us with a unique set of personal and education experiences. Some students have relatives or close friends who are teachers, and such students often have fairly well-informed views of what teaching physics entails. Others have more limited views. Determining where students are in their understanding of physics teaching is an important first step in recruiting students into the program. We have offered several opportunities for students to learn about what teaching physics entails prior to committing to completing their degrees.

Once prospective students express an interest in teaching physics, we take great care to establish a personal rapport with them and maintain constant contact, offering them several ways to learn more about the profession. As Mr. Selway is apt to say, we court prospective students the way a university would court potential student athletes. Interested students initially meet with Dr. Hermann, who advises physics majors in the secondary education concentration. Then, in order to help candidates make more informed decisions about physics teaching as a career, Mr. Selway and Ms. Rainey take students to visit local physics teachers, arrange for the students to attend a physics professor's class at TU, invite them to lunch, and engage them in numerous other personalized activities.

We have found that most students have not experienced this type of personalized attention elsewhere on campus. However, even after such attention, some students decide that teaching physics or our program are not the right fit. This is fine, as we are interested in both the number and the quality of the physics teachers we prepare, and we strive to ensure the success of those enrolled. Feedback from district science supervisors who hire our graduates suggests they are highly effective physics teachers. Indeed, it is common for students to be offered a teaching contract for the following year prior to completing student teaching, and supervisors often inquire about students in the physics teacher program to fill upcoming vacancies in their counties.

F. Realize that physics teachers can come from anywhere

Future physics teachers can come from anywhere, not just from the pool of physics majors. So be sure to welcome the diversity and avoid a "gatekeeper" mentality. Four studies on recruitment and retention of teachers have indicated that college graduates with the highest levels of measured academic abilities tend not go into teaching [18]. Physics majors are likely to be students with high levels of measured ability, and we haven't experienced too many switching to the secondary education track, though a few have. Our experience has shown that students are likely to migrate from other majors as well, including elementary education majors and even art majors. These may not be places where physics secondary education majors typically come from, but indeed some do. Students who are considering transferring from other majors do not always follow through and switch majors, but several have and are now competent and enthusiastic physics majors. Again, getting to know the students, exploring their transcripts, and discussing their experiences with physics are crucial to helping them make informed decisions. By doing so, we have tried to limit the gatekeeper mentality, which often leads faculty members to assume that certain students cannot do physics, or teach physics, well. Our approach has been, when reasonable, to give students who express interest in teaching physics the opportunity to do so, rather than trying to talk them out of the major.

III. RETENTION OF PHYSICS EDUCATION MAJORS

There is a natural overlap between activities to recruit students and activities to retain students. Many of the activities we will describe for retaining students are discussed with students as they consider physics teaching. Students naturally want to know they will be adequately supported during their course of study, and that seems to play a large role in their decision to enroll in the physics secondary education concentration.

Prior to our PhysTEC project, we didn't formally document which students were in a physics teacher education program. We therefore didn't keep track of which students remained in the program, which students left the program, why students left the program, or which students taught physics after graduating. This is consistent with the T-TEP finding that very few programs keep track of their graduates [1]. Anecdotally, we know of several students who were not retained, and, to a lesser extent, why they left. During our PhysTEC project, a few students were not retained despite the efforts outlined in this chapter. One student decided that teaching physics was not the right fit for him, one left for medical reasons, and two left the program due to inadequate academic performance. Again, for us it is important to retain students if they are committed to teaching physics. However, if they are no longer committed to teaching or if they are incapable of completing the coursework, we do not try to retain those students. All students who have completed the program since the inception of our PhysTEC project have obtained a physics teaching position and remain classroom physics teachers at the time of this publication. As a result of reflecting upon why students leave the major, we have identified a few effective practices that may help retain students.

A. Identify champions for physics teacher education

Identify an individual who will be the point person for communicating and advising physics education students and communicate that person's role across the university. The importance of this action is one of the most salient findings of the T-TEP report, and one that we have realized firsthand [1]. The report found that every active physics teacher education program has a champion, or multiple champions, who advise and mentor students, track graduates, recruit students, and arrange early teaching experiences. This individual or group can have the largest impact on the retention of students. A tremendous amount of effort and resources are required to recruit students into the program. Therefore, it is essential to retain students who enter the program, as the effort to retain a student can be much less than the effort required to recruit a student into the program. Put another way, the number of students graduating from the program is much more important than the number of students in the program at any point in time. The primary goal of champions of physics teacher education should be to have a high percentage of students graduating from the program in a reasonable timeframe. Ideally students will graduate in four years, though it may take longer for transfer students and for those who change their majors. As such, we help students identify and apply for scholarships to offset the cost of their education, especially when students are in a situation where they will take longer than four years to graduate.

Students may not involve themselves in physics education activities on campus, develop physics teacher identities, or feel a sense of belonging to the physics education community without the persistent and ongoing efforts of the champion(s). Indeed the authors of this chapter are the champions for physics education at Towson University, and had we not taken on the role of champion, the successes we are describing may not have come to fruition. To effectively recruit and retain students, there must first be a champion of this cause. Our hope is that those who read this chapter take on that role as they work towards preparing more physics teachers. Once other faculty and administrators are informed of your vision, they can help you achieve your goals.

B. Engage students in physics education activities

Once we have recruited students into the physics teacher preparation program, we encourage them to involve themselves in the physics teacher community and to participate in science education and physics education activities. As more students have entered the program, we have reached a critical mass that allows us to foster and support some physics education-specific initiatives in addition to more general science education initiatives. Students' pictures, names, and majors are placed on a poster in the department hallway, so that all students know who is in the physics teacher preparation program. We haven't noticed similar posters for other majors across campus, so the recognition that our students get seems to be unique to our program. We also display a smaller poster that includes alumni's pictures, names, majors, and current teaching locations. This poster helps us discuss career options with current and prospective students, because all of our majors in recent years hold teaching positions, several of which are in well-regarded local schools. Additionally, the poster helps current students develop contacts with TU alumni who are teaching physics at nearby schools. Current students can easily arrange to meet with an alum to discuss teaching physics, observe a physics lesson, or co-plan and co-teach a lesson, among a wide range of possible interactions. We also invite graduates to return to campus to speak to preservice physics teachers about their experiences and offer suggestions for enriching their physics teaching experiences.

In the last few years our students have attended and presented at science education conferences and American Association of Physics Teachers conferences, initiated and maintained an NSTA student chapter, participated in science-specific outreach activities with the community, served as Learning Assistants for physics faculty, and participated in early teaching courses, among other opportunities. In an effort to keep students engaged in teaching physics, we began offering three different early teaching courses. The first course consists of students teaching in an informal science education setting such as a nature center or the local science center. The second course consists of students teaching science at a local elementary or middle school.

After taking these two courses, a few physics students informed us that they would prefer to teach physics to high school physics students. They really wanted to teach content more closely aligned with their studies and to work with slightly more mature students. Our students did not feel the close age difference with high school students was an issue, and given the minimal time in the classroom, we decided a high school early teaching opportunity was a good fit for our students. So we initiated the third early teaching experience, which places physics majors in local high school physics classrooms working with physics teachers whom we hold in high regard. Indeed, the earlier potential candidates are partnered with master inservice teachers, the more likely they are to consider a career in teaching [14]. In addition, we have been able to support these students through Noyce Scholarships and the retention and induction activities conducted through that program.

C. Develop a sense of physics teacher identity

We have noticed that physics majors in the secondary education concentration seem to identify themselves more as physicists and less as physics teachers. That is to say, they seem to spend a lot more time thinking about and doing physics than thinking about how to teach physics. They are more likely to read articles about physics and physics research than about physics teaching or physics education research. We deliberately set out to include students in science education activities in an attempt to foster a stronger sense of teacher identity. Teacher identity has been characterized in numerous ways by researchers, but here we note that identity is a product resulting from influences on the teacher and a process of ongoing dynamic interaction within teacher development [19]. What is important here is that professional identify affects one's senses of purpose, self-efficacy, motivation, commitment, job satisfaction, and effectiveness [20].

We try to facilitate student development of a teaching identity through active participation in science education-related activities outside those required by the teacher education program. It is natural for students in a physics teacher education program to be interested in physics research, and at most universities opportunities to work with faculty are available and encouraged. Experience conducting physics research is invaluable for physics teachers, as it helps them articulate to students both physics content and the methodology of science, and, in most cases, the process of presenting research findings through written and oral communication. Equally important, but perhaps less natural, is for students to also explore the teaching and learning of physics. By engaging in research on the teaching and learning of physics, students may develop a better sense of themselves as teachers of physics. While our students are still interested in physics research, they have also become increasingly interested in reading about physics education research and conducting research with TU faculty, TIRs, and local high school physics teachers.

Over time, students have become more willing to participate in science education events, and student-initiated discussions more frequently center on the teaching of physics, whereas in the past the focus was more on their learning of physics and on physics content. Recently, Dana Molloy, a physics secondary education student and Learning Assistant, worked with Jim Selway, a TIR, and Dr. James Overduin, a physics professor, to develop a lesson that they subsequently implemented in high school and college-level classes. Student response to the lesson was so positive that the team submitted a manuscript that was published in *The Physics Teacher* [21].

TU students who have well-developed physics teacher identities have been much more involved in science education activities, have developed more robust relationships with the local and regional science education communities, and are more likely to implement student-centered, reform-based pedagogy once they graduate. For example, as part of his interest in physics teaching, one of our majors formed a student chapter of the National Science Teacher Association. Through his involvement with the student chapter, he has presented and attended local science education conferences, and is now on a first-name basis with science supervisors for a few of the surrounding counties. This student clearly has embraced the science education community and his place within that community, and has a well-formed identity as a preservice physics teacher. Our expectation is that students who become involved in the physics teaching community will be more likely to remain in a teacher education program than those who do not.

D. Remove or reduce programmatic barriers

The structure of our university, like that of most, does not always make it easy for students to smoothly navigate the physics teacher education program. There are both intentional gatekeepers (e.g., entrance requirements, application deadlines, and program personnel) and unintentional gatekeepers (e.g., institutional reputation, geography, and testing) that influence how individuals move through a teacher education program [15]. Students need an advocate to help minimize the effects of gatekeepers, and when they do, they turn to their advisors or TIRs.

Students often need help with issues that they perceive as barriers. Many times these barriers are not regarded as an issue for other disciplines that have many more students, where a "weeding-out" mentality exists. Due to the limited number of physics teacher candidates, we would rather remove or minimize barriers that could lead to students transferring or changing majors. These barriers include time overlaps of a few minutes with required courses, issues with course substitutions or class requirements, prerequisites in our department or in another department, issues with specific requirements (GPA, transfer credits, etc.) in the Department of Secondary and Middle School Education, and a host of other issues that vary from person to person. While these issues are often minimal, sometimes the frequency with which they occur is high enough that they could affect retention. So, as an advisor, it is necessary to take some time to meet with other faculty or administrators to try to resolve an issue before students become frustrated and explore other options.

Whereas faculty in other departments with numerous majors may not feel the need to do so, with our limited number of students we believe it is necessary to serve as advocates for students. We have seen faculty in other departments with signs on their doors stating, "A lack of preparation on your part does not necessitate an emergency on my part." We have tried to adopt the opposite mentality: If an issue is important to our students, it should be equally important to us. Through this mentality, we have been able to remove some barriers that students have met in the past. Bringing those issues to light was the first step in initiating a discussion with the parties involved and beginning work toward a solution. This process will pave a smoother path for future students and will ultimately improve the program and the reputation of the program and its faculty.

E. Maintain ongoing contact with students

Maintaining ongoing contact with students and alumni is a critical element in developing an effective physics teacher education program. We frequently disseminate information and opportunities to students through email, Facebook postings and direct messages, one-on-one advising, and group meetings. Whereas many other TU students are unsure who their major advisors are or where their advisors' offices are, our students are in constant contact with their advisors and frequently email, call, and visit their offices. It is important to have a few faculty members who know the students by name and know them on both an academic and personal level. Moreover, it is important for our students to feel that they are part of a community of like-minded people; they come to this feeling by interacting with their peers in and out of class. We also have several alumni teaching physics in local schools, and we encourage our students to visit with these physics teachers. Our current student teachers are sometimes placed in the same schools where our alumni teach, and eventually alumni may serve as mentor teachers. Developing this support network is an important goal and a way for us to stay in contact with current and former students.

We have also come to realize that it is difficult to determine if students are becoming dissatisfied with the program if we are not in contact with them. So we maintain contact through advising, social media, and email, and by encouraging students to attend local conferences. If asked, our students are pretty forthcoming about any areas of our program that are not well received. However, if they are not asked, they rarely volunteer information until it is too late and they are considering leaving the major or have already left. As a result of feedback we have received from students, we have been able to initiate discussions with other departments or within our own department to minimize or mitigate negative aspects of the program, so that students coming through the pathways are no longer met with persistent and ongoing challenges.

IV. DEPARTMENTAL AND COLLEGE SUPPORT

The emphasis above on the importance of champions notwithstanding, it is also very important to have a supportive department. The current and past PAGS chairs have initiated several support mechanisms for all students and have expanded those mechanisms specifically for students in the secondary education concentration. Physics faculty were generally supportive of the physics secondary education concentration prior to PhysTEC; however, through the efforts supported by our PhysTEC grant, departmental support for teaching as a profession is at an all-time high. In the past, students were mainly encouraged to pursue graduate degrees in physics, but more than ever, physics faculty are encouraging physics majors to consider teaching careers.

Physics faculty work with majors in the secondary education concentration who are Learning Assistants in the physics classes they teach, and some students conduct physics research with faculty as well. At TU, half of the funding for each LA is provided by the department in which the LA works, and half is provided by the college. LAs work with science education faculty and the TIR to develop interactive demonstrations, clicker questions, and other pedagogical methods for use during lectures. In this way, our students are supported by the department, which may retain them because they are perceived to be an integral part of the class. Additionally, the LAs support the physics faculty in learning more student-centered pedagogical practices. As a result, several physics faculty are incorporating more student-centered pedagogical practices in their teaching; such practices include clicker questions, interactive discussions, and demonstrations. All of these efforts have reduced the gap between the way we teach our students to teach physics and the way they experience physics teaching at the university level.

V. SUMMARY

Efforts to recruit more students into the physics teacher preparation programs at Towson University have been successful due to our ability to embrace a recruitment model that is flexible and adapts to student needs. Rather than try to maintain a static model with the same practices that we thought were working, we looked carefully at how our efforts to recruit students compared to how students actually entered the program. In doing so, we found several ways in which we could change our recruitment model. Most notably, we realized that high school physics teachers had a profound influence on several students' decisions to become physics teachers. So we began initiating discussions with local high school physics teachers about the physics teacher shortage and what they can do to help. Likewise, to retain students we reflected upon student feedback and realized that many students wanted an early teaching experience at the high school level, which we weren't offering at the time. As a result, we began offering an early teaching experience with local physics teachers.

We believe that efforts to recruit and retain students should be diverse, ongoing, and personalized, and should reflect the needs of the students. By having a champion or a group of champions who are in constant communication with students, efforts to recruit and retain students can be modified to reflect how students actually enter programs, and what students feel is necessary to stay engaged in the teacher preparation program.

---

[1] D. E. Meltzer, M. Plisch, and S. Vokos, eds., *Transforming the Preparation of Physics Teachers: A Call to Action. A Report by the Task Force on Teacher Education in Physics* (T-TEP) (American Physical Society, College Park, MD, 2012). http://www.ptec.org/webdocs/2013TTEP.pdf

[2] B. C. Clewell and L. B. Forcier, Increasing the number of mathematics and science teachers: A review of teacher recruitment programs, Teach. Change **8** (4), 331 (2001). http://eric.ed.gov/?id=EJ640140

[3] B. C. Clewell and A. M. Villegas, *Ahead of the Class: A Handbook for Preparing New Teachers from New Sources* (Urban Institute, Washington, DC, 2001). http://www.urban.org/url.cfm?ID=310041

[4] R. Ingersoll and D. Perda, *The Mathematics and Teacher Shortage: Fact and Myth* (Consortium for Policy Research in Education, Philadelphia, PA, 2009). https://www.csun.edu/science/courses/710/bibliography/math%20science%20shortage%20paper%20march%202009%20final.pdf

[5] S. Shugart and P. Hounshell, Subject matter competence and the recruitment and retention of secondary science teachers, J. Res. Sci. Teach. **32**, 63 (1995). doi:10.1002/tea.3660320107

[6] American Association for Employment in Education, Inc., *2010 Executive Summary: Educator Supply and Demand in the United States* (AAEE, Columbus, OH, 2010).

[7] C. L. Tesfaye and S. White, *High School Physics Teacher Preparation* (American Institute of Physics, College Park, MD, 2011). http://www.aip.org/statistics/trends/reports/hsteachprep.pdf

[8] National Science Teachers Association, *NSTA Releases Survey of Science Teacher Credentials, Assignments, and Job Satisfaction* (NSTA, Arlington, VA, 2000). http://www.nsta.org/publications/surveys/survey20000407.aspx, retrieved Sep. 8, 2013.

[9] R. Blank, D. Lanesen, and A. Petermann, *State Indicators of Science and Mathematics Education 2007* (Council of Chief State School Officers, Washington, DC, 2007). http://www.ccsso.org/Documents/2007/State_Indicators_of_Science_and_Mathematics_2007.pdf

[10] D. H. Monk and J. A. King, Multilevel teacher resource effects on pupil performance in secondary mathematics and science: The case of teacher subject-matter preparation, in *Choices and Consequences: Contemporary Policy Issues in Education*, edited by R. G. Ehrenberg (ILR Press, Ithaca, NY, 1994), pp. 29-58.

[11] D. D. Goldhaber and D. J. Brewer, Evaluating the effect of teacher degree level on educational performance, J. Hum. Resour. **32**, 3 (1997). http://nces.ed.gov/pubs97/97535l.pdf

[12] J. A. Luft, S. S. Wong, and S. Semken, Rethinking recruitment: The comprehensive and strategic recruitment of secondary science teachers, J. Sci. Teach. Educ. **22** (5), 459 (2011). doi:10.1007/s10972-011-9243-2

[13] V. Otero, S. Pollock, and N. Finkelstein, A physics department's role in preparing physics teachers: The Colorado learning assistant model, Am. J. Phys. **78**, 1218 (2010). doi:10.1119/1.3471291

[14] T. P. Scott, J. L. Milam, C. L. Stuessy, K. P. Blount, and A. Bentz, Math and Science Scholars (MASS) Program: A program for the recruitment and retention of mathematics and science teachers, J. Sci. Teach. Educ. **17** (4), 389 (2006). doi:10.1007/s10972-006-9026-3

[15] S. Abell, W. Boone, F. Arbaugh, J. Lannin, M. Beilfuss, M. Volkmann, and S. White, Recruiting future science and mathematics teachers into alternative certification programs: Strategies tried and lessons learned, J. Sci. Teach. Educ. **17** (3), 165 (2006). doi:10.1007/s10972-005-9001-4

[16] A. F. Artzt and F. R. Curcio, Recruiting and retaining secondary mathematics teachers: Lessons learned from an innovative four-year undergraduate program, J. Math. Teach. Educ. **11** (3), 243 (2008). doi:10.1007/s10857-008-9075-y

[17] T. Hodapp, J. Hehn, and W. Hein, Preparing high-school physics teachers, Phys. Today **62** (2), 40 (2009). doi:10.1063/1.3086101

[18] C. M. Guarino, L. Santibanez, and G. A. Daley, Teacher recruitment and retention: A review of the recent empirical literature, Rev. Educ. Res. **76** (2), 173 (2006). doi:10.3102/00346543076002173

[19] S. Chong, E. L. Low, and K. C. Goh, Emerging professional teacher identity of pre-service teachers, Aust. J. Teach. Educ. **36** (8), 50 (2011). http://eric.ed.gov/?id=EJ937005

[20] C. Day, A. Kington, G. Stobart, and P. Sammons, The personal and professional selves of teachers: Stable and unstable identities, Brit. Educ. Res. J. **32** (4), 601 (2006). doi:10.1080/01411920600775316

[21] J. Overduin, D. Molloy, and J. Selway, Physics almost saved the President! Electromagnetic induction and the assassination of James Garfield: A teaching opportunity in introductory physics, Phys. Teach. **52** (137), (2014). doi:10.1119/1.4865512

Building a thriving undergraduate physics teacher education program at the University of Wisconsin-La Crosse: Recruitment and retention

Jennifer Docktor and Gubbi Sudhakaran

Department of Physics, University of Wisconsin-La Crosse, La Crosse, WI 54601

The University of Wisconsin-La Crosse physics department is nationally recognized as a thriving undergraduate physics program, and routinely ranks among the nation's top graduaters of physics majors among departments that grant only bachelor's degrees. With the recent addition of a new secondary physics education degree and a new faculty member specializing in physics education, the very successful recruitment and retention practices of the physics department are now being used to increase the number of students pursuing careers in physics teaching. Effective practices identified include establishing a point person within the physics department to advise and mentor teacher candidates and collaborate with the School of Education, fostering a departmental culture that values teaching, providing undergraduates with early experiences as Learning Assistants or in outreach, and building a community of secondary education teacher candidates. This paper includes a detailed description of these practices and a summary of additional changes that have been made to further improve preservice physics teacher education at the university.

I. INSTITUTIONAL CONTEXT

A. Physics department

In order to get a comprehensive picture of physics teacher preparation at the University of Wisconsin-La Crosse (UW-L), it is essential to be cognizant of the university context as well as of the nature of the physics department. UW-L is a public university within the University of Wisconsin System, with approximately 9,000 undergraduate students and 1,000 graduate students [1]. La Crosse is located on the border between Wisconsin and Minnesota in a metropolitan area with a population of approximately 100,000, and most students at UW-L come from these two states. The physics department currently has 10 tenured or tenure-track faculty and three academic staff, and it boasts over 175 physics majors, making it one of the largest undergraduate physics programs in the state and one of the nation's top graduaters of physics majors among departments that grant only bachelor's degrees [2]. Recently, UW-L ranked number one in the nation by graduating an average of 31 physics majors per year from 2011 to 2013.

This was not always the case. In the late 1980s, the physics major was in danger of being phased out due to low enrollment. Key changes that have revived the program include the addition of new academic programs, the incorporation of undergraduate research into the curriculum, aggressive recruitment and retention efforts, and connections to the local community through outreach activities and lecture series. As a result of these innovations, the department received the 2004 University of Wisconsin System Regents Teaching Excellence Award, making it the first program at UW-L to receive this award. In 2013 the department won an American Physical Society Award for Improving Undergraduate Physics Education [3]. In addition, the department was selected for site visits by both the National Task Force on Undergraduate Physics [4] and the American Institute of Physics Career Pathways Project [5]. In both cases, the visiting teams sought to better understand the department's effective practices.

B. Secondary teacher education preparation

During the past few years, there have been several key changes that affected physics teacher preparation at UW-L. Starting in Fall 2011, all of the Secondary Teacher Education Preparation (STEP) program degrees were moved from the Department of Educational Studies into the content departments (physics, chemistry, biology, geography and Earth science, mathematics, English, and history). Each teacher candidate now receives a B.S. degree in 4.5 years from his or her content department and teaching licensure in Early Adolescence-Adolescence (grades 6–12). This is the only pathway available for UW-L students seeking to become secondary education teachers. Although this restructuring was instituted by the university's upper administration, it was also made in response to recommendations from the Wisconsin Department of Public Instruction. The rationale for reforming the program was to provide teacher candidates with discipline-specific advising, methods courses, and early field experiences, consistent with national recommendations for teacher preparation [6–8].

At UW-L, the School of Education is the teacher licensing body and the "umbrella" encompassing a variety of teacher education programs on campus. The programs in early childhood education and middle childhood education, and programs that span several grades (art, music, modern languages, and special education), are all housed in the Department of Educational Studies. There are also programs at UW-L for physical education teacher education (in the Department of Exercise and Sport Science) and school health education (in the Department of Health Education and Health Promotion).

The role of the School of Education is to interpret and communicate the requirements for obtaining a teaching license in the state. These requirements include licensing exams, field experiences and student teaching, documentation of classroom observations, and a portfolio process for teacher candidates. Each department sets the coursework required for the degree programs, with approval from the School of Education and the Wisconsin Department of Public Instruction. The College of Liberal Studies encompasses most teacher education programs (Departments of Educational Studies, English, and History) and the College of Science and Health includes the science, math, and health programs. Faculty involved in teacher preparation have an "affiliation" with the School of Education but a tenure-track home within their departments.

The secondary teacher education program restructuring brought with it a cluster of hires of discipline-based education research (DBER) faculty to direct the content-based degree programs. The physics department fully supported the change and was one of the first departments on campus to make a hire, bringing on board a physics education faculty member (Jennifer Docktor) in Fall 2011. In the following two years, there were three additional faculty hires in the sciences (one each in biology, chemistry, and Earth science), one in English, and one in history. The mathematics department already had three faculty members specializing in math education and hired two more. This has created a cohesive group of 11 faculty who are working together to improve the preparation of secondary teachers.

These discipline-based education faculty play hybrid roles, having responsibilities within their home departments for teaching, scholarship, and service, along with roles within the School of Education. Most notably, these faculty are responsible for recruiting students into the STEP program, for advising and mentoring students, and for supervising students during their clinical field experiences and student teaching experiences. The faculty also attend meetings and trainings with other teacher education faculty affiliated with the School of Education. These responsibilities have proved important for establishing relationships with local teachers and for staying up-to-date on issues affecting teachers.

The reformed secondary education programs also include streamlined degree requirements that balance content courses (required for a B.S. in a particular content area) with the professional education requirements and general education requirements of the university. Most programs require close to 120 credits and can be completed in 4 or 4.5 years, which is shorter than the average completion time in the old program (5 or 6 years).

The B.S. degree in physics education includes three main components: the physics major requirements, which include math courses (54 credits); the professional education requirements, which include education courses and student teaching (35 credits); and general education courses (48 credits). Since some courses in science, math, or education also count as general education courses, the total number of credits is near 120 (not 137). If an incoming student does not place into calculus for his or her first semester, this increases the total number of credits. The Appendix includes specific information about the physics courses required for the major.

Another key change was to move the secondary methods courses for English, history, science, and math into the content areas, so that these courses are taught by the new discipline-based education faculty. The structure of these courses was also changed to include an embedded clinical field experience, wherein students spend 100 or more hours in a local high school teacher's classroom. The field placement mentor teachers in science are coordinated by one of the science education faculty members, in collaboration with the School of Education's Office of Field Experience. The science course Teaching and Learning Science in the Secondary School is currently structured to meet for three hours per week on campus, and for one hour per week at a school site for a seminar with guest speakers from the school. The in-class portion is co-taught by two of the science education faculty members, and is structured into three main units: instructional planning, instructional strategies for teaching science, and assessment. There is an explicit focus on preparing students for the national edTPA teacher performance assessment, which was recently adopted by Wisconsin for assessing student teachers [9–11].

Additional changes to improve the quality of secondary teacher education preparation and to build a community of teacher candidates are ongoing. In response to perceived gaps in students' preparation, a new course, Curriculum and Assessment in Math and Science, was proposed and approved in 2013–2014; it was first offered in Fall 2014. This course is a highly recommended elective to prepare students for their teaching and learning (methods) course; it also provides students with an opportunity to meet other teacher candidates in secondary education. To help foster community, the faculty members directing secondary teacher education programs have successfully organized mentoring and networking "meet and greet" events each semester, beginning in Fall 2013.

In 2012, UW-L was selected as one of seven sites nationally to receive funding from the Physics Teacher Education Coalition (PhysTEC) to develop their physics teacher education programs. PhysTEC has further augmented the programmatic changes to teacher preparation at UW-L by funding students to serve as Learning Assistants in introductory courses, and by enabling UW-L to create a local network of physics teachers to mentor and support teacher candidates and build a community of students interested in science teaching. PhysTEC has also helped showcase the new physics education degree and bring visibility to preservice teacher education at UW-L.

II. RECRUITMENT AND RETENTION PRACTICES

Efforts to recruit and retain students in physics teaching fall into two key categories: (1) applying specifically to the physics education track practices already utilized by the physics department to attract students to UW-L; and (2) encouraging students already on campus (i.e., physics majors) to consider a teaching career.

A. Recruitment

1. Physics department recruitment activities

Six times each year, the physics department faculty participate in university-wide Campus Close-Up events, which bring high school students and their parents to campus for presentations and open houses. During these open house days, prospective students have the option to attend special sessions based on their interests. High school students and parents who come to the physics display booth receive brochures describing the different degree options, and can talk one-on-one with faculty to learn more about the program. There is also a designated time for interested students and parents to get a tour of the department; the tour includes some of the classroom and research laboratory facilities. Shortly after the restructuring of the secondary teacher education program, the department's recruitment materials were revised to highlight the new degree option in physics education.

Another recruitment strategy used by UW-L's physics department is to purchase from Educational Testing Service the names of ACT test takers who have expressed an interest in physics, engineering, or science teaching, and to send each test taker a letter and brochure about the department's programs along with a personal invitation to visit campus. This practice has been applied to the physics education option by expanding the search terms to include students who have expressed an interest in "science teaching." The recruitment letter and brochure were also revised to include information about the physics education track, and efforts are underway to create a physics education website and marketing posters about the program. It is difficult to determine to what extent the brochure mailings increase recruitment. Very few (if any) students enter UW-L with a declared major in physics education, yet several choose to major in physics, suggesting that UW-L has built a strong reputation for its physics program but not necessarily for its education programs.

The physics department further increases its visibility in the community via its annual Distinguished Lecture Series in Physics, which brings a Nobel laureate to campus for a public lecture. The first such lecture was held in the fall of 2000, and the 2014 event marked the 15th time a Nobel laureate in physics has visited UW-L. The department also hosts Physics and Laser Light Show Extravaganza presentations each spring, as well as frequent planetarium shows that attract more than 5,000 schoolchildren each year. The department has strong connections to local K-12 teachers and administrators through professional development workshops funded by a U.S. Department of Education Mathematics and Science Partnership grant, which has resulted in additional physics fairs and demonstration shows. Although there is no evidence directly connecting community-wide events or brochures to increased recruitment of physics majors, these events help increase the visibility of the department as a leading program for physics and physics education.

The department also recruits students internally through its introductory courses, which enroll 350 to 450 students each semester. Physics majors have the flexibility to take the introductory algebra-based course sequence, which makes it easier for students to switch into the physics major. Approximately 25% to 30% of physics majors take this route, and are able to catch up on their math by enrolling in Calculus I in their second semester and taking Calculus II in the summer or concurrently with their sophomore physics courses. Students meet with their academic advisors every semester to map out an individualized course plan. The physics department reformed its introductory calculus-based course sequence and adopted an integrated lecture-lab (studio) format in Fall 2011. The reformed course has been well received by students, and has aided in the recruitment of new physics majors and the retention of existing majors.

Successful recruitment is attributed (in part) to the range of degree options available to students. In the more "traditional" physics major, students can select an emphasis in astronomy, computational physics, optics, or a biomedical or business concentration. Dual-degree programs in physics and physical therapy or physics and engineering are also quite popular, and a new degree program in secondary physics education was added in Fall 2011. Students who are undecided on a specialization get to hear more about each area of study during the seminar for freshmen and sophomores, and the new degree in secondary physics education has been highlighted during at least one presentation each semester. In addition, the department has been very successful in placing students who receive their bachelor's degrees into science, technology, engineering, and mathematics careers [5]. Department faculty acknowledge that pursuing graduate studies is only one of many career paths available to students, and the faculty value teaching as a viable option.

2. Physics education recruitment activities

In addition to applying the physics department's existing recruitment activities to physics education, there are also several practices being implemented to encourage students such as physics or math majors, who are already at UW-L, to consider teaching physics. The department relies on academic advising and special events to publicize the STEP program.

Every student who has declared a physics major must meet with his or her academic advisor every semester to plan out a specific course sequence that aligns with the student's future career plans. These meetings typically last 15 to 30 minutes, and a registration "hold" placed on enrollment can be removed only by the academic advisor and only after the meeting. Students who have declared physics education as their majors are advised by the physics education faculty member, who provides program planning handouts and guides the student to complete licensure requirements such as applications and licensing tests. Faculty who advise the remaining physics students have a point person within the department to whom they can direct students for more information about physics teaching. Physics students who express some interest in teaching are quickly identified during an advising meeting and are often brought immediately to the physics education advisor to obtain more information.

In addition to identifying prospective teacher candidates during advising sessions, physics faculty also select particular physics majors to serve as Learning Assistants in the introductory calculus-based physics course and introductory laboratories. Faculty also select students to serve as university tutors. These opportunities present physics majors with early teaching experiences that may spur a broader interest in teaching. Physics majors are also selected to assist with outreach events to local schoolchildren; these include the annual laser light shows and physics demonstration shows.

In addition to efforts within the physics department, there is coordination across STEP program advisors to encourage teacher candidates to pick up additional licensure areas. For example, students who are majoring in one science education area or in math education are advised to pick up a certifiable minor in another area, such as physics, that would allow them to teach high school physics in Wisconsin. The STEP program faculty also organize special events each semester so that students can learn more about secondary teacher education at UW-L.

B. Retention

1. Physics department retention activities

As a primarily undergraduate institution, UW-L takes a very student-centered approach to teaching and mentoring. In addition to focused advising sessions, another outstanding effort by the department is its focus on involving all physics majors in undergraduate research. For example, during the 2012–2013 academic year, 44 unique students participated in undergraduate research in physics. Students now have the option to engage in physics education research (PER), and this experience is highly recommended for students in the physics education major. Some students who are early in their studies choose to investigate a particular topic, misconception, or curriculum in physics, and produce a summary paper from the PER literature. Students who are more advanced work on designing a study and/or analyzing existing data. Current projects focus on studying multiple representations (graphs, diagrams, equations, etc.) using eye-tracking technology; several students have taken advantage of the rare opportunity to use this equipment.

Community spirit pervades the department. This spirit is fostered through seminars and several student-led organizations that regularly sponsor campus events. These organizations include the Physics Club, a Sigma Pi Sigma chapter, and a Women in Physics club. In particular, the Physics Club hosts an annual fall picnic to welcome new physics majors to campus. This combination of activities yields high retention rates in the major (75% in the first year and 95% thereafter).

2. Physics education retention activities

Efforts to retain physics teacher candidates include discipline-specific advising and mentoring, and presenting students with opportunities to get involved in early teaching experiences and outreach events. During advising meetings, students receive formal mentoring regarding recommended courses and course sequences and informal mentoring on their decisions to pursue teaching. Physics education students are solicited in their first or second years to assist with specific courses, including the inquiry-based course Physical Science for Educators taken by future elementary and middle school teachers. This course gives physics education students teaching experience that they wouldn't otherwise get until much later in their education courses. Students who serve as Learning Assistants in this capacity have remarked that the experience validated their decisions to become teachers, and that they wish there were more classroom teaching experiences early in the program. Students also have the opportunity to participate in outreach events, including the laser light shows for elementary students each May and annual physics fairs and demonstration shows for local middle schools. In Spring 2014 the physics fair event was expanded to include 385 sixth graders from three middle schools over three days, with approximately 80 UW-L student volunteers.

III. EVIDENCE FOR SUCCESS

The STEP program and the physics education degree launched in Fall 2011, and many changes began to take place during the 2011–2012 academic year. These changes include academic advising and the new structure for the science methods course. A grant from PhysTEC began in 2012–2013, prompting the documentation and expansion of recruitment and retention efforts, and providing funds for Learning Assistant stipends. These two factors contributed to the increased number of students in the program in 2012–2013. Since there were five students in the program in 2011–2012 and only one of them graduated, four carried over into 2012–2013. In order to have a total of nine students, five *new* students must have been recruited during that year with higher than sophomore status. Three of the five new students in that year were recruited from the pool of physics majors, and their interest was identified during academic advising by physics faculty. In 2013–2014 there were six students who carried over from 2012–2013, and three physics majors who switched to the physics education program; however there were also two students who left the university or changed majors, leaving a total of seven students in the program. One of the recruited students in 2013–2014 had served as a PhysTEC Learning Assistant. In light of the challenges of tracking students entering and exiting the program, these preliminary counts indicate that UW-L is well positioned to increase the number of physics-certified teachers graduating in upcoming years.

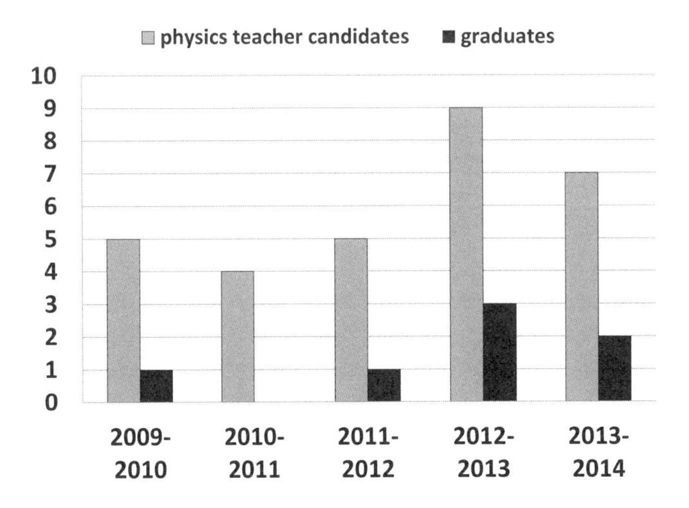

FIG. 1. Total number of physics teacher candidates and the number of those students graduating from the University of Wisconsin-La Crosse, by academic year. A student is considered a teacher candidate when he or she has completed at least one year of studies (i.e., the student has sophomore standing).

The quality of students recruited into physics teaching has remained high. In 2012–2013 one physics education student was selected to speak at graduation and another received the 2013 Strzelczyk Award in Science and Health, which recognizes an outstanding senior for academic achievement and service to the campus and community.

IV. DISCUSSION

UW-L recently made several changes to its physics teacher preparation program. These changes include aggressive recruitment and retention efforts. Although many of the activities were already part of physics department efforts to recruit students into physics, existing materials and activities have been applied or adapted for the physics education major. Additional advising, mentoring, and course reforms have been added to recruit and retain this particular group of teacher candidates. Preliminary results indicate that these practices have been effective at attracting physics majors into physics teaching. Key elements of UW-L's program have been summarized below.

UW-L's successful strategies for *recruiting* students into physics teaching include the following:

- Establishing a faculty member within the physics department to advise and mentor teacher candidates, and encouraging other physics advisors to identify potential teacher candidates from among the group of physics majors. This is consistent with reports indicating the need for a departmental "champion" who is committed to teacher preparation efforts [7].
- Establishing a departmental culture that not only values the undergraduate physics program, but also values teaching as a viable career option. Physics faculty at UW-L are actively involved in community outreach and course reform, and see this as a collaborative effort. The department is also known for having an open-door policy, which helps students feel comfortable having informal career-related discussions with faculty.
- Implementing active recruitment and retention activities to increase the number of physics majors in the program. This in turn increases the pool of potential physics teachers. There are many opportunities for students to obtain teaching experience, which may spur an interest in high school physics teaching. Opportunities include assisting with a course, tutoring, or helping with outreach events.

UW-La Crosse's successful strategies for *retaining* students in physics teaching include the following:

- Establishing a point of contact within the physics department to spearhead the secondary teacher education program and provide focused advising and mentoring to physics education students. This person also takes an active role in School of Education activities to foster collaboration between the physics department and the School of Education [7]. At UW-L this point person was a new hire in physics education. The job advertisement explicitly included teacher preparation responsibilities. Note that it is also critical to have a department chair who is willing to consider these activities as part of a faculty member's course load and to make accommodations as needed.

- Providing teacher candidates with early teaching experiences they wouldn't otherwise get until much later in the program, such as being a Learning Assistant for the physics for elementary teachers course, tutoring, and helping with outreach events for local schools. Engaging students in undergraduate research in physics education also presents them with additional opportunities for involvement and informal mentoring.
- Building a community of students interested in physics and physics teaching, both through departmental club events and by forming a cohort of secondary education teacher candidates. Similarly, a department should build a community of faculty committed to secondary teacher preparation. UW-L recently hired a cluster of DBER faculty, which makes teacher preparation a collaborative effort.

UW-La Crosse's challenges and lessons learned include the following:

- Very few (if any) freshmen enter the university with a declared major in physics education, and so far there is no evidence that expanding the recruitment letter to include science teaching or highlighting the program during campus open house events is having any effect on recruiting high school students into physics teaching. UW-La Crosse has a reputation for an exceptional physics program, but it does not have the same reputation in education.
- The process of selecting students to serve as Learning Assistants for a course needs to happen far in advance (prior to registration in the semester preceding the Learning Assistant experience), so that students block off the appropriate time commitments in their schedules prior to registering for courses. There is very little flexibility to change a student's course enrollment after the initial registration period. The Learning Assistants also need to be contacted and reminded about their commitments a few weeks before the start of the semester, in case their plans have changed and replacements need to be arranged.
- Some students who express an interest in physics teaching have the perception that it will be easier than pursuing a traditional physics degree, when in fact the GPA requirement for being admitted into the School of Education is quite high. This impacts the progress of some students through the teacher preparation program. It is important for students (and faculty) to be aware of this up front.
- The physics education major includes a substantial number of education courses—particularly a sequence of field experiences—that cannot be taken concurrently. This can lengthen the degree completion time for students who switch majors late in their programs. Identifying potential candidates early is important. Advising handouts need to be differentiated based on particular student situations, since most students do not begin as physics education majors in their freshman year.
- Before arranging early field experiences, UW-L requires approval from the Office of Field Experience and the school district(s) involved. This has presented some challenges when planning outreach events. It has been easier to hold physics fairs and demonstration shows at neutral or non-school locations (such as a local children's museum or environmental center), so that student volunteers do not need to submit paperwork several weeks in advance just to be able to enter a school building.

UW-L has a strong collaboration between content departments and the School of Education. However, there is a heavy time commitment on the part of content faculty involved in teacher education, especially to attend School of Education meetings and supervise student teachers (who can be placed anywhere within 50 miles of La Crosse). It is important for faculty to carefully document their activities related to teacher preparation and receive teaching credit when appropriate. It is crucial that these responsibilities are valued by the department and not just seen as additional "service."

V. CONCLUSION

The University of Wisconsin-La Crosse has made some substantial institutional changes to its secondary teacher education preparation program, including moving the degrees into the content departments in 2011 and hiring a cluster of DBER faculty. For the physics department, hiring a physics education specialist has meant that there is a champion to direct teacher preparation efforts, including academic advising, mentoring, acting as a liaison with the School of Education, and collaborating with other DBER faculty to develop a community of teacher education candidates. Beginning in 2012, as a result of PhysTEC funding, the physics department explicitly focused on recruiting and retaining future physics teachers. Students have several opportunities for early teaching experiences, including as Learning Assistants in introductory courses and through outreach to K-12 schools during physics fairs and demonstration shows. Counts of secondary teacher graduates and future physics teachers indicate that UW-L's physics teacher education program is showing signs of growth and has the potential to increase further in coming years.

[1] University of Wisconsin La Crosse Admissions, *Fast Facts about UW-La Crosse.* http://www.uwlax.edu/Admissions/Fast-facts/

[2] P. J. Mulvey and S. Nicholson, Physics bachelor's degrees: Results from the 2010 Survey of Enrollment and Degrees, AIP focus on (Sept. 2012). http://www.aip.org/sites/default/files/statistics/undergrad/bachdegrees-p-10.pdf

[3] See http://www.aps.org/programs/education/undergrad/faculty/awardees.cfm

[4] R. C. Hilborn, R. H. Howes, and K. S. Krane, eds., *Strategic Programs for Innovations in Undergraduate Physics: Project Report* (American Association of Physics Teachers, College Park, MD, 2003), pp. 136–138. http://www.aapt.org/Programs/projects/spinup/upload/SPIN-UP-Final-Report.pdf

[5] See http://www.uwlax.edu/uploadedFiles/Academics/Departments/Physics/Career%20Pathways%20Report%20-%20UW-L.pdf

[6] T. Hodapp, J. Hehn, and W. Hein, Preparing high-school physics teachers, Phys. Today **62** (2), 40 (2009). doi:10.1063/1.3086101

[7] D. E. Meltzer, M. Plisch, and S. Vokos, eds., *Transforming the Preparation of Physics Teachers: A Call to Action. A Report by the Task Force on Teacher Education in Physics* (T-TEP) (American Physical Society, College Park, MD, 2012). http://www.phystec.org/webdocs/2013TTEP.pdf

[8] D. E. Meltzer and V. K. Otero, Transforming the preparation of physics teachers, Am. J. Phys. **82**, 633 (2014). doi: 10.1119/1.4868023

[9] See http://www.edtpa.com

[10] See http://www.edtpa.aacte.org

[11] See http://www.pearsonassessments.com/teacherlicensure/edtpa.html

APPENDIX: COURSES REQUIRED FOR THE PHYSICS EDUCATION B.S. DEGREE PROGRAM

Physics Education Major

Physics Education Major Requirements (54 Credits):

Major Core Requirements (38 Credits)

PHY 103 (or 203) Fundamental (General) Physics I (4)
PHY 104 (or 204) Fundamental (General) Physics II (4)
PHY 250 Modern Physics (3)
PHY 302 Optics (3)
PHY 311 Experimental Physics (2)
PHY 321 Classical Mechanics (3)
PHY 334 Electrical Circuits (3)
PHY 469 Teaching and Learning Science in the Secondary School/Field Experience II (4)

Electives: Select 12 additional credits in Physics from the following and/or courses numbered 300 and above:

PHY 155 Solar System Astronomy OR PHY160 Stars, Galaxies, & the Universe (4)
PHY 332 Electrodynamics (3)
PHY 343 Thermodynamics (3)
PHY 356 Curriculum & Assessment in Math and Science (2)
PHY 401 Quantum Mechanics (3)
PHY 453 Topics in Physics & Astronomy (1-3 credits)
PHY 497 Physics & Astronomy Seminar (up to 2 credits)
PHY 498 Physics & Astronomy Research (up to 3 credits)

Additional Mathematics and DPI Requirements (16 credits)

MTH 207 Calculus I (5)
MTH 208 Calculus II (4)
MTH 310 Calculus III: Multivariate Calculus (4)
GEO 200 Conservation of Global Environments (3)

Professional Education Core (35 Credits):

EFN 205 Understanding Human Differences (3)
EDS 303 Foundations of Public Education in the United States (2)
EDS 309 Education in a Global Context (2)
EDS 319 Teaching with Integrated Technology (2)
EDS 351 Language, Literacy and Culture in the Secondary Classroom/Field Experience I (4)
PSY 212 Life-Span Development (3)
PSY 370 Educational Psychology (3)
SPE 401 Learners with Exceptional Needs (3)
EDS 492 Student Teaching Seminar (1)
EDS 494 Student Teaching – EA/A (12)

General Education Core (48 Credits):

STEP General Education Core Requirements: See STEP General Education Advising Sheet for additional General Education coursework.

- Literacy: ENG 110 or ENG 112 College Writing and CST 110 Communicating Effectively
- Minority Cultures or Multiracial Women's Studies: EFN 205 Understanding Human Differences
- International and Multicultural Studies: GEO 200 Conservation of Global Environments

Graduation Requirements: A minimum of 120 total credits are required.

Structuring Effective Early Teaching Experiences

Nationally scaled model for leveraging course transformation with physics teacher preparation

Valerie K. Otero
School of Education, University of Colorado Boulder, Boulder, CO 80309

All undergraduate physics instructors are involved in teacher preparation. Future physics teachers make up a fraction of the students in undergraduate physics courses, so the teaching and learning practices that take place in these courses serve as models for how future high school physics teachers will teach their students. It is therefore the responsibility of physics instructors to ensure that their methods for teaching, and the learning practices they promote, are aligned with their expectations for introductory students' high school preparation. However, making changes to a course is not always straightforward and is often time consuming. This article describes a model for making small to large transformations to undergraduate physics courses. Lessons learned from 10 years of implementation at the founding institution and from emulating institutions throughout the nation are discussed. Resources for making transformations, along with data to support claims about the effectiveness and adaptability of the model, are also described.

I. INTRODUCTION

A recent National Research Council report, *Adapting to a Changing World—Challenges and Opportunities in Undergraduate Physics Education*, points out that current practices in physics education are not serving students well, and that changes to undergraduate physics instruction are needed in order to improve the professional preparation of physics teachers [1]. Another recent national report demonstrates that few U.S. university physics departments are contributing substantively to the preparation of physics teachers, and therefore physics faculty contribute little to the preparation of the incoming freshmen whom they teach [2]. Both reports recommend that physics faculty begin to address these educational challenges by making small to large transformations to their courses, using instructional practices that have been shown to improve student learning. Such practices involve students in discussions and problem solving during class, and provide opportunities for instructors to engage directly with student groups. These instructional practices are the same teaching strategies that are promoted nationally for exceptional high school physics classroom experiences [3].

Implementing new teaching practices is challenging even for the best instructors and professors. What are the first steps? What should students do when they are in groups? How does one cover all the material and convince students that these novel formats are useful? These are just a few of the questions that emerge when thinking about making transformations to a physics course. In this manuscript, the Learning Assistant (LA) model is described as a practical and easy-to-adapt model for transforming undergraduate courses. Data to support claims about the efficacy and scalability of the model are also discussed, along with lessons learned and tips for implementation. Finally, support structures and materials that are accessible to physics faculty members are described.

II. THE LEARNING ASSISTANT MODEL

The Colorado Learning Assistant model integrates goals of teacher recruitment and preparation, course and curriculum transformation, discipline-based educational research, and departmental and institutional change. With multiple facets appealing to different stakeholders (i.e., deans, chairs, physics faculty, and students), the LA model has spread rapidly throughout our own campus and throughout the nation, and is beginning to spread internationally as well [4]. Since the national introduction of the model in *Science* in 2006 [5], over 60 U.S. universities report using the LA model. The LA model is one of few national-scale, university-based efforts that involves physics research faculty in making modifications to their courses specifically to improve the educational experiences of future teachers as well as those of students aiming for graduate school or industry.

The Colorado LA model [6] is recognized nationally as a hallmark model for both course transformation and teacher recruitment and preparation [7,8]. The full program was launched at the University of Colorado Boulder (CU Boulder) in 2003 to engage science and math research faculty in making small to large changes to courses. These changes promote teaching practices that are aligned with education research as well as with the methods of instruction promoted by teacher preparation programs for future K-12 teachers (discussed in more detail in Section III). Undergraduate LAs at CU Boulder are hired by math, science, and engineering faculty (and paid for by provost and departmental funds) to help transform large-enrollment courses so that enrolled students have ample opportunities to work in small groups to articulate, defend, and modify their ideas about a relevant problem or phenomenon. This strategy of active learning through scientific modeling of phenomena has been shown to facilitate learning and increase engagement in class [5,9–12].

While instantiations of the LA model differ across departments and programs, the main role of the LA is to work with small groups of three to six students as they solve challenging conceptual or mathematical problems. Small-group settings take place in the actual lecture hall during the regularly scheduled lecture, or in additional recitation or laboratory sections of the course (see Section III). Physics courses transformed with LAs show conceptual learning gains nearly twice the national average of gains in traditional courses [13]. In a longitudinal study, Pollock demonstrated that students who had previously served as LAs in introductory courses outperformed their peers in upper-division electricity and magnetism (E&M) [14]. There were two groups of students in the study who were enrolled in upper-division E&M but did not previously serve as LAs: those who had taken an LA-supported E&M course as freshmen and those had taken traditional introductory E&M courses as freshmen. Pollock's study showed that those who had taken an LA-supported E&M course as freshmen outperformed those who did not. In the same study, Pollock demonstrated that students' performance on traditional, instructor-generated, problem-solving exams mirrored their performance on the conceptual exams—that is, former LAs outperformed the students who, as freshmen, had an LA-supported E&M course, who outperformed those who did not, as freshmen, have an LA-supported course.

The Colorado Learning Assistant model is based on the philosophy that learning involves actual participation in the specific practices of a discipline. Just as the learning of physics requires students to investigate and make inferences about real physical phenomena, learning new ways of teaching requires teachers to use the instructional methods they are trying to learn. Both LAs and their faculty supervisors practice using methods involving small groups of students discussing their thinking with one another. Some of the strategies that LAs use include open questions that do not have one specific right answer, "wait time" wherein the instructor allows for a few (sometimes uncomfortable) minutes of silence after asking a question (giving students a chance to gain the confidence they need to express their answers), and "probing questions" that encourage students to discuss their thinking even when they are uncertain.

LAs support large-enrollment courses by effectively increasing the teacher-to-student ratio and allowing for multiple collaborative working groups, either in a large lecture hall with the lead instructor present, or in LA-led, small-group meetings that take place at various times and places throughout the week. In these settings, LAs support students as they analyze data and make inferences about phenomena—practices necessary for understanding science.

A subset of the students who serve as LAs go on to obtain certification to teach high school physics and become high school physics teachers. Since the first physics LA was recruited to K-12 teaching through the LA program in 2006, the graduation rate of physics and astronomy majors certified to teach has increased from less than one per year to an average of three per year (see Figure 1).

III. USES OF LEARNING ASSISTANTS: FROM RECITATION TO FLIPPING THE CLASSROOM

Many different models of transformation can be used to structure learning environments to support LAs as they work with students in small groups [12,15]. However, a necessary component of the program is that groups work on *group-worthy problems* in these settings. One type of group-worthy problem has been described in the literature as a "*context-rich problem*" [16,17], and examples are available on the University of Minnesota website [18] or in the online PhysPort resource [19].

The LA-supported calculus-based introductory physics courses at CU Boulder primarily use the University of Washington *Tutorials in Introductory Physics* [20]. The *Tutorials* use problems that are context rich and that facilitate group engagement through challenging and carefully sequenced questions. Implementing the *Tutorials* requires a setting in which groups of three to four students work through the questions, which are carefully sequenced conceptual probes into students' understanding of equations, relations, and phenomena. The *Tutorials* contain different types of questions, such as those listed below:

- Ranking questions: These questions require that students rank characteristics within a physical system (e.g., the brightness of two bulbs in an electric circuit), between two different physical systems (e.g., the brightness of bulbs in two different circuits), or before or after something is done to a physical system (e.g., adding a bulb to a circuit).
- Modeling questions: Students are asked to formulate a rule on the basis of various observations they have made (or that are described to them).
- "Predict, observe, explain" questions: Such questions ask students to predict, observe, and resolve differences in their predictions and observations.
- Mathematical relationships questions: Students are asked to establish relationships by writing mathematical equations.
- Diagramming initial or final conditions: Students are asked to draw diagrams of various inputs or outcomes of a phenomenon.

These questions are presented to students in a workbook that is used as a foundation for small-group discussion. Each major concept that is developed through the *Tutorials* comes with pre- and post-assessments that allow the teacher to assess the effect of the tutorial on student learning. In these

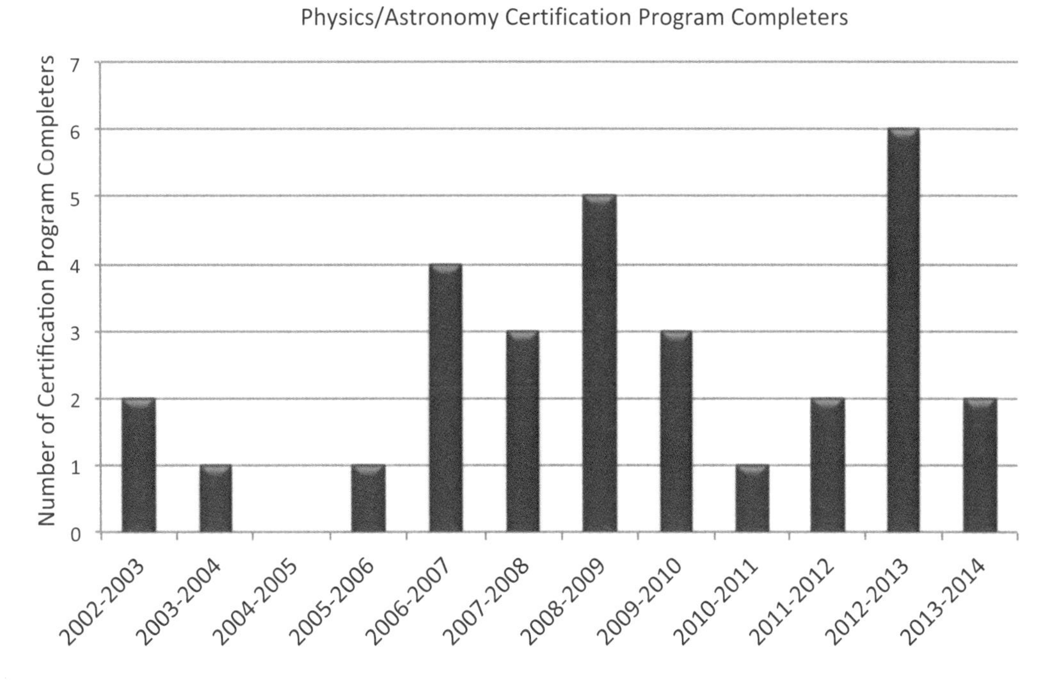

FIG. 1. Physics and astronomy majors at CU Boulder certified to teach high school science.

settings, the LA's role is to ensure that students interact with one another, that each group member has an opportunity to express his or her ideas, and that groups are working toward a solution and making appropriate assumptions and progress. In some cases this requires that LAs assess group progress and ask probing and guiding questions, but it rarely requires that LAs tell students answers.

At CU Boulder, one or two LAs and one graduate TA are typically assigned to each tutorial session; approximately 24 students attend each session. The *Tutorials* are used in recitation sections that run throughout the day on Tuesdays and Thursdays to accompany the Monday, Wednesday, and Friday lectures. In other departments, professors have cancelled one lecture per week in their introductory courses and replaced it with a large number of small-group sessions (with 12 to 24 students per session) that are headed by LAs, LAs and TAs, or LAs and the lead professor of the course. Group sessions in other departments follow the same tutorial model used in the physics department, only using different materials, such as *Lecture-Tutorials for Introductory Astronomy* [21]. In certain CU science courses, including some physics courses, the *full lecture* is transformed and LAs are used to help the instructor "flip" the classroom so that lecture activity takes place as homework: Students obtain the lecture online, and class time that was traditionally used for lecture is now used for context-rich problem solving, as described above.

Although there is a diverse range of manifestations of the LA model throughout CU and throughout the nation, one of the important elements of the model is that LAs facilitate groups of students in group-worthy activities. Unlike many other peer mentoring models, the LA model is not designed to promote one-on-one tutoring, but instead to help the physics instructor manage the conversion of his or her large lecture course into (some) smaller settings in which students have opportunities to articulate, defend, discuss, and modify their ideas as a group.

In addition to facilitating group activities, LAs engage in various forms of preparation. During a typical week, LAs engage in three core activities: (1) They lead *Tutorials* sessions consisting of approximately 20 students that are organized into smaller groups of three to four students (most LAs run three sessions weekly); (2) they meet weekly with the lead faculty member of the course to plan for the upcoming week, reflect on the previous week, and analyze assessment data; and (3) they attend a science, math, and engineering education course once per week. In this education course, which is required for all first-time LAs, LAs are introduced to practical techniques that include:

- using a variety of questioning strategies, such as asking open, closed, synthesis, evaluation, and application questions;
- facilitating group participation by, for example, asking one member to evaluate or comment on another group member's idea;
- diagnosing and attending to common conceptual difficulties encountered in specific content areas and topics (this is not "fixing misconceptions" but rather helping students reconcile their intuitions with physics principles and their observations); and

- developing pedagogical responses that can promote progress in reasoning without directly giving the answer to a problem.

The reason LAs are encouraged to avoid giving answers directly is that the philosophy of the LA model is to represent math and science principles as solutions that make sense rather than as dictums from authority. In fact, one of the major additions of the LA model to the traditional lecture environment is that it helps students see scientific knowledge as evidence-based claims that are generated through careful research and sensemaking, rather than as answers found in the back of a textbook. The strategy of engaging students in generating and being skeptical about ideas, rather than simply memorizing them without question, is also central to the practices promoted for high school teaching and learning [19].

The education course also uses readings from education research and cognitive science to introduce empirical studies and theories of teaching and learning that support the value of the instructional techniques discussed above. As course assignments, LAs complete weekly online teaching reflections on how their teaching (and learning) integrates with the educational research they are reading. Through discussions with other physics LAs and LAs from other departments, and through their weekly reflections and readings, LAs begin to develop their own principles of teaching and learning. Thus, this model for LAs' learning about effective instructional techniques follows the same model LAs are being asked to use with their students, as their students learn scientific principles. This is also the model for learning promoted in national reports [1,2] and in the Next Generation Science Standards for K-12 science and engineering [19].

IV. APPLICATION AND SELECTION PROCESS

The LA program at CU Boulder runs in 12 departments in the College of Arts and Sciences, the College of Engineering and Applied Science, and the School of Education. Approximately 260 LAs are hired to serve in 40 to 80 courses each year. Many LA programs throughout the nation, however, are smaller and run in only one or two departments. It is nevertheless helpful to have a central organizing structure for LA program activities, including the recruiting, hiring, and preparation of LAs.

At CU Boulder, an information session is held four to five weeks into each semester, in which current LAs and faculty make brief presentations about the LA-supported courses and answer questions from students who are interested in the LA program. The LA application system is then open for two weeks following the information session. The session runs early in the semester so that the faculty interested in using LAs can interview and hire students far in advance of the upcoming semester, and also so newly hired LAs are aware of any time restrictions before registering for their own courses for the upcoming semester.

Because the CU Boulder LA program has grown so large, faculty must apply to receive "Provost's LA Awards." The provost's award assigns a certain number of LAs to faculty members who have applied. Not all faculty applicants receive the provost's awards, and not all applicants who receive awards receive the number of LAs they applied for. In some cases it is recommended that faculty applicants use more LAs (for example, when they apply for only one LA), and in other cases faculty members receive only a fraction of the LAs requested. From some applications it is clear that the course is not "LA ready," for reasons that include no real evidence of transformation and few opportunities for LAs to work with groups of students. In such cases some consultation is provided, and applicants are encouraged to apply for a Chancellor's Award for education research offered through the Center for STEM Learning on campus.

At CU Boulder we have an online application and evaluation system that allows faculty to apply for the Provost's LA Awards and students to apply to be LAs. In their applications to obtain LAs, faculty members must state specifically how they intend to use LAs, in what ways the course will undergo transformation, and anticipated procedures and content of weekly planning sessions with LAs. Other application questions have to do with anticipated course enrollments, whether the applicant has attended a workshop session on using LAs, the curricular materials the applicant is planning to use (e.g., *Tutorials*), roles other than facilitating group work that LAs might have, and how LAs might benefit from the experience.

When students apply online for an LA position, they are allowed to make three course selections and rank them in order of preference. Faculty can then enter the system, rank selected candidates, make a short list, interview top candidates, and log their final selections. When final selections are made, the LAs are contacted, their paperwork is processed, and they are hired through the provost-funded "LA Central." LA Central is the central coordinating body for the campus and currently consists of a program director, a technology and data specialist, a project administrator, and an executive director who is a tenured faculty member. (Staff needs at other institutions might be larger or smaller depending on program size and scale.) LA Central interacts with a single liaison from each department who coordinates the department's LA activities and applications. Each coordinator has access to the faculty applications for LAs in his or her department so that the quality and number of these applications can be monitored.

All faculty and student data is fed into a database to keep track of the program and maintain statistics on how and by whom LAs are used, in which courses and departments they are used, and how many students are impacted. For comparative purposes, we also keep records of LAs who are not hired.

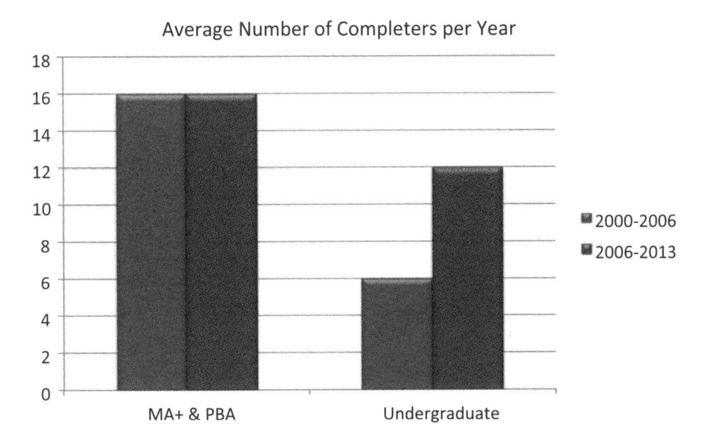

FIG. 2. Average number of math and science teacher certification completers per year at CU Boulder: Before the start of the LA program (2000–2006) and after the start of the LA program (2006–2013).

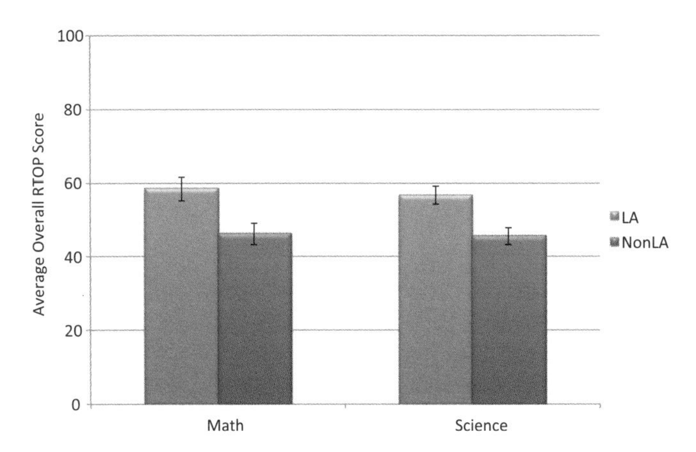

FIG. 3. RTOP scores for former LAs and a matched sample of teachers who did not serve as LAs.

V. IMPACT OF THE CU BOULDER LA PROGRAM

A. Teacher recruitment and preparation

Students apply to become LAs for many different reasons. Some students are interested in finding out more about teaching (at both the K-12 and college levels), others are interested in getting a different perspective on the science content, and still others are looking for a rewarding way to earn extra money. (LAs are paid $1,500 per semester to work approximately 10 hours per week.) Regardless of their reasons for their initial interest, approximately 12% of LAs decide to get their K-12 teaching certifications and become teachers. Previous research at CU Boulder showed that the two most frequently stated reasons for LAs deciding to become teachers were (1) encouragement and support from science and mathematics research faculty, and (2) recognizing teaching as a challenging, problem-solving activity [22].

LAs who decide to become teachers are eligible for National Science Foundation (NSF)-funded Noyce Scholarships [23,24] as they participate in the CU-Teach certification program. As Noyce Scholars, CU-Teach students (prospective teachers) can participate in the Teacher Research Team program [25-27] with master teachers, veteran teachers, and early-career teachers. Teacher research teams are professional learning communities within which teachers conduct publishable, discipline-based educational research in K-12 classrooms. Teachers and prospective teachers publish their work in journals [e.g., 28–30] and present at national research and practitioner conferences. These teams also provide workshops for other teachers locally and throughout the nation. As CU-Teach students move on to careers in teaching, they frequently opt to serve as early-career representatives on research teams, and these early-career teachers have found the team experience to be a valuable form of induction. In the future, as teachers gain even more experience and move toward becoming leaders, they will eventually be able to serve in the more experienced roles on the teacher research teams.

The LA model at CU Boulder tripled the number of talented physics majors entering teaching careers, and it has positively impacted the quality of high school instruction in the Boulder area [31]. Figure 2 shows the effect of the LA program on math and science teacher graduation rates at CU Boulder; the figure compares the undergraduate certification program (into which LAs are recruited) to the master's (MA+) and post-baccalaureate (PBA) programs. As shown, the average annual number of undergraduate secondary math and science certification program completers has doubled, while the average number of master's and post-baccalaureate program completers per year has remained relatively constant. The LA model recruits LAs into the undergraduate program so that they can obtain certification while completing their bachelor's degrees in math, science, or engineering through the CU-Teach program.

A longitudinal study of LAs who became K-12 teachers (N=15) was conducted from 2006–2011. A matched sample of teachers (N=14) who went through the same teacher preparation program, had similar GPAs, taught in the same schools, and had similar majors, but did not serve as LAs, was also studied. All participants were certified to teach either math or science. K-12 teaching practice was investigated through interviews and observations. The research team made 178 observations using the Reformed Teacher Observation Protocol (RTOP). The RTOP is made up of 25 statements scored from zero to four, with zero indicating the practice was not observed and four indicating that the practice was characteristic of the class period [32].

Figure 3 shows that the average RTOP scores for former LAs were significantly higher than those of non-LAs in both math ($p=0.005$) and science ($p=0.001$). This indicates that students who served as LAs were more likely to engage in evidence-based teaching practices than were similar colleagues who did not serve as LAs. At the time of the study, the sample size of physics teachers was too small to investigate separately from the entire group of science teachers.

B. Enrollment, retention, and cost

The LA program officially began at CU Boulder in 2003. From 2003 to 2014, a total of 1,627 LA positions were filled and more than 120 individual faculty members used LAs in 73 different courses in 12 departments within three colleges and schools. At this rate of participation, the program costs approximately $690,000 per year and impacts approximately 15,000 CU Boulder students per year. This amounts to a cost of about $46 per impacted student.

Increased retention results in tremendous financial savings for the university. The program has therefore captured the interest of the university's higher administration. In a recent retention study, it was found that 95% of LAs had graduated from CU Boulder within six years, compared to 84% of a matched sample of students who did not serve as LAs (see Table I). Students were matched on the basis of gender, predicted GPA in their freshman year (a measure based on high school GPA, ACT scores, and SAT scores), class level, and college. Table II shows average time to degree and cumulative GPA at graduation.

TABLE I. Comparison of LAs against a matched sample of peers (non-LAs): Percentage of students graduating in four, five, and six years.

Enrollment status as of mid-Spring 2014	LA (count)	LA (%)	non-LAs (count)	non-LAs (%)
Graduated in 4 years	83	48.0	4,790	51.1
Graduated in 5 years	143	82.7	7,283	79.0
Graduated in 6 years	164	94.8	7,773	84.4

TABLE II. Comparison of LAs against a matched sample of peers (non-LAs): Time to degree and cumulative GPA.

	LA (mean)	LA (SD)	non-LAs (mean)	non-LAs (SD)
Time to degree (years)	4.28	0.84	4.23	1.08
Cumulative GPA at graduation	3.52	0.404	3.28	0.43

C. International Learning Assistant Alliance

The LA model has had a broad impact across the CU Boulder campus and across the nation. On campus, the LA model has catalyzed weekly discipline-based educational research (DBER) meetings and has led to a professional learning community consisting of the faculty coordinators from the different participating departments. Because of the broad adoption, the financial and human resource commitment by departments to the program, and the alignment with our university's strategic plan, the LA model has garnered full financial support from the university.

Nationally, the LA model has been emulated from coast to coast. As shown on the census diagram at the national Learning Assistant Alliance website [4], over 60 universities have adopted the LA model, and many others, both within and beyond the U.S., are interested (see Figure 4). Over 220 science, mathematics, and engineering departments currently use an LA program; 68 of those are physics departments and 14 are astronomy departments [33-35]. These programs are funded by a variety of entities, including universities, the National Science Foundation (through the Noyce, MSP, STEP, CONTENT, and DRK-12 programs), the APS/AAPT Physics Teacher Education Coalition program, the Howard Hughes Medical Institute (HHMI), and other foundations such as Ball Aerospace, the Gill Foundation, Amgen, and Walmart.

As a result of large-scale adoption and interest, an international Learning Assistant Alliance has been established to help institutions using LAs share and adapt materials and develop a research agenda. The alliance has been funded by HHMI and NSF [36]. Currently available resources include census data about the usage and growth of the LA model, contact information for departments throughout the world using LAs, and relevant grant funding agencies. Resources are continually added and updated on the Learning Assistant Alliance website, which is organized into three areas: implementation, research, and scaling/sustaining. For implementation, an updated LA manual and content-specific materials (and links) are available, along with downloadable acceptance and rejection letter templates and materials for running faculty workshops and the pedagogy course. The research section of the site contains a collection of assessment instruments and interview protocols for measuring learning gains and assessing other impacts of the LA experience. The goal is for a large number of programs to use the same instrument, so that we can pool data to make claims about the program generally, and about how our own programs compare. The sustaining and scaling section of the site includes slides with data for making the case for an LA program and communicating with colleagues, administrators, and funding agencies. Talks, tips, funding sources, and other useful information are also included.

CU Boulder has run a national LA workshop since 2006, at first semi-annually and now annually; a total of 171 faculty members representing 86 universities have attended. An additional 153 faculty attended regional LA workshops that were held in the Mid-Atlantic (at University of Maryland,

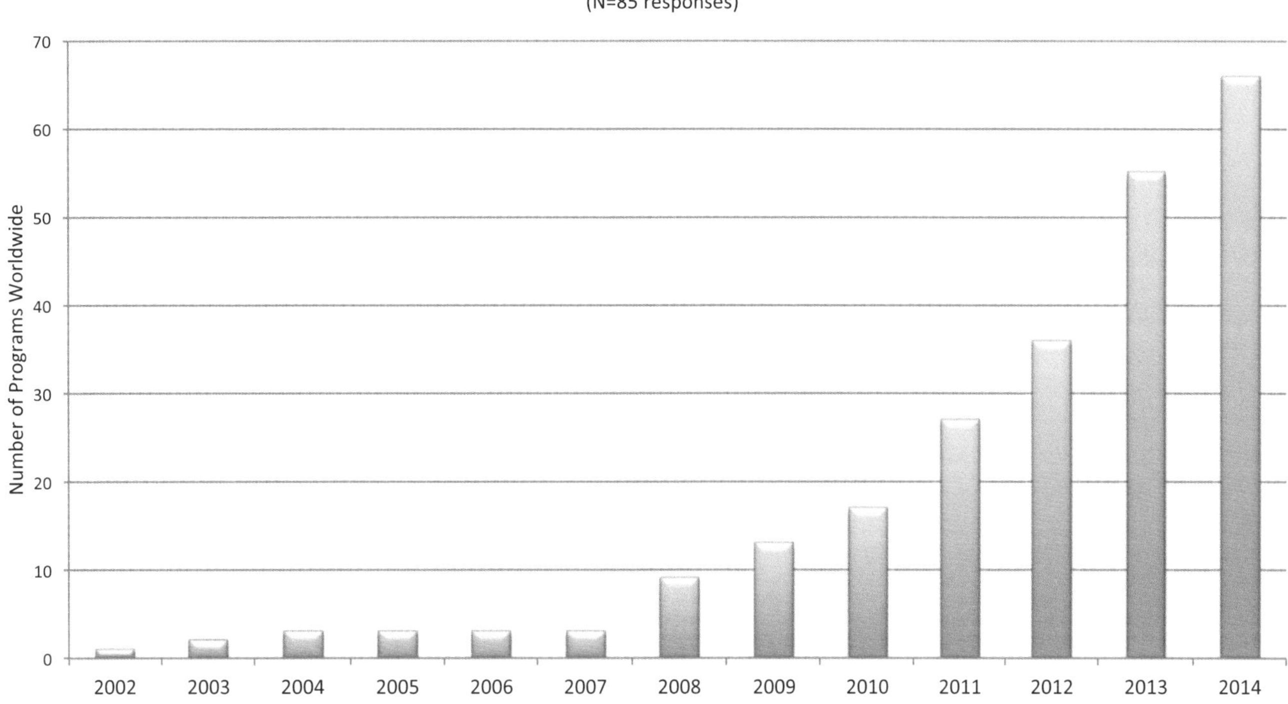

FIG. 4. Total number of LA programs globally, by year.

College Park), the Midwest (at Chicago State University), the Northeast (at Boston University), the Southeast (at Florida International University), and the Southwest (at Texas State University in San Marcos) in Spring 2014. For more information and data about the Learning Assistant model or the Learning Assistant Alliance, please see [4].

VI. ENACTING AND SUSTAINING THE LA MODEL

In this section, practical advice is provided for readers interested in enacting the LA model. This advice is based on CU Boulder's decade of experience with the model and in assisting other universities in starting, expanding, and sustaining the model.

A. Getting started

(1) Start in one or two departments and expand from there. Expansion is helpful for sustainability purposes.

(2) LAs should be positioned as elite among undergraduate math and science majors. This will help to establish departmental and university norms of valuing evidence-based teaching for faculty and for students. We often use the challenge: "Oh, so you understand math and science…but can you teach it?"

(3) Find another faculty member, assistant, or graduate student to help as soon as possible. As is the case with any mission, a partner is needed. Whether you have access to a small percentage of the departmental administrator's time or work with another faculty member, graduate student, master teacher, or postdoctoral scholar, it is useful to have someone to work with.

(4) Institute all three main LA activities (LA-led small groups, faculty-LA weekly planning meetings, and the science, math, and engineering education course) at the start of the new LA program. Also, decide early whether stipends, wages, and/or credit will be used as compensation for LAs' time. Once norms are set they are hard to change.

(5) Be intentional about how LAs will use their time. Remember that this experience should be of value to all who participate (faculty, LAs, and enrolled students). In order for the model to take hold, participants must feel valued by others and they must feel that their experiences are of personal value.

(6) LA-supported sessions should have direct and obvious value to enrolled students. Either recitation sessions or other sessions that are run by LAs should be required, or content covered (or uncovered) in these sessions should be clearly included on exams. Students often measure what instructors value by what is on the exam.

B. Gather data early and often

(1) Take baseline data at least one semester prior to implementing changes to courses, so that you can compare transformations to the status quo. These comparative data will be helpful for getting buy-in from faculty and administrators and for sustaining and measuring change.

(2) Maintain electronic records for tracking LAs. These data will be helpful for recruiting teachers, making claims about recruitment, and doing follow-up surveys.

(3) Keep track of the faculty using LAs and the courses in which LAs are used. Each administrator is interested in the welfare of a different group of stakeholders. Having data on the value of the experience to multiple stakeholders is useful for growing and scaling up the program. Keeping track of enrollments of these courses is also useful when you seek to make claims about the numbers of students impacted and the cost per impacted student.

C. Growing and sustaining

(1) Identify "live ones" in any departments of interest. "Live ones" are faculty who seem to be particularly interested in the education of undergraduate students or in course development and transformation. It is helpful to have more than one live one for garnering enthusiasm initially and for expanding to other departments.

(2) Report to your administrators regularly, especially when you are not asking for anything. Pre- and post-course learning gain data, LA experience data, and faculty experience data are all of value to your administrators. Each type of administrator is attending to different sets of stakeholders, so highlight those in your conversations. For example, if reporting to the department chair, be sure to demonstrate the value of the experience to faculty and the buy-in of faculty in your department. If reporting to the dean of the science college, highlight the welfare of the enrolled students; for the dean of the education college, include information about recruitment and preparation of teachers. If reporting to the provost, highlight the value of the program to goals set forth by your university's strategic plan and to faculty development, as well as the economic benefits of the model. If reporting to the president or chancellor, highlight the potential of the model for local, national, and international recognition and fundraising.

(3) Become familiar with the strategic plans of higher administrators and make sure to utilize language that is aligned with these plans. Since the LA model is multifaceted, it can often fit into a variety of different financial and strategic plans. Therefore, find the areas of the plans into which the LA model fits, and ensure that the appropriate language is used when talking with administrators. Any new idea that is worthy of investment should appear as a stepping stone toward meeting the stakeholder's existing goals. If a new idea does not appear to meet existing goals, it will appear to the stakeholder to be a resource sink.

(4) Attend the international Learning Assistant workshop in Boulder or a regional workshop near you. Consider holding a regional workshop in your area.

(5) Join the Learning Assistant Alliance. The alliance provides periodic updates and surveys, and will maintain a large database to help you implement, research, and scale up your program.

D. Additional advice

(1) Always ensure that there are at least two new LAs in each course. LAs need a partner with whom to share ideas, from whom get feedback, and with whom to work on projects from the science, math, and engineering education course. The course requires peer observation and other group projects, so it is critical that more than one LA per course are hired.

(2) Do not "market" LAs as cheap replacements for graduate TAs. LAs are not cheap TAs; they are specially trained and supported to help to foster enhancements to undergraduate education on campus. Use the LA model as a catalyst for upgrading the graduate TA experience.

(3) Garner some type of support for "department coordinators" early on, so that norms can be set. Try to ensure that the department coordinator is a respected, high-ranking faculty member. This helps establish the legitimacy of the LA model. Find ways to establish norms among faculty and commitments from the administration that will allow the associated efforts of faculty development, undergraduate education, course transformation, and program leadership to be valued as contributions toward tenure and promotion. For research universities, this may amount to legitimizing the publication of discipline-based educational research.

(4) Take efforts to ensure that the term "LA" has a specific meaning. Faculty will start referring to any

undergraduate teaching assistant as an LA; yet the LA experience involves much more than serving in a teaching role. The LA assists in transforming the way learning and teaching is viewed, they receive a synchronous pedagogical education in the science, math, and engineering education course, and they reflect on their experiences as teachers and learners.

(5) Begin to engage in retention studies as soon as possible. These data could support arguments that the LA model is income generating rather than a resource sink.

VII. CONCLUSION

The LA experience is integrated, involving multiple aspects of teaching and learning. As undergraduate math, science, and engineering students, LAs experience content in their own courses. They also draw on relevant education literature and work with content experts to plan instruction, teach their carefully planned lessons and activities, and reflect on their teaching. LAs thus have repeated opportunities to connect theory to practice. This integrated experience has been shown to be a crucial element of successful professional development for even the most experienced teachers [37].

The LA model also offers physics faculty a starting point for making small to large transformations to their courses. The presence of LAs increases the teacher-to-student ratio and enables a broader range of students to participate in class activities. We often hear faculty who use LAs talking about how much they "love" their LAs and how much more they enjoy teaching since they started using LAs. The LAs themselves often comment on how valuable the experience is for their own learning about themselves and about the content of the courses in which they teach. These may be some of the reasons that the LA model has spread so quickly and so broadly throughout the nation. As national-scale, collaborative research projects begin to produce results, we will know more about exactly which elements of the program are critical for various aspects of its success.

ACKNOWLEDGMENTS

I am grateful to Laurie Langdon, the director of the LA program, for her work on campus. She sustains the Departmental Coordinators group, maintains the program operations, and continues to manage and innovate with the International Learning Assistant Alliance. Laurie also facilitated the implementation of the first set of regional workshops. This work would also not be possible without the collaboration of Michael Oatley, the designer and engineer of the electronic data management system that allows us to manage LA and faculty applications and track all data. Olivia Holzman has been instrumental in managing and administering all aspects of the program and maintaining communication with colleagues throughout campus and the world. I am also grateful for the data collection and analysis conducted by Amy Biesterfeld for the ongoing retention studies. Her diligence and communications with University Records has been invaluable. Kara Gray and David Webb carried out the intensive longitudinal RTOP study and maintained strong communication with K-12 schools. Finally, without the generous contributions of funds and the intellectual guidance of Dean Lorrie Shepard, Provost Russ Moore, and Chancellor Phil DiStefano, the work reported here would not be possible.

---

[1] National Research Council, *Adapting to a Changing World—Challenges and Opportunities in Undergraduate Physics Education* (National Academy of Sciences, Washington, DC, 2013). http://www.nap.edu/openbook.php?record_id=18312

[2] D. E. Meltzer, M. Plisch, and S. Vokos, eds., *Transforming the Preparation of Physics Teachers: A Call to Action. A Report by the Task Force on Teacher Education in Physics* (T-TEP) (American Physical Society, College Park, MD, 2012). http://www.ptec.org/webdocs/2013TTEP.pdf

[3] NGSS Lead States, *Next Generation Science Standards: For States, By States* (National Academies Press, Washington, DC, 2013).

[4] See http://www.learningassistantalliance.org

[5] V. Otero, N. Finkelstein, R. McCray, and S. Pollock, Who is responsible for preparing science teachers?, Science 313, 445 (2006). doi:10.1126/science.1129648

[6] V. Otero, S. Pollock, and N. Finkelstein, A physics department's role in preparing physics teachers: The Colorado learning assistant model, Am. J. Phys. **78**, 1218 (2010). doi:10.1119/1.3471291

[7] See http://www.phystec.org/components/learning-assistants/index.php, retrieved Sep. 9, 2013.

[8] See https://www.aplu.org/SMTI/PP, retrieved Sep. 9, 2013.

[9] L. S. Vygotsky, *Thought and Language* (MIT Press, Cambridge, MA, 1962).

[10] E. Wenger, *Communities of Practice: Learning, Meaning, and Identity* (Cambridge University Press, New York, NY, 1998).

[11] M. P. Jiménez-Aleixandre, A. B. Rodrígues, and R. A. Duschl, "Doing the lesson" or "doing science": Argument in high school genetics, Sci. Educ. **84** (6), 757 (2000). doi:10.1002/1098-237X(200011)84:6<757::AID-SCE5>3.0.CO;2-F

[12] D. Meltzer and R. Thornton, Resource letter ALIP—1: Active-Learning instruction in Physics, Am. J. Phys. **80** (6), 478 (2012). doi:10.1119/1.3678299

[13] S. Pollock and N. Finkelstein, Sustaining educational reforms in introductory physics, Phys. Rev. ST Phys. Educ. Res. **4**, (2008). doi:10.1103/PhysRevSTPER.4.010110

[14] S. Pollock, Longitudinal study of student conceptual understanding in electricity and magnetism, Phys. Rev. ST Phys. Educ. Res. **5** (2), 020110 (2009). doi:10.1103/PhysRevSTPER.5.020110

[15] See http://perusersguide.org, retrieved Sep. 9, 2013.

[16] P. Heller, R. Keith, and S. Anderson, Teaching problem solving through cooperative grouping. Part 1: Group versus individual problem solving, Am. J. Phys. **60** (7), 627 (1992). doi:10.1119/1.17117

[17] P. Heller and M. Hollabaugh, Teaching problem solving through cooperative grouping. Part 2: Designing problems and structuring groups, Am. J. Phys. **60** (7), 637 (1992). doi:10.1119/1.17118

[18] See http://groups.physics.umn.edu/physed/Research/CRP/crintro.html, retrieved Sep. 9, 2013.

[19] See https://www.physport.org/guides/ for information on cooperative group problem solving and context-rich problems.

[20] L. C. McDermott, P. S. Shaffer, and the Physics Education Group at the University of Washington, *Tutorials in Introductory Physics* (Prentice Hall, Upper Saddle River, NJ, 2002).

[21] E. E. Prather, T. P. Slater, J. P. Adams, and G. Brissenden, *Lecture-Tutorials for Introductory Astronomy*, 3rd ed. (Addison-Wesley, Boston, MA, 2012).

[22] V. K. Otero, Recruiting talented mathematics and science majors to careers in teaching: A collaborative effort for K-16 educational reform, in *American Association for the Advancement of Science Annual Conference Proceedings* (Washington, DC, 2005). http://www.colorado.edu/physics/EducationIssues/Science_Supp/Otero_AAAS_05.pdf

[23] V. K. Otero, W. Wood, J. Curry, and R. McCray, *Colorado STEM/Noyce Fellowship Program*, NSF DUE-0434144, January 2005–September 2010. http://www.nsf.gov/awardsearch/showAward?AWD_ID=0434144

[24] V. K. Otero, D. Webb, N. Finkelstein, L. Langdon, M. Klymkowsky, and S. Pollock, *STEM Colorado/Noyce Teacher Scholarship Program*, NSF DUE-0833258, January 2009–December 2012. http://www.nsf.gov/awardsearch/showAward?AWD_ID=0833258

[25] V. K. Otero, N. Finkelstein, and L. Langdon, *STEM Colorado Teaching to Learn Program*, NSF DUE-1240073, September 2012–August 2016. http://www.nsf.gov/awardsearch/showAward?AWD_ID=1240073

[26] V. K. Otero, S. Pollock, L. and Langdon, Streamline to Mastery Phase II: Teacher-Led Professional Partnerships, NSF DUE-1340083, September 2013–August 2018. http://nsf.gov/awardsearch/showAward?AWD_ID=1340083

[27] V. K. Otero, N. Finkelstein, and L. Langdon, *STEM Colorado's Streamline to Mastery*, NSF DUE-0934921, September 2009–August 2015. http://nsf.gov/awardsearch/showAward?AWD_ID=0934921

[28] S. Belleau and V. K. Otero, Critical classroom structures for students to participate in science, AIP Conf. Proc. **1513**, 11 (2013). doi:10.1063/1.4789639

[29] S. Severance, What are the effects of self-assessment preparation in a middle school science classroom?, AIP Conf. Proc. **1413**, 351 (2012). doi:10.1063/1.3680067

[30] A. Grimes and V. K. Otero, The effects of student input on homework completion and student performance, in *Proceedings of the 2013 Physics Education Research Conference* (2014), pp. 19-22. doi:10.1119/perc.2013.inv.003

[31] K. Gray, D. Webb, and V. K. Otero, Effects of the learning assistant experience on in-service teachers' practices, AIP Conf. Proc. **1413**, 199 (2012). doi:10.1063/1.3680029

[32] D. Sawada, M. D. Pilburn, E. Judson, J. Turley, K. Falconer, R. Benford, and I. Bloom, Measuring reform practices in science and mathematics classrooms: The Reformed Teaching Observation Protocol, *School Sci. Math.* **102** (6), 245 (2002). doi:10.1111/j.1949-8594.2002.tb17883.x

[33] V. K. Otero and L. Langdon, From Undergraduate STEM Major to Enacting the NGSS, NSF DRL 1317059, August 2013–July 2015. http://www.nsf.gov/awardsearch/showAward?AWD_ID=1317059

[34] V. K. Otero, HHMI grant #54107045

[35] V. K. Otero, HHMI grant #1547162

[36] See http://www.learningassistantalliance.org/stats.php, retrieved Sep. 9, 2013.

[37] H. Borko, Professional development and teacher learning: Mapping the terrain, Educ. Res. **33** (8), 3 (2004). doi:10.3102/0013189X033008003

The Teacher Immersion course model: A reform-oriented early teaching experience that capitalizes on collaborations between high schools and universities

Mel S. Sabella
Department of Chemistry, Physics, and Engineering Studies, Chicago State University, Chicago, IL, 60628

Amy D. Robertson
Department of Physics, Seattle Pacific University, Seattle, WA, 98119

Andrea Gay Van Duzor
Department of Chemistry, Physics, and Engineering Studies, Chicago State University, Chicago, IL, 60628

We describe the Teacher Immersion course model, in which university physics departments and inservice high school teachers collaborate in the recruitment and preparation of future physics teachers. The immersion model creates opportunities for early teaching experiences by engaging prospective teachers in the planning, implementation, and assessment of a high school science lesson. Students in the immersion course define objectives; consult literature; assess learning needs; develop, practice, and implement lessons; and then reflect on the experience. The course supplements the existing science teaching methods courses and student teaching experiences that are typical in many secondary education programs. The Teacher Immersion course can be implemented at colleges and universities for a relatively small cost.

I. INTRODUCTION

Despite the best intentions of teacher preparation programs, novice teachers (and teachers, in general) rarely adopt reform-oriented teaching practices once they enter their classrooms [1,2]. This fact is often connected to specific, systemic challenges: Few preservice candidates teach in high school classrooms until their internship years, and many do so in contexts in which authentic reform-oriented teaching is countercultural and, in some cases, discouraged [3]. Additionally, the traditional teaching that is often the cultural norm in high schools aligns with novice teachers' own accumulated experiences of teaching (as learners), reinforcing a vision of teaching as information delivery [4,5].

If our nation wants competent, confident reform-oriented novice teachers, preservice teacher education programs need to offer the following [6–8]:

Early opportunities for preservice teachers to practice reform-oriented teaching. When such opportunities are embedded in courses on educational theory, they provide authentic integration of theory and practice, situating learning in the practice of teaching [9–13].

Relevant, discipline-specific pedagogical instruction. Not only do novice teachers need to situate their teaching in general, reform-oriented pedagogical instruction; they also need to develop pedagogical content knowledge via discipline-specific pedagogical instruction [14–16] . Ideally, such instruction would integrate knowledge about existing research-based instructional materials, research on children's ideas, epistemological resources, and accumulated wisdom of experience from teachers who have taught science for many years.

Mentorship from teachers who have intimate knowledge of the local context and who have successfully implemented reform-oriented teaching. Research increasingly acknowledges the importance of local context on one's learning and actions [17,18]. In this sense, learning to teach involves *learning to teach a specific topic to a specific population.* Because the majority of teachers accept their first positions within a 50-mile radius of the teacher preparation programs they attend, learning from local teachers is a critical part of a novice teacher's professional preparation [19]. It is especially important that preservice teachers be mentored by inservice teachers who have successfully implemented reform-orientated teaching within the local context.

Using the Teaching Immersion (TI) course model, Chicago State University (CSU) and Seattle Pacific University (SPU) have implemented courses that respond to this need. These courses, the Teacher Immersion Institute (TII) at CSU and the Science Teaching Immersion Experience (STIE) at SPU, provide scaffolded, reform-oriented teaching experiences for individuals considering a career in science teaching. This is accomplished through collaboration and joint leadership between high school and university faculty [20]. The TI model does not require extensive resources, avoids removing the teacher from the high school classroom (as in the Teacher-in-Residence model [21]), and is designed to utilize local teaching resources and contexts. Committed and prospective preservice secondary teachers who enroll in the course benefit from connections to experienced inservice teachers, subject-specific pedagogical content knowledge that blends the expertise of scientists and practitioners, and access to high school students prior to the student teaching experience. This type of early teaching experience capitalizes on local strengths and resources and responds to local challenges.

In this chapter, we will describe the TI course model by introducing the shared content and structure of the TII and STIE courses; by discussing how different players—including university faculty, mentor teachers and university students—play key roles in the course; and by describing the affordances and challenges of implementing a TI course. Throughout the paper we will use TI to refer to the immersion courses at both institutions, while the TII and STIE refer to the specific courses at CSU and SPU, respectively. We will refer to the university science students enrolled in the TI courses as interns (to distinguish them from the high school students with whom they are working). The chapter is aimed at teacher educators who wish to implement a similar course, or components of this course, in their local contexts.

Although there are a number of differences between CSU and SPU, with CSU being a public, majority-black institution and SPU being a private, majority-white institution, the universities share a number of goals and objectives. Students at both institutions share a commitment to service and often come from nearby communities. Both SPU and CSU have science faculty who are committed to high-quality teaching and who recognize the importance of effectively preparing physics majors to teach in area high schools.

Both institutions offer multiple routes to becoming a teacher; these routes are known as "certification" or "licensure," depending on the state. At CSU, students can enter the teaching profession through any one of three tracks: an undergraduate degree in physics with licensure in science, a licensure-only program for students who already have an undergraduate degree, and a Master of Arts in Teaching with initial licensure. The state of Illinois licenses students broadly in science, with designations in the disciplines. At SPU, students can become certified as undergraduates or via post-graduate programs, which include a Master of Arts in Teaching, a Master in Teaching Mathematics and Science degree, or an alternative route to certification program.

As a standalone course in one of these tracks to certification or licensure, the Teacher Immersion course can support a science teaching methods course or a student teaching experience by providing early, small-scale, focused activities that may make the interns' transition into these courses easier. Components of a TI course could also be added to existing courses, such as science teaching methods, although we have found that many methods courses have little extra time available for additional activities.

II. TEACHER IMMERSION COURSE CONTENT AND STRUCTURE

In both the TII and the STIE, our science, technology, engineering, and mathematics (STEM) interns plan, implement, and assess lessons in high school science courses. The purpose of the TI course is to immerse each intern in the experience of planning, assessing, and implementing a single science lesson under the mentorship of an experienced high school teacher, rather than immerse the interns in the day-to-day experiences of a high school physics teacher. TI courses are designed to be taken by individuals with no background in educational theory, as a means of introducing these individuals to the field of education; the courses are also designed for those who have already chosen teaching as a career and are enrolled in a teaching certification or licensure program.

Teaching a topic in a way that effectively builds content understanding, incorporates inquiry, and establishes positive attitudes and expectations toward science is incredibly challenging. Because of this, in the TI course model, we spend the entire semester focusing on a single topic that is taught in a single class period at a local high school toward the end of the semester. While this may seem limited—we complete in a semester what an inservice teacher completes in a day—focus on a single topic allows novices to engage in deep reasoning and reflection.

In this section, we offer the reader an overview of the content of the TI courses at CSU and SPU, and of some of the structural logistics. We emphasize the similarities between the courses in order to illustrate the model generally, and in some cases we provide context-specific details to demonstrate ways in which one might adapt the model to meet local needs. First we briefly describe the recruitment and role of the mentor high school teachers in the course, as a way of framing the remainder of the section.

A. Participation of university faculty, mentor high school teachers, and university interns

The TI course is designed to capitalize on the expertise of both university faculty and high school teachers. Both contribute relevant content knowledge and familiarity with research on student ideas and reform-oriented pedagogies. In addition to this, high school teachers have intimate knowledge of local contexts and experience planning and implementing lessons in ways that meet local and national standards. At both CSU and SPU, university faculty recognize that the mentor teachers bring this specialized professional knowledge. Teachers play important leadership roles in the courses at both institutions, although the specific roles taken on by faculty and high school teachers differ.

At CSU, two high school teachers, Jennie Passehl and Kara Weisenburger, have mentored TI interns. Passehl and Weisenburger were participants in professional development programs at CSU and were originally selected as mentor teachers based on their commitments to reform-oriented teaching. Mel Sabella (CSU physics professor), Andrea

Gay Van Duzor (CSU chemistry professor), Passehl, and Weisenburger initially met and discussed the course syllabus and activities. During the first year, Sabella led the course by introducing assignments and choosing readings; consequently, interns directed questions to him. In the second, third, and fourth years of the TII, Passehl and Weisenburger assumed greater leadership roles and were the contact points for the interns, drawing on classroom experiences and issues relevant to physics teaching in Chicago Public Schools.

At SPU, Amy Robertson (physics professor) planned assignments and structured the course, and mentor teachers Linda Anderson and Nina Christensen, as well as Physics Teacher Education Coalition (PhysTEC) Visiting Master Teacher "B." Lippitt, worked with small groups of university interns during course meetings to plan and assess the interns' lessons. Anderson and Christensen shepherded the implementation of TI interns' lessons in their high school classrooms.

B. Course content

Interns in the TI courses at CSU and SPU spend several weeks planning a lesson on a single topic, implementing the lesson in a high school science course, and then assessing the lesson. During this process they engage in the following pedagogical processes: forming groups, choosing topics, defining objectives, consulting literature, assessing learning needs, developing lessons, practicing lessons, implementing lessons, and reflecting on lesson implementation.

1. Planning the lesson

Interns begin by *forming collaborative groups* and *choosing a lesson topic*. The lesson topic is negotiated between the mentor teacher and the TI interns. TI interns are encouraged to choose topics they have extensively covered in their own university science courses (e.g., electric circuits, energy, forces and motion). Often the final choice is influenced by timing (i.e., when the TI interns are available and ready to implement the lesson within the mentor teacher's curriculum structure). In some cases, TI interns choose to adapt an existing lesson or activity rather than construct their own. Existing high school course materials may also influence their topic choice.

After they have chosen a topic, TI interns and mentor teachers work together to *articulate lesson objectives* while *consulting the literature* and *assessing learning needs*. Lesson objectives are what the TI students hope that the high school students will learn by participating in the lesson. TI interns are encouraged to connect their lesson objectives to:

- *Relevant state and national standards*. Mentor teachers support TI interns in understanding how to navigate existing standards documents (e.g., the Washington State K-12 Learning Standards, the Benchmarks for Science Literacy, and the Next Generation Science Standards (NGSS)) [6,22,23]. Interns are asked to earmark those standards and benchmarks that are related to their topics of choice, and the interns are held accountable for articulating the connection between the earmarked standards and their lesson objectives.
- *Research on children's ideas about related science topics*. Interns are assigned one or more published articles or book chapters that describe research on children's ideas about their chosen topic. (Two useful sources of relevant articles are the books *Children's Ideas in Science* and *Making Sense of Secondary Science* [24,25].) Interns write a short literature review and revise their lesson objectives on the basis of this review. They are required to articulate the connection between their lesson objectives and the existing literature.
- *Existing high school classroom norms*. Interns observe their mentor teachers' classrooms at least once prior to implementing their lesson, preferably toward the beginning of the quarter or semester in which the interns are enrolled in the TI course. Interns reflect on the existing classroom structures and norms immediately following their visits, and they describe how their lesson objectives are appropriate for the mentor teachers' classrooms.
- *Personal teaching values*. Interns at SPU are also asked to connect their lesson objectives to their teaching values. In particular, they are asked to reflect on what they value instructionally and why. For example, several STIE interns articulated that they value high school students being engaged and having fun, and they value high school teachers who clearly demonstrate that they care about their students. After their teaching values are articulated, STIE interns reflect on the alignment of their lesson objectives with their values. SPU interns are asked to be transparent about values for several reasons. For one, these values undoubtedly affect their choice of lesson objectives and their delivery of the lesson; however, these effects are often invisible to TI interns [17]. Making these effects visible offers interns an opportunity to assess their values and how they affect teaching practice. In addition, consciously choosing to align one's teaching with one's values grounds teaching practice in passion and identity.

Although it is rare for practicing high school teachers to explicitly connect their lesson objectives to all four of these—standards, research, norms, and values—our goal is to introduce TI interns to a variety of sources that might inform their lesson objective development.

After articulating lesson objectives, TI interns *plan and develop their lessons* or *adapt existing materials*. At CSU, interns collectively pick a topic such as energy or electric circuits, and each intern develops his or her own lesson. At SPU, students work together in groups of two or three to plan and collaboratively teach lessons. In each case, shared lesson topics foster collaboration among the interns.

Because the TI courses are staffed by reform-oriented mentor teachers and faculty, lesson plans primarily focus on opportunities for high school students to actively participate, with minimal formal lecturing. To supplement the mentorship from the teachers, interns also read papers that focus on inquiry instruction. For example, at CSU, interns read an overview paper on inquiry, "Inquiry Learning: What is it? How do you do it?" [26].

When TI interns construct their own materials, they tend to focus on lab stations or experiments. For example, TI interns in the SPU course planned a lesson on electricity that involved four lab stations: (1) a hand-crank generator, (2) a potato-powered electric circuit, (3) a Stirling engine, and (4) current induced by changing magnetic fields. TI interns developed worksheets to go with each station; the worksheets asked students to record their observations and make sense of the phenomena they observed, as well as to brainstorm possible real-world applications. At CSU, lessons tend to be class activities that incorporate interactive lecture and small-group lab experiments.

Aligning lesson content with lesson objectives requires multiple levels of coordination. For instance, one must consider the order of questions and activities (e.g., is the order logical?); the coherence of questions and activities (e.g., what is the big-picture message, and does this group of activities communicate that message?); and the alignment between specific objectives that, as noted above, coordinate standards, research on children's ideas, classroom norms, and instructional values (e.g., does this set of activities support learners in accomplishing the lesson objectives? Does it attend to possible pitfalls and/or capitalize on existing learner resources? Is it appropriate for this classroom?). This coordination is one of the most challenging tasks of the course for TI interns, not only in terms of execution, but also in terms of understanding what it means to coordinate objectives and lesson plans. For example, interns are sometimes tempted to use experiments that are exciting or novel but that are not tightly coordinated to the objectives.

Prior to implementing their lessons in the high school classroom, interns *practice their lessons* with their peers in the TI course. Each intern or group of interns is given the opportunity for a "dress rehearsal," treating their peers as though they were the high school students for whom the lesson was designed. Their peers are instructed to ask questions as though they were high school students experiencing the lesson for the first time. After each intern (or team) has taught his or her lesson or a portion of the lesson, peers, mentor teachers, and faculty instructors offer feedback on the content and structure of the lesson as well as on the delivery. At CSU, the practice lesson happens early in the semester, so interns can receive initial feedback. Interns teach the first 10 minutes of their lessons to the others in the course. This helps set a tone that focuses on inquiry and active student engagement, and aids our interns by structuring the challenging task of lesson development into manageable increments.

The final step in lesson planning is to submit a summary of the lesson topic and objectives, a copy of the lesson that includes sufficient detail to enable someone who was not a part of the class to replicate it, a copy of the assessments (described below), and a detailed description of how the lesson addresses the lesson objectives. At SPU this is called a lesson portfolio.

2. Implementing the lesson

Interns then *implement their lesson in their mentor teacher's high school science courses.* Typically, the lesson spans a single class period rather than multiple days, and interns may teach the lesson to more than one class of their mentor teacher's students. Because the interns have observed their mentor teacher's classroom before lesson implementation, the high school students know who the interns are, and introductions have already taken place. As the interns facilitate the lesson, the mentor teacher observes and assists as needed. For example, during small-group activities, the mentor teacher may act as an additional instructor, supporting group work and answering questions. During larger class activities, the mentor teacher will mostly observe and offer support (e.g., classroom management or clarification) as requested by the TI intern(s). Mentor teachers offer feedback to interns immediately following lesson implementation. At CSU, mentor teachers often use portions of a popular observation rubric, the Reformed Teaching Observation Protocol (RTOP), to organize and structure feedback [27].

3. Assessing the lesson

Interns *assess and reflect on their lesson implementation* via three mechanisms: (1) traditional pre- and post-assessments completed by high school students, (2) written reflections on their implementation of the lesson, and/or (3) video records of implementation.

CSU requires the interns to do the first two of the above; SPU requires all three and asks the interns to synthesize the results in an assessment portfolio. Each assessment is described in detail below. As with the formulation of lesson objectives, we expect that a typical (single) high school science lesson would rarely call for all three modes of assessment; our goal is to introduce the interns to different forms of assessment.

Pre- and post-test analysis. TI interns' lesson planning includes the design of pre- and post-assessments that match lesson objectives. As with the coordination of lesson objectives and lesson content, the design of assessments that are intended to measure a shift in student understanding and that are aligned with lesson objectives requires multiple levels of coordination. Assessments must not only include well-crafted questions, but the questions on the pre- and post-assessments must also provide similar enough information that one can determine whether student performance has improved. In addition, the information that these questions provide must be relevant to the lesson objectives, so that the TI interns can assess whether or not the objectives have been met.

TI interns are offered a variety of options for pre- and post-assessment design, including written quiz questions, whiteboard questions, and targeted whole-class discussion. The first of these is usually the most popular choice; pre-assessment quizzes are typically given by mentor teachers in advance of lesson implementation, and post-assessment quizzes are given immediately following implementation.

Reflection on experience. Immediately following lesson implementation, TI interns write personal reflections. They are prompted to answer questions about their impressions of lesson success, what they learned about student understanding of their lesson topic, how well they think their lessons accomplished their objectives, the alignment of their teaching with their instructional values, and any modifications they would make to their lessons based on their impressions. At CSU, interns also describe the processes by which they chose their objectives, explain the challenges they faced in developing these objectives, and articulate how the experience shaped their views of the high school teaching profession.

Analysis of classroom video. At SPU, as TI interns implement their lessons, the interns' interactions with high school students are videorecorded. There are two cameras in the room: One captures a wide-angle view of the entire class, and one points at a single group of high school students. TI interns are trained in video equipment use prior to implementation and do the recording themselves. Permission from students, parents, and school administration is acquired early in the quarter.

TI interns at both institutions then analyze, synthesize, and coordinate the assessment data they collected. Each SPU group must submit an assessment portfolio at the end of the quarter that contains:

- A copy of each group member's personal reflections.
- An analysis of pre- and post-assessments. The analysis includes a copy of the assessments, a discussion of the learning objectives the assessments are designed to assess, a discussion of student responses to the pre- and post-assessments, a comparison of pre- and post-assessment results, and an analysis of what this comparison implies about student learning and assessment design.
- A transcribed video clip (one per group member) and analysis. TI interns reflect on what they noticed while watching the clip but not while in the classroom, specific instructional choices they made, and student ideas about the science topic they chose. They offer an assessment of their teaching and/or their lessons in light of their lesson objectives.
- A synthesis of these three analyses. Interns are asked to identify any consistent themes or conflicting messages and, based on their data, to make recommendations for lesson modification.

At SPU, TI interns present their assessment results in the undergraduate research symposium at the end of the quarter. CSU interns submit post-test analysis and personal reflections at the end of the semester and engage in an informal group discussion about the experience with their peers, their mentor teachers, and the university faculty.

The combination of overview material on effective instruction and inquiry-based science and specific studies on student understanding of a focused topic provided a useful background for TI interns. They are presented with big-picture ideas that can be applied to a wide variety of subjects that they may teach and to focused, discipline-specific pedagogical content knowledge.

C. Course structure and specifics

1. Length of course and timing of course elements

At both CSU and SPU, the TI course meets in the evening to accommodate mentor teachers' schedules. The TI course structures at CSU and SPU differ somewhat due to local environments and available resources. The TII course at CSU is two credits and spans one semester (Spring); the STIE course at SPU is three credits split across two quarters (one credit in Winter, two credits in Spring). Approximate timings for the different course elements are summarized in Table I.

TABLE I. Approximate timings for course elements in CSU and SPU Teaching Immersion courses.

Course Element	CSU (15 weeks)	SPU (19 weeks)
Lesson Planning	*11 weeks*	*10 weeks*
Discussion about literature and effective teaching	2 weeks	
Choose topic	1 week	1 week
Develop lesson objectives	2 weeks	3 weeks
Plan or adapt lesson	3 weeks	3 weeks
Develop assessment	2 weeks	1 week
Participate in lesson as learners	1 week	2 weeks
Lesson Implementation	*2 weeks*	*1 week*
Lesson Assessment	*2 weeks*	*8 weeks*

2. Grading and assignments

At CSU, grades are based a series of papers in which interns reflect on the lesson planning and implementation process, as well as on participation and attendance. These assignments include a literature review of student understanding of a specific physics topic, development of pre- and post-assessment instruments, observation notes and reflection on high school visits, development of an activity for the mock lesson, development of the full activity for use in the high school classroom, implementation of the activity, written analysis of pre- and post-assessment results, and a written reflection about implementation.

At SPU, grades are based on participation in course activities, weekly written assignments, and lesson (Winter) and assessment (Spring) portfolios (the latter discussed above). Weekly readings introduce TI interns to contemporary science education reform efforts (e.g., AAAS's Project 2061 and the National Science Education Standards), research on children's ideas, and best practices in lesson design and assessment [7,28].

III. AFFORDANCES OF PARTICIPATION IN TEACHING IMMERSION COURSE

The Teaching Immersion model provides early exposure to the teaching profession, immerses interns in lesson planning and assessment practices, promotes reform-oriented teaching, illustrates the complexity of teaching, supplements course offerings in the School of Education, and empowers inservice high school teachers to participate in the preparation of future teachers. While some of these positive outcomes overlap with a standard methods course, some are specific to the Teaching Immersion model.

A. TI model provides early exposure to the teaching profession

Early teaching experiences have been shown to be an important component of teacher preparation programs [8]. These experiences provide individuals considering a career in science teaching an opportunity to decide whether they can see themselves in this challenging role *early* in their undergraduate careers, rather than much later during their student teaching experiences. In addition, the course provides a scaffolded experience of reform-oriented teaching to students who have not yet committed to teaching. It thus serves as a recruitment tool.

B. TI model promotes reform-oriented teaching

Research has consistently demonstrated the importance of situating *learning to teach* in the *experience of teaching* [9]. Interns in the TI courses at CSU and SPU are not only exposed to educational theory and research-based pedagogical techniques; they are also encouraged to try on these theories and techniques in high school classrooms, under the support and guidance of reform-oriented mentor teachers.

One interesting illustration of implementing educational theory in the classroom comes from the TI course at CSU. An intern who is planning to be a high school science teacher and who had completed the majority of her education courses was challenged by her mentor teacher to engage students in exploring a phenomenon first, rather than explaining the phenomenon directly.[1] The mentor teacher noticed that the intern had provided significant background and explanation on the topic in her lesson:

> TI Teacher: So I'm wondering...what was your thought process for why you did this picture and reading before the lab?
>
> Lisa: ...Just to refer to the text so they will have a place to go back to and look at the relationship...
>
> TI Teacher: What do you think is some feedback I gave to [Becky] that I could also give you?
>
> Lisa: To reverse the order of what I did...the hands-on part first.
>
> Pablo: Yeah.
>
> Lisa: So basically we are giving the answer away...OK.
>
> Pablo: I think the both of you guys did like a reverse order type of thing...

[1]Pseudonyms are used in the dialogue.

Lisa: I guess we're just so used to where…I just thought about it just now; I just did a cookbook lab. I gave them the point first and then we did the lab to confirm what I said…

TI Teacher: Returning to the 5 'e's methodology –

Lisa: Yeah so we just explained before we engaged or explored…

Here, the TI Teacher supported Lisa in applying her theoretical knowledge about the structure of inquiry teaching to her practice. Support from reform-minded teachers is especially important given the "apprenticeship of observation" that interns acquire over their lifetime of experience with traditional instruction [5]. Despite extensive exposure to reform-oriented teaching at the college level,[2] ongoing support and repeated experiences with inquiry-based instruction is often required to break away from the traditional lecture. The limited content focus of the TI model, paired with observations of high school classrooms and participation in peer TI interns' lessons as learners, offers multiple opportunities for discussion of effective pedagogies for science instruction and fosters TI interns' understandings of the function of inquiry in student learning.

C. TI model illustrates the complexity of teaching

Effective science teaching requires the successful coordination of content knowledge, pedagogical knowledge, pedagogical content knowledge, curricular knowledge, and interpersonal skills [14,15]. Faculty at CSU and SPU recognize, appreciate, and emphasize the challenging, professional nature of teaching, elevating the status of the profession and encouraging students to pursue certification/licensure. This recognition is especially important in light of research findings suggesting that one reason the United States is experiencing a shortage of high-quality physics teachers is the relatively low status of secondary education programs versus positions in industry or graduate school [29].

As interns participate in the TI course, read papers about student learning, observe mentor teachers' classrooms, develop assessments and lesson activities, implement their lessons, and assess their lessons' success, they increasingly understand the complex sequence of activities and careful reflection that leads to effective instruction. One SPU intern acknowledged this understanding explicitly in her final course reflection:

I used to believe that teaching was a relatively "easy" profession. I thought that you showed up, taught a lesson, and graded some papers. However, I am realizing that it is a highly adaptive occupation. You can never "master" it the way you can other jobs. Teaching becomes about learning to use formative assessment in-the-moment to help those around you. Lesson planning is a delicate process which involves a lot of preparation. This class has made me really appreciate the teachers I have had in my life because they undoubtedly put a lot of work into the lessons that they taught and the methods that they used to implement it.

This intern describes her evolving understanding of the professional nature of teaching through her experiences in the TI course.

D. TI model supplements course offerings in schools of education

1. TI model immerses university students in lesson planning and assessment practices

By design, university students in the TI courses at CSU and SPU spend between 15 and 19 weeks planning, implementing, and assessing a single science lesson. Throughout those weeks, interns are exposed to multiple lenses for viewing lesson planning, research-based teaching practices, and various means of assessing student learning. Because there is only one topic in the class, there are many opportunities for iterative refinement of lesson plans and for exploring different aspects of lesson development; these opportunities include observing experienced teachers, teaching a topic, and reflecting on learning. The goal is for interns to leave the course with a "toolbox" of best practices for planning, implementation, and assessment.

One intern at SPU reflected on the expansion of her repertoire of assessment tools via the TI course, writing:

Prior to taking this class, I was a strong believer in end of the unit assessments or homework as the means to gauge student success. I now see that this is not the only way of doing this. In developing our lesson, we decided to [use] proximal formative assessment, …assessing students['] understanding by their work on their whiteboards during their lab. I believe that this type of in the moment assessment is valuable because you can see how students develop their ideas and see the actual learning process. I think that this may be just as valuable of an indicator of student understanding as requiring students to recall information during a test and crunch numbers using equations. One benefit…is that teachers…assess students['] understanding right in that moment and alter or change the curriculum if he or she sees necessary. The teacher does not have to wait for a number of students to fail a test to see if the material was taught well. I think that utilizing a number of assessments is more successful than the typical assessment we are used to.

The intern came to see the pedagogical benefits of formative, in-the-moment assessment, in addition to summative

---

[2]Both SPU and CSU use research-based curricula in our introductory physics courses. Curricula include: SMART Physics (https://www.smartphysics.com); C. J. Hieggelke, D. P. Maloney, S. Kanim, Newtonian Tasks Inspired by Physics Education Research: nTIPERs (Addison-Wesley Series in Educational Innovation, 2001); L. C. McDermott, P. S. Shaffer, Tutorials in Introductory Physics (Prentice Hall, 2001); Tutorials in introductory physics and homework package (Prentice Hall, 2001).

assessment at the end of a unit. Likewise, one CSU College of Education master's candidate who was observing the TI course for her research commented on the affordances of the in-depth focus on a single lesson, and particularly on the opportunities to revise and improve the lesson. She compared these affordances to those in courses where many lessons must be planned and implemented, mirroring a high school teacher's lesson planning schedule. Thus, the TI course provides an experience that complements, rather than replaces, important College of Education courses in which students develop integrated lessons across a unit.

We note that although there is a significant amount of pressure on the interns for their lessons to go well, instructors in the TI courses also emphasize that the lesson planning and development process is extremely challenging. All lessons are viewed as learning experiences, in which interns must think about how to improve the lessons for the future. Most interns are proud of the materials they develop, and if the lessons do not go as planned, they see opportunities to modify their lessons to address unexpected concerns. Faculty and mentor teachers emphasize the fact that lessons do not always go as intended for even the most experienced teachers, and that this is part of the profession.

2. TI model offers discipline-specific, culturally relevant methods instruction that capitalizes on the experiences of multiple communities

Interns in TI courses have at their immediate disposal disciplinary experts, discipline-based science education researchers, and experienced high school science teachers. Each of these mentors contributes unique, discipline-specific expertise. Together with the interns, the TI course team brings together knowledge of:

- science-specific pedagogical skills and methods;
- the content and epistemology of the discipline;
- research on how children learn science;
- secondary teacher preparation programs;
- the local community, including the realities of high school, the culture of the students, student life outside of school, and administrative constraints on curriculum;
- what is fun and exciting to learn; and
- state and national standards.

In particular, the presence and mentorship of Chicago and Seattle high school science teachers offer TI interns culturally relevant methods of instruction. Because society and policy play such an important role in the development of individual teachers and students, discussion of student learning focuses on the sociocultural and educational policy frameworks of the city. Passehl, Weisenburger, Anderson, and Christensen share their knowledge about what it means to be not just a physics teacher but a *Chicago* or *Seattle* physics teacher. One CSU intern reflected on the importance of this mentorship, writing:

> It also helped that they are currently teaching in an urban setting, so they are familiar with the struggles that teachers will face in an urban setting. I feel like I will really use Kara and Jennie as professional contacts in the future as I begin my first years teaching.

This intern appreciated mentorship in the context of science teaching in Chicago because she seeks to become a Chicago science teacher.

3. TI model promotes deep study of a science topic

TI interns have often been exposed to their lesson content as learners. However, lesson planning requires that interns choose and order relevant content in a way that attends not only to the content itself but also to the means by which students understand it (i.e., process skills). Implementation requires that TI interns listen and understand alternative solution paths, and assessment requires discerning *progress* toward canonical or epistemological understanding (i.e., the nature of scientific knowledge and how to learn it), even if the product does not match the teacher's existing understanding [30]. The TI course's extended focus on a single science topic promotes the kind of deep understanding—and the application of that understanding—that is required of practicing teachers. A number of interns have reflected positively on this opportunity to study one topic in depth and to grapple with the content in the way that their students will [31]. One CSU intern wrote:

> At first, I thought working on one lesson all semester would be redundant and that I would lose interest. That was not the case. In fact, I came to appreciate the time spent on each component of the lesson. We came up with ideas, they evaluated them, we revised ideas, they evaluated them, we finalized ideas, they approved them…etc.

An SPU intern reflected:

> I have become aware and more confident in my own knowledge. Throughout the class, I have learned things that I already "knew" in more depth. Planning a lesson makes you really think about why a mechanism works the way it does. This quarter I have really learned to appreciate the way that electricity and power generation works. I have gained the confidence to be able to explain the system to someone with very little physics understanding. I think this is part of the very definition of being a teacher, becoming so invested in a subject that you want to pass it on to someone who doesn't know anything about it–this can only be done by intimately knowing the subject yourself.

Interns were explicitly aware of how focusing on one topic enabled them to generate a much deeper understanding of the content and of how a learner approaches the content.

4. TI model familiarizes interns with state and national standards and the edTPA

With the advent of the Next Generation Science Standards and with institutions in 34 states—including Illinois and Washington—and the District of Columbia participating in the teacher performance assessment edTPA, the TI model may help secondary education programs address new teaching and learning standards and assessments [6,32]. The NGSS promote the mastery of fewer content standards in greater depth, with more cross-disciplinary connections and an emphasis on scientific processes and the nature of science. During the TI course, interns examine standards documents, including the NGSS, and try to focus on core content and process standards in their lesson development. The extensive time the interns devote to one topic allows them to view the NGSS as a guide and to make the advocated-for application connections.

The edTPA was designed to be educative and predictive of effective teaching and student learning. In this assessment, teacher candidates must provide, explain, and reflect on evidence of teaching effectiveness for a three- to five-day lesson sequence. The targeted competencies include planning, instruction, assessment, reflection, and academic language. CSU and SPU, like many universities, are finding that they must implement curricular changes to teacher education programs to prepare candidates for the deep reflection on practice required by edTPA. Rich early teaching experiences like the TI course can provide early scaffolding for edTPA core competencies. Interns are required to think deeply and reflect on planning, instruction, and assessment. The video assessment component of the TI at SPU also mirrors the type of video analysis required by edTPA.

E. TI model empowers inservice high school teachers to participate in the preparation of future teachers

The TI course structure empowers inservice high school teachers to participate in the preparation of future physics teachers. One CSU teacher explained, "This experience has changed…how I view my role in the science education community…I have a lot more confidence in my abilities and I now believe I have a lot of ideas to offer the community of physics teachers."

CSU mentor teachers have become very invested in the physics education program. The CSU collaboration has led to presentations by the high school teachers, co-authorship on a paper published in *The Physics Teacher* [20], co-led workshops at the national PhysTEC conference and a local conference for science and math educators in Chicago, and invitations to give talks on science teacher preparation. The high school teachers we work with continue to lead the TI course and assist in efforts to recruit physics education students.

Mentor teachers have also described the role of the TI course in their own professional development. One SPU teacher wrote:

> The mentor teacher can engage in rich dialogue with the STIE students about physics concepts, assessment strategies, and the challenges of classroom management. Revisiting these ideas is important for veteran teachers, and can have a rejuvenating effect on our own instructional practices.

IV. CHALLENGES OF IMPLEMENTING AND SUSTAINING A TEACHER IMMERSION COURSE

A. Funding for mentor teachers

One challenge related to the sustainability of the Teacher Immersion course is funding for mentor teachers. (University faculty can be paid by course load.) Early on, both SPU and CSU benefited from PhysTEC funding. Currently, both CSU and SPU physics teacher preparation program leaders have negotiated or are in the process of negotiating for institutional funds to pay mentor teachers. CSU has found university support through a commitment from our university president. SPU physics faculty are collaborating with their School of Education colleagues to transform the TI course so that it is also appropriate for students obtaining a Master of Arts in Teaching. The course would then be cross-listed in the Department of Physics and in the School of Education, and mentor teachers would be listed as co-instructors of the course and paid by the university for their time and expertise. Teachers at CSU are hired for approximately $1,200 a semester, and teachers at SPU are hired for approximately $1,200 split over two quarters.

Although funding for the TI course presents a challenge, the cost is relatively small compared to other models that employ inservice teacher experts. For example, many successful teacher preparation programs incorporate a Teacher-in-Residence (TIR) [21,33,34]. A full-time TIR is often unaffordable for university departments, particularly those with smaller secondary education programs. The TI course is an alternate model that includes many of the mentoring elements of a TIR model, and thus allows for relatively easy adoption.

B. Course enrollment

An additional challenge to the sustainability of a TI course is low course enrollment [35].[3] Both CSU and SPU are small universities with proportionally small physics departments, and hence there are few potential teacher candidates. This means that there may be insufficient enrollment to offer the course on a yearly basis. CSU has mitigated this challenge by opening the course to students already committed to science teaching (in addition to those considering it) and by encouraging chemistry majors and licensure candidates with a strong interest in physics to enroll. As we say above, SPU physics faculty are currently negotiating with the university's School of Education to transform the TI course so that it is suitable for students obtaining a master's degree in STEM teaching.

C. Incorporating additional courses into existing secondary education programs

An additional challenge that may be encountered when implementing a TI course is how to fit additional coursework into a secondary education program that is already at capacity in terms of the number of credit hours required. A course based on the TI model may substitute for a previous requirement, or the content of an existing education class could be altered to incorporate elements of the Teacher Immersion model and its links to the edTPA.

Furthermore, many physics departments recognize the importance of discipline-specific education courses to quality science teacher preparation. At CSU, the TI course fills the same place in the physics secondary education degree option as does an upper-level physics elective in the general physics degree option. Similarly, SPU has a "teaching physics" degree track, and the TI course counts as an upper-division physics elective for students pursuing that path.

[3]At CSU, years one and two of TI implementation involved only those students considering careers in physics teaching, while in year three the course was opened to students already committed to science teaching. There were three interns in year one, four in year two, six in year three, and six in year four. At SPU, six interns participated in the first and only implementation of the TI course: five who were simultaneously participants in SPU's Learning Assistant Program and one who was enrolled in SPU's master of arts in teaching program.

D. Mentor teacher recruitment

Choosing mentor teachers whose objectives and goals are aligned with those of your teacher education program, who use research-based best practices, and who adhere to the ideas put forth by the Next Generation Science Standards is crucial, so that students are not receiving mixed messages about how to teach effectively [6]. Because TI interns are new to the teaching profession, any presentation of conflicting ideas by the high school teachers, the college faculty, and outside sources such as NGSS can derail interns' ideas, which may still be somewhat tenuous. CSU and SPU physics faculty have established strong relationships with local teachers who are experienced in reform-oriented pedagogy and who have been in the classroom for more than five years. All teachers were originally involved in professional development courses at CSU or SPU before becoming involved with the TI courses.

When a university hires a high school teacher, it sends an important message to interns in a secondary education program. The message is that the university respects and values these teachers and believes they bring something unique to the university. This endorsement of the skills, current experiences, and knowledge of high school teachers impacts how new teachers seek mentorship from experienced teachers in the field. Hiring a high school teacher also sends a message that mentorship from experienced teachers is valuable and that novice teachers should seek out mentoring relationships [8]. This emphasis on mentorship aligns with the TIR model, in which a master teacher is brought to the university to offer the secondary education program his or her unique, critical wisdom and expertise [21].

V. SUMMARY

In this chapter we describe a course that immerses university students in physics teaching and provides an early teaching experience for prospective and committed preservice teachers. The course capitalizes on the expertise and mentorship of inservice high school teachers, as interns in the course focus on a single topic and develop lessons that they then implement in their mentors' high school classrooms. The course supplements the existing methods courses and student teaching experiences typical in many secondary education programs. It provides a recruiting tool, at a low cost, for students who want to consider, in a low-stakes way, a career in teaching; it also provides exposure to the profession for students who may have committed to teaching but who have limited experience in the classroom.

[1] G. H. Roehrig and R. A. Kruse, The role of teachers' beliefs and knowledge in the adoption of a reform-based curriculum, School Sci. and Math. **105** (8), 412 (2005). doi:10.1111/j.1949-8594.2005.tb18061.x

[2] L. Cuban, The hidden variable: How organizations influence teacher responses to secondary science curriculum reform, Theor. Pract. **34**, 4 (1995). doi:10.1080/00405849509543651

[3] D. M. Levin, D. Hammer, and J. E. Coffey, Novice teachers' attention to student thinking, J. Teach. Educ. **60** (2), 142 (2009). doi:10.1177/0022487108330245

[4] S. Feiman-Nemser and M. Buchmann, Pitfalls of experience in teacher preparation, Teach. Coll. Rec. **87** (1), 53 (1985).

[5] D. C. Lortie, *Schoolteacher: A Sociological Study* (University of Chicago Press, Chicago, IL, 1975).

[6] See http://www.nextgenscience.org/

[7] American Association for the Advancement of Science, *Project 2061: Science for All Americans* (AAAS, Washington, DC, 1989).

[8] D. E. Meltzer, M. Plisch, and S. Vokos, eds., *Transforming the Preparation of Physics Teachers: A Call to Action. A Report by the Task Force on Teacher Education in Physics* (T-TEP) (American Physical Society, College Park, MD, 2012). http://www.ptec.org/webdocs/2013TTEP.pdf

[9] H. Borko and R. T. Putnam, Learning to teach, in *Handbook of Educational Psychology,* edited by D. C. Berliner and R. C. Calfee (Prentice Hall International, New York, NY, 1996), pp. 673-708.

[10] T. P. Carpenter, E. Fennema, and M. L. Franke, Cognitively guided instruction: A knowledge base for reform in primary mathematics instruction, Elem. School J. **97** (1), 3 (1996). doi:10.1086/461846

[11] S. B. Empson and V. R. Jacobs, Learning to listen to children's mathematics, in *Tools and Processes in Mathematics Teacher Education*, edited by D. Tirosh and T. Wood (Sense Publishers, Netherlands, 2008), pp. 257-281.

[12] S. Grossman and J. Williston, Teaching strategies: Strategies for helping early childhood students learn appropriate teaching practices, Childhood Educ. **79** (2), 103 (2002). doi:10.1080/00094056.2003.10522780

[13] R. T. Putnam and H. Borko, What do new views of knowledge and thinking have to say about research on teacher learning?, Educ. Res. **29** (1), 4 (2000). doi:10.3102/0013189X029001004

[14] L. S. Shulman, Knowledge and teaching: Foundations of the new reform, Harvard Educ. Rev. **57** (1), 1 (1987).

[15] D. Loewenberg Ball, M. Hoover Thames, and G. Phelps, Content knowledge for teaching: What makes it special?, J. Teach. Educ. **59** (5), 389 (2008). doi:10.1177/0022487108324554

[16] R. K. Blank, N. de las Alas, and C. Smith, *Analysis of the Quality of Professional Development Program for Mathematics and Science Teachers: Findings from a Cross-State Study*, Report for the Council of Chief State School Officers (Washington, DC, 2007).

[17] F. Erickson, Qualitative methods in research on teaching, in *Handbook of Research on Teaching*, edited by M. C. Wittrock (Macmillan, New York, NY, 1986), pp. 119-161.

[18] J. A. Maxwell, Causal explanation, qualitative research, and scientific inquiry in education, Educ. Res. **33** (2), 3 (2004). doi:10.3102/0013189X033002003

[19] P. Mulvey, C. L. Tesfaye, and M. Neuschatz, *Initial Career Paths of Physics Bachelor's with a Focus on High School Teaching* (American Institute of Physics, College Park, MD, 2007). http://www.phystec.org/phystec/webdocs/status/aip_careerpaths.pdf

[20] M. S. Sabella, A. G. Van Duzor, J. Passehl, and K. Weisenburger, A collaboration between university and high school in preparing physics teachers: Chicago State University's Teacher Immersion Institute, Phys. Teach. **50** (5), 296 (2012). doi:10.1119/1.3703548

[21] J. Anderson, What can a TIR do for your teacher preparation program?, American Physical Society Forum on Education Newsletter (Fall 2012). http://www.aps.org/units/fed/newsletters/fall2012/anderson.cfm

[22] M. McClellan, C. Sneider, R. I. Dorn, and K. Kanikeberg, *Washington State K-12 Science Learning Standards* (2009).

[23] American Association for the Advancement of Science, *Benchmarks for Science Literacy* (AAAS, Oxford University Press, 1993).

[24] R. Driver, E. Guesne, and A. Tiberghein, *Children's Ideas in Science* (Open University Press, Philadelphia, PA, 1985).

[25] R. Driver, A. Squires, P. Rushworth, and V. Wood-Robinson, *Making Sense of Secondary Science: Research Into Children's Ideas* (Routledge, New York, NY, 1994).

[26] L. Trout, C. Lee, R. Moog, & D. Rickey, Inquiry learning: What is it? How do you do it?, in *Chemistry in the National Science Education Standards* (American Chemical Society, Washington, DC, 2008).

[27] D. MacIsaac and K. Falconer, Reforming physics instruction via RTOP, Phys. Teach. **40,** 479 (2002). doi:10.1119/1.1526620

[28] National Research Council, *National Science Education Standards (*National Academy Press, Washington, DC, 1996).

[29] V. H. Shipp, Factors influencing the career choices of African American collegians: Implications for minority teacher recruitment, J. Negro Educ., 68 (3) 343 (1999). http://www.jstor.org/stable/2668106

[30] D. Lewenberg Ball, Bridging practices: Intertwining content and pedagogy in teaching and learning to teach, J. Teach. Educ. **51** (3), 241 (2000). doi:10.1177/0022487100051003013

[31] R. Bjuland, Student teachers' reflections on their learning process through collaborative problem solving in geometry, Educ. Stud. in Math. **55,** 199 (2004). doi:10.1023/B:EDUC.0000017690.90763.c1

[32] See http://www.edtpa.com/

[33] J. Hehn and M. Neuschatz, Physics for all? A million and counting! Phys. Today **59,** 37 (2006). doi:10.1063/1.2186280

[34] N. Finkelstein, M. Dubson, C. Keller, S. Pollock, S. Iona, and V. Otero, CU physics education: Recruiting and preparing future physics teachers, American Physical Society Forum on Education Newsletter (Spring 2005). http://www.aps.org/units/fed/newsletters/spring2005/cuphysics.html

[35] V. Otero, N. Finkelstein, R. McCray, and S. Pollock, Who is responsible for preparing science teachers?, Science, **313** (5786), 445 (2006). doi:10.1126/science.1129648

Early teaching experiences at Towson University: Challenges, lessons, and innovations

Cody Sandifer, Ronald S. Hermann, Karen Cimino, and Jim Selway
Department of Physics, Astronomy and Geosciences, Towson University, Towson, Maryland 21252

The Department of Physics, Astronomy and Geosciences at Towson University offers two types of department-led early teaching experiences (ETEs) for majors: (1) a three-semester sequence of one-credit early teaching courses, and (2) an undergraduate Learning Assistant program. Each ETE course focuses on a different instructional setting: an informal setting (semester 1), elementary and middle school (semester 2), and high school (semester 3). In the Learning Assistant (LA) program, faculty hire undergraduate physics majors to assist with "active learning" instruction in lectures, labs, group tutoring sessions, and test review sessions. Towson's ETEs target physics majors who have not considered high school teaching as a possible career as well as those majors already leaning toward a teaching career. Our course innovations include the reduction in the length of the informal and elementary/middle school courses to 10 weeks, and a teaching format in the elementary/middle course in which multiple interns teach at the same time in the same classroom. Our most important innovation in the LA program is having the undergraduate LAs meet regularly with a Teacher-in-Residence to plan and implement Interactive Lecture Demonstrations. Evidence for the effectiveness of these ETEs is provided. We also provide a list of challenges, suggestions, and lessons learned to help faculty successfully implement ETEs at their own institutions.

I. INTRODUCTION

One method that can help a university increase its physics teacher graduation rate is to offer early teaching experiences (ETEs) to physics majors, so that they might discover the joy of teaching firsthand and increase their awareness of and interest in teaching as a possible career. The philosophy underlying this recruitment strategy is that students cannot truly develop a passion for teaching until they are given the opportunity to teach [1,2].

The national Physics Teacher Education Coalition (PhysTEC) leadership has identified ETEs as one of the 10 key components of successful secondary teacher education programs [3]. A particularly important aspect of thriving physics teacher education programs is that their associated ETEs are led by the physics department itself rather than an outside organization or department [4].

Given the importance of ETEs, the purpose of this article is to provide an overview of physics department-led ETEs at Towson University, in order to highlight innovative or unique aspects of these experiences. We also describe challenges, lessons learned, and implementation tips, so that departments wishing to offer their own ETEs can benefit from our successes and mishaps.

II. BACKGROUND

The definition of an early teaching experience depends on the teacher population in question. For a secondary education major, an ETE is any teaching experience that occurs before student teaching [5]. For a non-education physics major, an ETE is any teaching experience that occurs early in that student's academic career. "Early field experience" is a closely related term that applies when the teaching experience is off-campus in a formal school setting [6].

ETEs can be formally integrated into coursework, but this is not always done. Some undergraduates engage in ETEs as part of their employment, such as when undergraduates are paid to tutor or assist students in a lecture or lab. Other ETEs are unpaid, informal activities in which students volunteer their time to participate in early teaching programs simply for the experience.

Examples of ETEs include teaching visits to local schools and science centers, undergraduate Learning Assistant (LA) programs, and group tutoring sessions. Generally speaking, care needs to be taken to differentiate early teaching experiences from other types of observation-based school internships in which participants primarily (or exclusively) observe experienced teachers rather than engage in teaching activities themselves.

Although increased recruitment into the secondary physics education major is a primary focus of ETEs, these programs provide other benefits, including the fact that the students doing the teaching improve their physics content knowledge and successfully contribute to course redesign [7]. ETEs also serve as self-evaluation tools to help physics majors determine whether a teaching career is genuinely appropriate for them. The opposite effect can also occur, in that ETEs can help secondary education majors recognize early on that teaching is not what they want to do for the rest of their lives, rather than have this unfortunate realization during student teaching (just before graduation), by which time it is too late to select a different major.

III. INSTITUTIONAL CONTEXT

Towson University (TU), a member of the University System of Maryland, began as a teacher's college, and the tradition of developing qualified teachers continues as TU graduates more education majors than any other institution in the state. Currently 2,382 undergraduate and 1,698 graduate students are enrolled in teacher education programs at TU.

TU is one of the few academic institutions nationwide to have a significant number of education faculty housed in its science content departments. For example, the Department of Physics, Astronomy and Geosciences (PAGS) includes six science education faculty members, one of whom specializes in high school education and the rest of whom specialize in elementary or middle school education.

The PAGS department currently has 98 physics majors distributed across concentrations in general physics, applied physics, astrophysics, and physics secondary education. The most recent incoming class size was 26 majors (all concentrations).

In 2010, TU joined the national PhysTEC project as a PhysTEC Supported Site, and worked with a full-time high school Teacher-in-Residence to increase the recruitment and retention of physics secondary education majors. Since then, the total number of physics secondary education majors at TU has grown from three to 13.

IV. GOALS AND DESIGN PRINCIPLES

A. Overview

As related to secondary education, the primary purposes of Towson's ETEs are to promote and maintain student interest in teaching, and to expose physics majors to inquiry-based science instruction.

A unifying design principle of TU's science teacher education courses is a focus on inquiry-based instruction [8] and other "active learning" pedagogical methods that promote reflection, discussion, and student engagement [9]. At TU, science instruction is considered to be inquiry-based if it is directed by one or more focus questions, engages students directly with scientific phenomena, has the teacher act primarily as a guide rather than as a source of information, promotes deep thinking, and emphasizes evidence-based reasoning [10,11].

ETEs at Towson are consistent in their adherence to active learning principles, although the instruction takes on a slightly different flavor and appearance depending on the teaching context. In particular, the activity structure and teacher guidance can differ depending on the whether the ETE occurs in a lecture, laboratory, outreach visit, or group tutoring session. Table I provides examples that illustrate this important idea.

TABLE I. Active learning examples in different ETE contexts.

Teaching context	Active learning example
Lecture	The instructor periodically breaks up her lecture by having students discuss and vote on answers to "clicker" questions. As pairs of students discuss possible answers, undergraduate Learning Assistants circulate around the room and pose questions to help students clarify their ideas and reasons. After the initial vote, the instructor holds a whole-class discussion and asks students to share the reasons for their answers. During this time, the undergraduate assistants occasionally interject interesting ideas that were raised by the student pairs.
Laboratory	A lab instructor replaces a scripted cookbook lab with an inquiry lab, in which students are exposed to a new concept for the first time. During the lab, students share their initial predictions with the class and think deeply about possible experimental procedures to test their ideas. Once the class has come to consensus on an acceptable experimental procedure (with the instructor's help), undergraduate assistants answer questions about equipment, draw students' attention to potential sources of error, and promote further group discussion.
Outreach	Interns in an "early teaching" course are led through an inquiry physics lesson by the university instructor, after which they discuss the types of activities and teacher-student interactions that make the lesson successful. Later that week, the interns teach a version of the lesson in a local middle school classroom. The lesson starts with a whole-class discussion, but then individual interns perform lesson activities with their own small groups of students. To conclude the lesson, the groups come together again as a class to discuss what was learned and make connections between the lesson content and real-life phenomena.
Group tutoring	An undergraduate tutor makes a conscious effort to follow the suggestions from this semester's teaching and learning seminar, and therefore tries not to dictate mathematical procedures to students when they need help with homework. Instead, the tutor focuses on using questioning techniques to guide the tutored students through their conceptual difficulties and the steps of the problem-solving process.

B. Ideal sequences of early teaching experiences

Each ETE at Towson University can be treated as a standalone experience, as it is possible for an undergraduate student to become involved in a single type of ETE without becoming involved in the others. However, some undergraduates participate in more than one type of ETE, and in those cases there are ideal sequences of ETEs with respect to recruiting and retaining physics secondary education majors, as depicted in Figure 1.

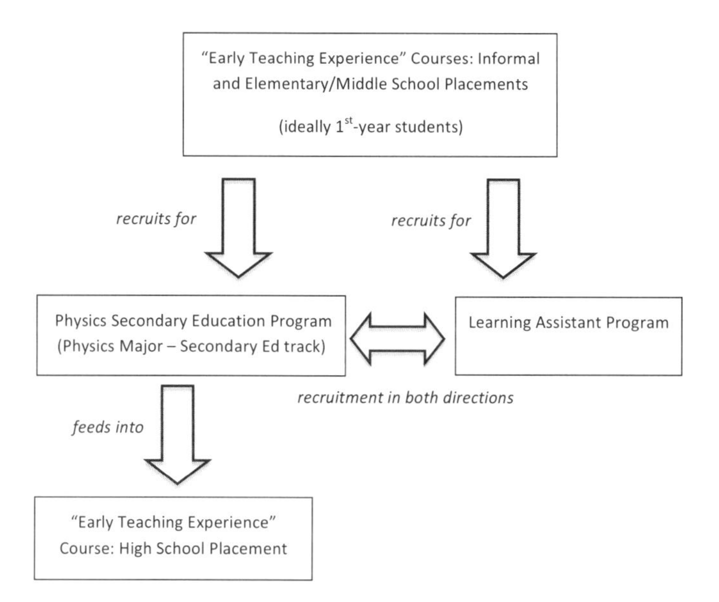

FIG. 1. Ideal sequence(s) of ETEs for secondary education recruitment and retention.

In our experience, one or two first-year physics majors per year have a strong interest in pursuing a high school teaching career, with the rest imagining themselves in industry jobs or in graduate school after earning their degree. The purposes of the informal and elementary/middle "starter" ETE courses are therefore twofold: (1) to give non-education physics majors firsthand experiences that foster positive attitudes toward teaching (e.g., that teaching is fun and fulfilling), and (2) to help education-focused physics majors maintain their interest in teaching and establish social connections to other teaching-oriented students and faculty. For both populations, initial ETE courses can also help some students realize that teaching is not what they want to do with their lives, which is a valuable outcome that saves those students time, effort, and money.

Having completed the starter ETE courses and their own calculus-based physics courses, the next logical step is for students to join the secondary education program (if interested) and/or join the LA program. In the LA program, undergraduate assistants support student learning in the algebra- and calculus-based physics courses, even assisting with the calculus-based physics courses that they recently completed. The purposes of the LA program are to: (1) help education-focused physics majors maintain their interest in teaching and further develop their pedagogical skills, (2) support and sustain course improvement efforts, and (3) allow undergraduates (regardless of teaching interest) to earn money in a job that is more intellectually stimulating than the usual fast food or retail options.

The lateral recruitment shown in Figure 1 can also occur. In some cases, faculty members actively recruit sophomore- and junior-level LAs from the pool of secondary education majors; in other cases, LAs become sufficiently excited about their college peer-teaching experiences that they join the secondary education program. Our experience has been that LAs are more likely to be recruited from the secondary education program, rather than the reverse, but recruitment in either direction is possible.

The last step in the sequence is the high school ETE course, for which enrollment is limited to secondary physics education majors only. The purpose of this course is to allow secondary education interns to interact with high school students in various teaching contexts—one-on-one, in small groups, and as a whole class—in a comfortable, novice-friendly environment. Once the secondary education majors enter formal student teaching, their high school placements become much more high-pressure in terms of evaluation and performance, which is why an early experience in a more relaxed high school setting is especially worthwhile.

V. INNOVATIONS IN STRUCTURING EFFECTIVE EARLY TEACHING EXPERIENCES

A. Early teaching coursework

TU offers early teaching courses inspired by the STEP courses in the UTeach program. UTeach is a nationally recognized program, first developed at the University of Texas at Austin, that has been extremely successful at secondary teacher recruitment and preparation [2]. In the UTeach-based one-credit, low-stakes courses, science majors ("interns") learn about effective teaching methods and plan and teach science lessons under the close supervision of the course instructor. At TU the lessons are chosen by the instructor, although there is some room for interns to individualize their lessons in ways that they find personally interesting. Required assignments in the ETE courses consist of response papers in which interns reflect on articles on science teaching and learning, and reflection papers that focus on the interns' teaching experiences.

Our three-course series places interns in informal settings (fall only), elementary and middle schools (spring only), and high schools (as needed). The justification for placing potential secondary education majors in pre-high school classrooms is that new university students only recently graduated from high school themselves, and are typically uncomfortable teaching students who are very close to their own age. For these interns, a younger student audience is naturally less intimidating than an older student (or adult) audience.

Since our ETE courses were created, 40 science and mathematics majors have enrolled; this number includes 12 physics majors. Students are recruited into the ETE courses via email announcements, academic advising, orientation sessions for new students, brief announcements in introductory science courses, a "teaching career" session in our department seminar for first-year physics majors, and word of mouth from former interns. Our goal is to recruit first- and second-year undergraduates into these courses, since we want science majors to consider teaching as a possible career very soon after their entry into the university. But undergraduates at all levels are allowed to enroll.

Full-time science education faculty, the PAGS department chair, and, most often, our elementary science internship coordinator (Karen Cimino) teach the informal and elementary/middle ETE courses. The high school course is taught by a former PhysTEC Teacher-in-Residence (Jim Selway), who still teaches and does volunteer work for the department. Mr. Selway provides additional support for Towson's informal and elementary/middle ETE courses as he helps the course instructors and interns plan and implement physics-specific lessons.

TU has implemented various innovations with respect to its ETE courses. Originally, the two-course UTeach sequence focused on elementary school and middle school placements, respectively. We modified this sequence to instead focus on informal and then elementary/middle school settings, as mentioned above, with the additional modification that each course meets for only 10 weeks. Our third-semester high school course is a new 15-week (i.e., full semester) course that is not part of the original UTeach sequence.

TU has a history of outreach and public involvement, so it was natural for UTeach's first-semester ETE course to be modified to take advantage of the community connections that had already been established by the department, college, and university. Informal teaching placements for the first-semester course have included publicly- and privately-funded nature centers, the Maryland Science Center at Baltimore's Inner Harbor, an on-campus Saturday science program, an after-school elementary science club, and National Engineers Week activities hosted by a regional airport.

These sites were particularly attractive as teaching placements because they already had well-established educational programs for the general public, which meant that the course instructor was relieved of the burden of creating teaching opportunities at these sites. Informal sites are also enjoyable as a change of pace from a structured school setting. Both the interns and public visitors find informal, hands-on environments to be creative and exciting.

1. First-semester course: Informal placements

The informal ETE course meets once per week (two hours per class meeting) for the first 10 weeks of the semester, for a total of 20 hours. The first meeting is an orientation, and the other nine class meetings are dedicated to site orientations, teaching preparation, and actual teaching. The course was reduced to 10 weeks because meeting for fewer weeks allows us to increase the meeting time per class session, which provides additional time to properly prepare interns for their teaching sessions. The shorter session also means that the course finishes a full month before the end of the semester, which allows interns to focus their time on end-of-semester obligations such as projects and final exams.

The informal ETE course has been structured two different ways. The original structure had interns teach three different lessons at three different informal sites (e.g., nature center, after-school science club, airport), but more recently interns have been teaching three different lessons at the same site (the Maryland Science Center). The multiple-site structure has two advantages: (1) it provides the interns with a variety of enjoyable teaching experiences with different populations in different contexts and (2) in the unfortunate instance when interns have a bad or marginal experience at a particular site, there are still two other sites that can provide positive experiences.

Under either course structure, the teaching settings are spaced out across the semester so that interns have time for site orientations and planning before implementing each lesson. Specifically, each teaching rotation adheres to a three-week cycle, as follows:

- Week 1: Interns attend a site orientation (at placement).
- Week 2: Interns participate in science activities related to the lesson topic and engage in lesson planning (on campus).
- Week 3: Interns implement science lessons (at placement).

Table II illustrates a sample three-week cycle for a school rotation.

Sample lessons for the three-site version of the course include a science-oriented puppet show (nature center), building a simple hovercraft (National Engineers Week), and investigating the scale of astronomical bodies and distances (Saturday science program). Please refer to the Appendix for a sample syllabus for an informal ETE course, which includes details on the reflection and response paper assignments and class meeting schedule. Similar syllabi are used in our other ETE courses, described below.

An interesting aspect of the Maryland Science Center site is that interns are broken into subgroups that rotate through three different types of teaching activities. During the three-week orientation/preparation/teaching cycle, one subgroup

TABLE II. Class activities during a three-week orientation/preparation/teaching cycle at an informal placement.

	Purpose of meeting	Activities
First meeting (week one)	Orientation (in lecture hall and adjoining rooms)	The interns and university instructor meet with the director of the TU on-campus "Saturday Science" public lecture series for families. The meeting (15 min) occurs on a Saturday just before a lecture begins. The director explains the structure of each session: a one-hour lecture for all visitors, followed by hands-on activities for those families who sign up in advance. He also explains the goals of the program and takes questions. The interns then experience one of the lectures firsthand (60 min) and observe the hands-on session that follows the lecture (30 min). Afterwards, the instructor and interns discuss the upcoming meeting schedule and other course logistics (15 min).
Second meeting (week two)	Preparation (in assigned course classroom)	With the interns, the university instructor reviews the engineering design process (30 min) and implements an engineering activity about disassembling and assembling ballpoint pens (30 min), since that topic happens to coincide with the topic of next weekend's Saturday Science lecture. The instructor then helps the interns prepare to implement this same activity in the hands-on session that follows next week's Saturday lecture (30 min). The instructor also engages the interns in a teaching methods activity that addresses teaching issues such as engaging the entire family—since both adults and children will be present—and teacher guidance (30 min).
Third class period (week three)	Teaching (in lecture hall and adjoining rooms)	The interns engage in last-minute preparation of their hands-on activities before the Saturday Science lecture begins (10 min). The interns watch the public lecture on engineering principles and procedures (60 min), and then implement their engineering activities in the hands-on session that follows (30 min). The session is introduced by the university instructor, and then pairs of interns are assigned to small groups of three to four families. Finally, the university instructor holds a post-teaching discussion with the interns concerning activity successes and challenges, participant reactions, teaching methods, and guidelines for the written teaching reflection (20 min).

focuses on interactive demonstrations in the museum lobby; one subgroup focuses on implementing lessons for visiting school groups in the museum classroom; and one subgroup focuses on helping families interact with exhibits on the exhibition floor. It is further arranged so that the subgroups are rotated at the end of each cycle: The lobby demonstration subgroup becomes the school visit subgroup, and so forth. Over the 10-week duration of the course, this allows each intern to experience each type of teaching activity.

Given the small number of students involved (and less than 100% response rates), it is challenging to assess with any statistical certainty the degree to which the informal ETE course has been effective in accomplishing its course goals. We wish to provide evidence of our ETE successes, however, so we report here a sample of the quantitative and qualitative course evaluations, pre/post surveys, and summative course reflections that we have available.

One item on the TU end-of-semester course evaluation is the degree to which students agree with the following statement (1–5 scale, with 5 being the greatest agreement): "Course learning objectives were met." The average score on that item was 4.14 ($N = 7$ respondents), illustrating that interns felt that they had enjoyable teaching experiences, had a more positive outlook on teaching as a career, and successfully explored active learning and effective teaching methods, which were our stated course goals. The qualitative course evaluations and reflections provide insight into precisely how the interns benefited, as seen below:

> The course is a great way to be exposed to STEM outreach and it's a lot of fun! (course evaluation)

> This course helped me know I want to be a teacher. (course evaluation)

> It allowed us to reach outside of the classrooms and use the skills we learned in college to make an immediate change. We were given the opportunity to teach and share our love for STEM-related subjects with others. (course evaluation)

> The course was great in increasing my enjoyment of teaching about STEM activities. All in all this was a great course that I found very enjoyable and that is why I plan on continuing

> to learn about STEM outreach and teaching. (summative reflection)

> Teaching should not just be lecturing and giving notes to people. There should be activities involved, and the lecturing should be used to support the lesson. (summative reflection)

Finally, on pre/post surveys, the first-semester interns ($N = 4$ respondents) rated their knowledge of and comfort with certain science education activities, using a 1–5 scale (5 being the greatest). "Knowledge of" and "comfort with" were measured separately, and are therefore reported separately.

On the pretest, interns had average scores of 1.5 and 1.5, respectively, in their knowledge of and comfort with "National STEM education reform efforts"; on the posttest, interns had average scores of 4.75 and 4.75, respectively, for those same items. On the pretest, interns had average scores of 2.75 and 3.0, respectively, in their knowledge of and comfort with "Engaging others with science-related activities"; on the posttest, interns had average scores of 4.75 and 4.75, respectively, for those same items.

2. Second-semester course: Elementary and middle school placements

The school-based course follows the same meeting schedule as the informal ETE course: one two-hour class meeting per week for 10 weeks. Like the informal course, the school-based course also has interns teach a total of three lessons, with the first two teaching visits held at an elementary school and the third teaching session held at a middle school. The same cyclical three-week orientation/preparation/teaching structure is also employed. Table III illustrates a sample three-week cycle for a school rotation.

The teaching experiences are 45-to-60-minute guided inquiry lessons that allow students to explore physical science concepts (e.g., magnetism or light reflection), occasionally in the context of the engineering design process. In these lessons, students develop science concepts via hands-on investigations, reflection on prior knowledge and everyday experience, explorations of physical and mental models, and small-group and whole-class discussions. Lesson guidance is provided both by the teachers (i.e., the interns) and the activity handouts. The lesson content is often, though not always, tied to state or local standards. Our elementary placements are most commonly in the upper elementary grades (third through fifth).

A key innovation in our elementary/middle early teaching course is that multiple interns teach in the same classroom at the same time, with each intern or pair of interns teaching science to his or her own small group of four to six students. For some teaching sessions, interns in the same classroom teach the same lesson in parallel to their own groups; for other teaching sessions, the interns plan a stations-based lesson in which students rotate from station to station (i.e., from intern to intern) to participate in different activities related to the same general topic (e.g., forces). To foster collaboration and the creation of high-quality lessons, all interns sharing a classroom plan their science lessons together with the help of the course instructor.

The benefits of a multiple-interns-per-classroom teaching structure include the following:

- It is only necessary to recruit a single elementary school and a single middle school per course section, and even then only three to five classrooms are needed at each school. Example: There are 20 interns enrolled in the ETE course, and our host elementary school has 24 children per class. We want each intern to teach science to four to six children, so we place five interns in four separate classrooms, giving each intern a small group of four to five children to work with (24 children ÷ 5 interns).
- Since all interns are placed at the same school site, it is possible to mentor and observe every intern on teaching days. Typically, a planning space is provided for the entire group of interns; in that space, the instructor is available to provide last-minute assistance (answer final questions, help with setup, etc.) to help the teaching go smoothly. Once the interns disperse and the lessons begin, the instructor is free to wander through the different classrooms to make observations.
- Interns who are new to teaching are less intimidated when instructing a small group of students (as opposed to an entire class).
- By the end of the course, most interns have learned that collaborative lesson planning is a useful process that results in better lessons than what any one intern could produce individually. Ideally, the interns continue to plan lessons collaboratively during student teaching and after graduation.
- In a small-group setting, the interns get to know their students' thoughts, strengths, and personalities in a deep and meaningful way, and are able to focus squarely on the pedagogical issues that are relevant to inquiry-based science teaching (idea development, questioning, etc.). In a traditional ETE model, with only one or two interns per classroom, instruction takes on a more impersonal tone, and the teaching sessions often end up focusing on classroom management rather than the principles of inquiry.

Note that, when appropriate, the small-group teaching model is also used in the first-semester informal ETE course.

The effectiveness of the second-semester ETE course can be seen in the interns' qualitative evaluations and end-of-semester reflections, which indicated that the course

TABLE III. Class activities during a three-week orientation/preparation/teaching cycle at a school placement.

	Purpose of meeting	Activities
First meeting (week one)	Orientation (at school site)	The interns and university instructor meet at the school site with the school principal, who outlines volunteer training requirements and responsibilities, provides background on the school and population, and answers questions (30 min). The interns observe their host classrooms to get a general idea of the students, their knowledge, and the classroom contexts (60 min). Afterwards, the instructor and interns discuss the upcoming meeting schedule, past and future assignments, and other course logistics (30 min).
Second meeting (week two)	Preparation (on the university campus)	The university instructor implements an inquiry-based science lesson with the interns (60 min). The instructor then helps the interns prepare to teach this same lesson at the school site (in week 3) by providing a lesson plan outline and guiding the interns as they discuss lesson implementation (required materials, lesson timing, etc.) in small groups (30 min). The instructor also engages the interns in a teaching methods activity (or discussion) that addresses inquiry teaching issues such as wait time, posing questions, and evidence-based reasoning (30 min).
Third class period (week three)	Teaching (at school site)	The interns engage in last-minute preparation in a school-assigned planning room (30 min). The interns then divide into their assigned classrooms and teach their lessons (60 min). Multiple interns are placed in each classroom, and each intern teaches to a small group of four to six students. Lastly, the university instructor holds a post-teaching discussion with the interns concerning lesson successes and challenges, student reactions, teaching methods, and guidelines for the written teaching reflection (30 min).

generally had a positive impact. (Quantitative evaluation scores were not available.)

> It was fun and provided a good insight into the world of teaching. (course evaluation)

> This semester has definitely increased my desire to be a STEM educator because, unlike last semester—where it was all community based—this semester we were actually in schools and there is nowhere I feel more comfortable than in a school and around children. I just absolutely love it. Being with the older students helped too; last semester we worked with a variety of ages but it was really nice to be around the older students. (summative reflection)

> My eyes were opened to the scope of discovery-based learning through this course because I thought it was solely linked to laboratory activities rather than many aspects of teaching. Over the course of the semester we explored different avenues of presenting material that widened my perspective in how to use it. This course has certainly increased my desire and passion for becoming a teacher. (summative reflection)

> I found being in front of students exhilarating when they are using inquiry-based learning. What I really enjoyed from this course was the freedom of collaborating on lessons with fellow peers, which I think should be done more often in the education system. (summative reflection)

The pre/post surveys also revealed positive outcomes. On the pretest, interns ($N = 3$ respondents) had average scores of 3.0 and 3.0, respectively (out of 5), in their knowledge of and comfort with "National STEM education reform efforts"; on the posttest, interns had average scores of 4.3 and 4.0, respectively, for those same items. On the pretest, interns had average scores of 3.3 and 3.3, respectively, in their knowledge of and comfort with "Engaging others with science-related activities"; on the posttest, interns had average scores of 4.7 and 4.7, respectively, for those same items.

3. Third-semester course: High school placements

Our newly-created high school ETE course does not follow the structure of the informal and elementary/middle school courses, but instead has its own format. Logistically, as a first step, one high school teacher is recruited for each intern. During the semester, the teacher makes his or her classroom available to the assigned intern for teaching and observing purposes, and serves as an informal mentor.

The course itself consists of an initial orientation meeting, five two-hour seminars, 10 to 15 hours of volunteer time in a high school physics classroom, and a final class discussion. Each of the five interactive seminars deals with a practical topic related to physics teaching: questioning techniques, problem-solving strategies, inquiry physics lessons, classic physics experiments, and resource materials for new teachers. The high school volunteering starts one month into the semester, and the five on-campus seminars are completed by mid-semester. During the school visits, the interns meet with their mentor teachers, conduct classroom observations, tutor students (if needed), and assist with lectures, interactive demonstrations, labs, and problem-solving sessions.

Interestingly, the creation of the new high school ETE course came at the request of the secondary majors themselves, who appreciated the informal and elementary/middle school courses, but felt that, as future high school teachers, they should have more high school teaching experience.

As far as evidence for course effectiveness, on their end-of-semester evaluations students' average rating was 5 out of 5 ($N = 3$ respondents) in their agreement with the statement, "Course learning objectives were met." Qualitative comments from the course evaluations and reflections are also overwhelmingly positive; samples are provided below:

> The fact that the instructor Jim [Selway] advocated for a course like this to be designed and offered was great. It really helped us going into secondary education to get a taste of what's in store for us and what we're getting into. The other [ETE] courses were very effective as well, but I think it's definitely important to have a secondary level course for this kind of program. (course evaluation)

> It was a great opportunity for student interns to see firsthand what teaching is like and being able to participate in small ways. It was an invaluable class for me and I wish I would have pushed myself to visit the high school more than I did. (summative reflection)

> Going in the classroom was a good experience to learn about what a teacher can really accomplish in a standard 45-minute class. What concepts will take a week to teach? Two weeks? Watching the teachers move about the classroom gave me a few ideas about what I might want to do in my own classroom. Also, the on-campus [seminar] was great. Having the small class allowed for discussion and group activities [about teaching]. You also gave us a variety of research materials on inquiry-based learning that supplemented our field experience and application. (summative reflection)

> I really liked my experience at the high school. I didn't feel pressured to teach the class but was free to help out where I felt comfortable. I think a lot of people feel thrown into a classroom during student teaching, so opportunities like this class give potential teachers the chance to "test out the waters" before running down the deck and doing a cannonball. Being in a classroom is intimidating, especially if you have little experience. Including these extra hours in a school where there isn't a ton of pressure to perform is beneficial to easing us into the craziness that is inevitable. (summative reflection)

Pre/post surveys were not administered in this course, so those results are not available.

B. Learning Assistant program

1. Program description and innovations

In Towson's LA program, undergraduates are paid to assist faculty in implementing "active learning" teaching strategies in large lectures, small lectures, laboratory classrooms, and various out-of-class contexts (e.g., test review or group tutoring sessions). TU's program is similar to other LA programs around the nation in that LAs meet regularly with their supervising faculty to prepare for their teaching activities, and new LAs attend an activity- and discussion-based teaching and learning seminar to explore pedagogical issues [7]. Minor logistical differences between TU's LA program and LA programs at other institutions are that our LAs are paid hourly ($10 per hour for new LAs and $12 per hour for experienced LAs) rather than via a semester-long stipend; our teaching and learning seminar is not a course taken for credit, but is instead a part of an LA's paid duties; and our seminar meets biweekly rather than weekly.

Overall, our LA program has been successful in its ability to redesign and improve lower-division physics courses, provide teaching internships for physics and secondary physics education majors, and establish an expanded teaching community that includes faculty, undergraduates, and Teachers-in-Residence. Thirty-one College of Science and Mathematics faculty and 78 Learning Assistants have participated in TU's undergraduate LA program since its inception in 2008. The project has impacted 5,042 students enrolled in 20 different courses. Fifteen of the faculty participants have been from the PAGS department, and 22 LAs have been physics majors.

The unique aspect of our LA program is that our Teachers-in-Residence have become actively involved in the mentoring of LAs regarding pedagogical reforms, particularly the planning and implementation of Interactive Lecture Demonstrations (ILDs) [12]. Periodically, during a physics lecture class, LAs (or pairs of LAs) are given an allotted time to briefly "teach" the class with an ILD, possibly with help from the resident teacher. These ILDs are jointly planned by the LAs and Teacher-in-Residence, with input from the supervising faculty member.

Resident teachers at other institutions are frequently involved in course reform efforts, but it is unusual for resident teachers to work one-on-one with LAs to the extent that it has been possible at TU, where teachers frequently work with LAs on ILDs for as many as 100 hours per semester. This teacher participation has resulted in instructional modification that would not otherwise have been possible, primarily due to the fact that faculty have difficulty finding the time to meet extensively with LAs and plan these innovations.

2. Program impact

For the first three years of the Learning Assistant program, the LAs reflected on their experiences as part of their paid duties. The following are excerpts of those reflections, by category.

Understanding and appreciation of "active learning" modes of instruction:

> I learned that providing a student with a straightforward answer is not always necessary. In fact, it is better to ask them questions so that they can come to a conclusion on their own.

> I used to think that there was only one dimension to teaching. This program has introduced me to a couple of other dimensions of teaching, which includes planning making lectures dialogic, encouraging students to think beyond class, etc.

> Sometimes hands-on activities are more helpful for some people than others, but overall most people benefit greatly from this type of learning.

Gaining a better understanding of and appreciation for teaching as a career:

> I have realized how much time and preparation goes into teaching. This makes me value the way my teachers teach and what they do to make sure that I learn and understand more.

> I am starting to lean more toward doing teaching after graduation in addition to my current goal. Except for the fact that high school students today can be very difficult I would love to teach in that setting.

> I now realize I want to become a STEM educator in a college.

> I have learned that I might want to be a teacher...!

> I am beginning to seriously enjoy being in a classroom in this capacity and am thinking about entering this field as a career.

General assessment of the program:

> I have really enjoyed my experience as an LA. Although I may have gotten stressed out at times because of it, I'm really glad I stuck with it throughout the entire semester. It has really been a rewarding experience.

> Overall, I truly enjoyed my experience as a Learning Assistant and would recommend it to other students.

> This is a very good project as it helps both the student and me.

> This is by far the most beneficial program that I have had the opportunity to be a part of, and I look forward to participating again.

VI. CHALLENGES, LESSONS LEARNED AND SUGGESTIONS

Throughout our program's development, we have been careful to document any ongoing successes, problems, and solutions regarding ETEs.

A. Early teaching courses

1. Challenges and lessons learned

Transportation is a familiar obstacle in all ETE courses, though there are also challenges specific to ETEs with elementary, middle, and high school placements. It is important to consider and address these issues in advance of implementation—rather than during implementation, when it is too late—to ensure that a new ETE runs smoothly.

Almost all ETE courses require off-campus placements. In these courses, factors such as a lack of transportation, parking expenses, travel time, and the occasional need to schedule teaching sessions outside of normal class time (including on weekends) can be enrollment deterrents for some undergraduates.

When at elementary and middle school placements, secondary education majors can be uncomfortable because they are unsure how to interact with students below high school age. The secondary interns also have little idea as to the level of science content that is appropriate in pre-high school classrooms. Therefore, instructor guidance is critical in ETEs in elementary and middle schools.

An important decision regarding a school-based ETE is whether the interns will be observing any science lessons in the school, or whether all course activity will be dedicated to giving the interns opportunities to teach. In our combined elementary/middle school ETE course, we avoid having interns observe science lessons, reasoning that any time spent observing could instead be spent on additional teaching experiences. The elimination of observations also sidesteps the potential problem of interns observing traditional science instruction (i.e., lecture, reading, worksheets, or the memorization of vocabulary words) instead of the inquiry-focused science lessons.[1]

In an ETE course with high school placements, we have discovered that the host teachers are key to the course's success or failure. These teachers are in frequent contact with the interns, which in principle allows the teachers to model effective teaching practices, engage in deep discussion of lesson planning and physics instruction, encourage the interns to continue to pursing teaching as a career, and provide general advice about the school system's hiring practices, physics curricula, and local politics.

---

[1]We allow for a limited amount of observation time in our high school ETE course, but that is because there are many more teaching opportunities in that class, and also because the mentor teachers have been carefully selected such that the observed physics lessons should represent the type of instruction that we would like the interns to observe.

2. Suggestions

- As described above, we place multiple interns in each classroom to teach in parallel to their own small groups of four to six students. This multiple-interns-per-classroom model is an effective model for school-based ETEs, and is frequently useful in ETEs in informal settings as well.
- Care needs to be taken to recruit host teachers for ETEs who have teaching philosophies that align with the program goals (e.g., a focus on inquiry). In our program, this is more critical for the high school ETE course because: (a) the interns in our high school placement (compared to our elementary/middle placement) are more likely to observe the mentor teacher's instruction, and so the high school teachers have a strong influence on the secondary education majors in terms of instructional modeling, and (b) the high school teachers play an important mentoring role, so teaching discussions between the interns and teachers are commonplace.
- Providing transportation (e.g., buses or vans) promotes intern attendance at off-campus placements, because some interns do not own a vehicle and therefore cannot easily travel to the site. Even interns who own a vehicle frequently prefer to utilize shared transportation to save money on parking and fuel.
- To increase enrollment in ETE courses, the department should encourage the course graduates to communicate their positive early teaching experiences to other physics majors. This might be done informally in a Society of Physics Students meeting, or formally in a department seminar jointly given by physics faculty, former and current interns, and the course instructors. The best form of advertisement for an ETE course is the joy and excitement that an enthusiastic undergraduate shares with a captive audience. As much as faculty sing the praises of ETE courses, it is the words of fellow undergraduates that are perceived as being the most valid and sincere.
- Host teachers for high school ETEs can be recruited by soliciting suggestions from county science coordinators or by working with the university's intern placement offices, but in our case our recruited teachers were either members of our department's Teacher Advisory Group or personal acquaintances of our Teacher-in-Residence.
- Local high school teachers should be encouraged to structure ETEs for their physics students so that a possible career in teaching is encouraged even before the students enter college. There are numerous approaches to structuring ETEs for high school students, such as having them teach physics to other students at their school (e.g., ninth-grade physical science students), getting them involved in after-school science clubs at local elementary and middle schools, or having them solve and explain problems on the board in their own physics classes.

B. Learning Assistant programs

1. Challenges and lessons learned

The impact of Towson's Learning Assistant project on the "assisted" students, the supervising faculty, and the LAs themselves is connected to attitudes toward teaching, the degree to which faculty and LAs genuinely understand the project goals, and faculty workload and staff support.

Like all instructors, faculty and LAs have preconceptions about effective ways of teaching and learning physics, including preconceptions about student effort, student preparation, and the types of students who "can" and "can't" learn physics. Given these cognitive predispositions, it can be difficult to move faculty and LAs away from focusing on student weaknesses or shortcomings—a mentality that tends to be unproductive—toward a focus on instructional or curricular improvement. Faculty and their LAs also tend to focus on their students' mathematical understanding of physics rather than their deeper conceptual understanding; this can be an obstacle to supporting student learning.

It is particularly important for the project participants to understand the goals and philosophies that underlie the course reform efforts, including our focus on active learning and inquiry instruction, yet helping participants develop these understandings is not an easy task. TU's teaching and learning seminar for new LAs has been well received, for example, but the seminar has not been as effective as desired in terms of changing LA attitudes toward teaching and learning.

Despite the eagerness of faculty and LAs to participate in the program, faculty are frequently unable to invest as much time as they would like due to their full workloads. Meeting with LAs takes a significant amount of effort, as do the development and implementation of new course activities. In Towson's case, it was only when a Teacher-in-Residence began to mentor and support the faculty and LAs that the majority of project participants became truly excited about the instructional changes that were occurring. The resident teacher's vast knowledge of physics lessons, physics demonstrations, pedagogical techniques, and student thinking became a tremendous motivational resource for faculty and undergraduates alike.

Finally, in terms of project outcomes, even if the LA experience does not persuade non-education physics majors to become high school teachers, the program still has the effect of encouraging undergraduates to continue with teaching in some fashion in their careers. As a result of their program participation, for example, some LAs have expressed strong interest in becoming university instructors, even if they have no interest in becoming high school teachers.

2. Suggestions

- To adequately support LA-based course reform, it is important to schedule multiple kinds of meetings: (1) small-group faculty meetings centered on curricular improvement, (2) opportunities for the faculty members to observe and discuss each other's classrooms and use of LAs, and (3) classroom visits by the project director. Recruiting the assistance of another pedagogical expert, such a resident master teacher, can also help an LA program become successful.
- Decisions about recruiting and retaining LAs should be made far in advance of the upcoming semester. Highly effective LAs often have multiple external demands on their time, so these LAs need advance notice to plan their schedules (if possible) around the next semester's teaching responsibilities. For example, current LAs might be notified in March (before registration for the fall semester begins) that faculty members hope to continue their employment in the fall.
- With practice and support, new LAs are typically comfortable tutoring or working in a lab class, as well as helping with classroom demonstrations. LAs may need additional support for a large-group teaching experience without a faculty member present, such as an after-class test review session, and in these contexts LAs sometimes prefer to work in pairs. It is rare for a new LA to be willing to give a lecture, although toward the end of a semester (or in future semesters) an LA may be willing to give a small portion of a lecture (10 to 15 minutes) with sufficient guidance from the class instructor.

VII. CONCLUSION

Towson's participation in the national PhysTEC project has been a rewarding experience that has helped increase the recruitment and retention of physics secondary education majors in our department. A portion of this success is related to the institutionalization of multiple forms of ETEs that have been carefully planned and implemented. Our intent is that the ETE programs described above, along with our accompanying lists of challenges, suggestions, and lessons learned, are useful to faculty who are interested in creating or redesigning ETEs at their own institutions. Please contact the authors if you would like further information about planning ETE programs that may be effective in your own institutional context.

ACKNOWLEDGMENTS

The early teaching programs at Towson University were graciously funded by grants from the University System of Maryland and the Physics Teacher Education Coalition (PhysTEC). The support of our department chair, Dr. David Schaefer, and college dean, Dr. David Vanko, has also been instrumental in the success of these programs.

[1] B. C. Clewell and L. Forcier, Increasing the number of mathematics and science teachers: A review of teacher recruitment programs, Teach. Change **8** (4), 331 (2001). http://eric.ed.gov/?id=EJ640140

[2] A. J. Petrosino and G. Dickinson, Integrating technology with meaningful content and faculty research: The UTeach natural sciences program, *Cont. Iss. Tech. Teach. Educ.* **3** (2003). http://www.citejournal.org/vol3/iss1/general/article7.cfm

[3] Physics Teacher Education Coalition (PhysTEC), *Key Components* (2012). http://www.phystec.org/keycomponents, retrieved Jan. 5, 2013.

[4] D. E. Meltzer, M. Plisch, and S. Vokos, eds., *Transforming the Preparation of Physics Teachers: A Call to Action. A Report by the Task Force on Teacher Education in Physics* (T-TEP) (American Physical Society, College Park, MD, 2012). http://www.ptec.org/webdocs/2013TTEP.pdf

[5] H. C. Waxman and H. J. Walberg, Effects of early teaching experiences, in *Advances in Teacher Education Volume 2*, edited by J. D. Raths and L. G. Katz (Ablex Publishing, Norwood, NJ, 1986), pp. 165-185.

[6] J. R. Cannon and L. C. Scharmann, Influence of a cooperative early field experience on preservice elementary teachers' science self-efficacy, Sci. Educ. **80** (4), 419 (1996). doi:10.1002/(SICI)1098-237X(199607)80:4<419::AID-SCE3>3.0.CO;2-G

[7] V. Otero, S. Pollock, and N. Finkelstein, A physics department's role in preparing physics teachers: The Colorado learning assistant model, in *Teacher Education in Physics: Research, Curriculum and Practice*, edited by D. E. Meltzer and P. S. Shaffer (American Physical Society, College Park, MD, 2011), pp. 84-90.

[8] National Research Council, *Inquiry and the National Science Education Standards: A Guide for Teaching and Learning* (National Academy Press, Washington, DC, 2000).

[9] J. A. Michael and H. I. Modell, *Active Learning in Secondary and College Science Classrooms: A Working Model for Helping the Learner to Learn* (Lawrence Erlbaum Associates, Mahwah, NJ, 2003).

[10] C. Sandifer and P. Lottero-Perdue, Delving deeper into science teaching: An early childhood magnetism lesson as a context for understanding principles of inquiry, Connect **24** (2), 11 (2010).

[11] C. Sandifer and P. Lottero-Perdue, *Teaching Science in Early Childhood*, 2013 (unpublished course text).

[12] D. R. Sokoloff and R. K. Thornton, *Interactive Lecture Demonstrations: Active Learning in Introductory Physics* (Wiley, Hoboken, NJ, 2004).

APPENDIX: SAMPLE COURSE SYLLABUS

SCIE 170: Experiences in STEM Outreach, Teaching and Learning

Course Description:

This one-credit introductory course provides students with experiences in teaching and an understanding of effective techniques for communicating STEM topics to visitors of local science museums, nature centers, after-school programs, and other informal science education settings in a supportive learning environment. There are no course prerequisites.

Course Objectives:

This course is designed to help students:

- Distinguish science from non-science
- Develop an understanding of the nature of science
- Explore various methods of active learning through scientific inquiry
- Learn strategies and techniques for inquiry-based science teaching, lesson planning and assessment
- Use resources to relate science to local and regional communities
- Engage members of the pubic in science-related activities to locally important issues

Course Activities:

For the first class session, we will discuss the nature and philosophy of STEM disciplines, STEM inquiry, and effective STEM instruction. Thereafter, the class will participate in three-week rotations related to different outreach programs: Towson University's Saturday Science lecture/activity series, Engineering Day at a local airport, and Marshy Point Nature Center's environmental program.

For each three-week site rotation, students will:

- Week 1 (outreach site): attend an orientation at the site and observe an outreach activity
- Week 2 (TU campus): participate in STEM learning activities related to the outreach site (guided by the course instructor) and engage in activity planning
- Week 3 (outreach site): implement similar STEM learning activities with small groups of visitors

In this manner, students will become familiar with the outreach sites, experience inquiry activities related to those outreach sites, and return to the sites to teach similar activities themselves. Our weekly meeting schedule (including time and location) is attached to this syllabus. Occasionally, students will read short papers related to STEM teaching and learning and provide a brief written response to those papers. Students will also provide brief written reflections focusing on their teaching sessions at the three different outreach sites.

Grading Criteria:

Grades for the course will be based as follows:

	Points
Response papers	20
Teaching reflections	30
Final course reflection	50
Professionalism	50
Total Points	**150**

Response papers:

Response papers related to the brief course readings (there will only be a few) must be complete before the start of the class period and will be collected during the first few minutes of class. There is no credit for late assignments. If you must miss class, please send me your homework to me via email on the day that the homework is due. Guidelines for response paper assignments are attached to this syllabus.

Teaching reflections:

The purpose of the teaching reflection is to have you reflect each STEM outreach lesson in terms of overall effectiveness, possible areas of improvement, your thoughts on STEM teaching and STEM teaching careers, and so forth. **Teaching reflections, which are to be double-spaced and a minimum of 1.5 pages long, are due the week of each teaching session by 12 noon on Friday.** (The earlier the better, since you don't want to forget important details about your teaching and the activity.) The reflections may be sent to me by email, or they may be put in my mailbox in the main physics office. Reflections will be graded on completeness and adherence to the reflection guidelines. Reflections turned in on Friday after 12 noon will be marked down 10%; if they are turned in Saturday or later, they will be marked down 20% per day that they are late. Guidelines for teaching reflections are attached to this syllabus.

Professionalism:

Students will lose 10 points from their professionalism grade each time that they do not attend one of the on-campus class meetings. Since the outreach sites are partly relying on your participation for the success of their educational programs, students will lose 15 points from their professionalism grade each time that they do not attend one of the off-campus (outreach site) class meetings.

Cell phones are disruptive – especially in inquiry courses - and should not be visible in class at any time. Each time a student is seen text-messaging (even "out of view" under the table) or talking on the phone, that student will have 5 points deducted from her professionalism grade. Class will not be interrupted to inform the student of this loss of points. Other distracting behaviors (e.g., reading the newspaper) will result in a similar loss of points.

Final Course Reflection:

The purpose of the final course reflection is to have you reflect on all of your activities in the course and consider what you have learned about STEM topics, STEM teaching, and STEM outreach. Also, since we want SCIE 170 to be as useful and effective as possible, a portion of the reflection will be dedicated to your thoughts on course improvement.

This final reflection assignment is due by 5:00 pm on Wednesday, November 17th. The reflections may be sent to me by email, or they may be put in my mailbox in the main physics office. Reflections turned in on Wednesday, November 17th, after 5:00 pm will be marked down 10%; if they are turned in Thursday, November 18th, or later, they will be marked down 20% per day that they are late. Guidelines for the final course reflection are attached to this syllabus.

Grading Scale:

Final grades will be based on overall course percentage, as follows:

Final Course %	Grade
92 - 100	A
90 - 91	A -
88 - 89	B +
82 - 87	B
80 - 81	B -
78 - 79	C +
70 - 77	C
60 - 70	D
< 60	F

SCIE 170 meeting schedule

Date	Time	Activity	Location
Wed, Aug 25	3:30 pm	Course orientation	Smith 420
Sat, Sep 11	9:30 am	Rotation 1 orientation: *Saturday Science*	Smith 326
Wed, Sep 22	3:30 pm	Rotation 1 activities and planning	Smith 420
Sat, Sep 25	9:30 am	Rotation 1 outreach session	Smith 326
Wed, Oct 6	3:30 pm	Rotation 2 orientation: *Science and Engineering Festival* at Martin Airport's Aerospace Exploration Education Gallery	Smith 420 (carpooling to Middle River, MD)
Wed, Oct 13	3:30 pm	Rotation 2 activities and planning	Smith 420
Wed, Oct 20	10:00 am	Rotation 2 outreach session	Martin Airport's Aerospace Exploration Education Gallery
Sat, Oct 23	10:00 am	Rotation 3 orientation: *Marshy Point*	Smith 420 (carpooling to Marshy Point Nature Center)
Wed, Nov 3	3:30 pm	Rotation 3 activities and planning	Smith 420
Wed, Nov 10	10:00 am	Rotation 3 outreach session	Marshy Point Nature Center
Wed, Nov 17	5:00 pm	Final course reflection due	By email

Response Paper guidelines

The purpose of this assignment is for each student to read an assigned article related to STEM outreach and provide a written response to the article. A response paper is not simply a summary of the article that you have been assigned to read. Rather, the response paper is an opportunity for you to reflect upon and share your reactions to the article.

Assignment guidelines:
Your written response should be 1.5 to 2 pages long (double-spaced). This is approximately 475-600 words. Reflections that are the proper length, but not the proper number of words - due to top or side margins that are much too large, etc. – will be marked down accordingly.

Your response should be broken down into five sections:
1. How does this article relate to STEM-related outreach programs or activities that you experienced yourself as a K-12 student?
2. What questions did the article raise in your mind?
3. What portions of the article did you find particularly interesting or relevant? Why?
4. What did the article help you better understand about effective STEM outreach?
5. What else, if anything, would you like to share in your response to the paper?

Each response paper is worth 10 points, and is graded on clarity, effort, and adherence to the assignment guidelines.

Teaching Reflection guidelines

Your written reflection should be 1.5 pages long (double-spaced). This is approximately 475-500 words. (Reflections that are the proper length, but not the proper number of words - due to top or side margins that are much too large, etc. – will be marked down accordingly.)

Your reflection should be broken down into six sections:
1. What the teaching experience helped you better understand about STEM outreach
2. How the participants responded
3. What went well in the activity, and why
4. Changes that you would make in the activity if you taught it again
5. What you are learning about yourself as a STEM teacher
6. Anything else that you would like to share in response to your outreach teaching experience

Each response paper is worth 10 points, and is graded on clarity, effort, and adherence to the assignment guidelines.

Final Course Reflection guidelines

Please review your previous class assignments (teaching reflections and reading responses) before completing this assignment, as the purpose of the final course reflection is for you to summarize what you learned from your entire SCIE 170 experience.

Your written reflection should be at least 2 pages long (double-spaced). This is approximately 500-600 words. Reflections that are the proper length, but not the proper number of words - due to top or side margins that are much too large, etc. – will be marked down accordingly.

Your reflection should answer the following questions:

1. The three different outreach sites had different audiences, physical locations, and activities.
    a. What did SCIE 170 help you understand about making STEM outreach/teaching effective for different audiences?
    b. What did SCIE 170 help you understand about the types of physical locations that support effective STEM outreach/teaching?
    c. What did SCIE 170 help you understand about the types of activities that support effective STEM outreach/teaching?
2. What are other aspects of effective STEM outreach/teaching that you understand now, but you didn't understand at the beginning of the semester?
3. Was SCIE 170 effective in increasing your enjoyment of STEM teaching and outreach? Why or why not? (Please be specific.)
4. How could SCIE 170 be improved? (Please be specific.)

Note: The final course reflections are graded on clarity, effort, and adherence to the assignment guidelines.

High school physics placements for undergraduate Learning Assistants

Doug Steinhoff, Linda M. Godwin, and Karen E. L. King
Department of Physics and Astronomy, University of Missouri, Columbia, MO 65211

The University of Missouri has implemented a Learning Assistant (LA) program that places university students in local high school physics classrooms every time the class meets during one semester. The goals of the program are to increase participants' interest in becoming future physics teachers; to educate participants in best teaching practices, including modeling-based methods; and to build a community of preservice teachers that will support participants in persisting in the teaching profession. Surveys show that 16 of the 20 LAs who completed the program were either "very interested" or "interested" in being a high school physics teacher after their LA experiences. The University of Missouri anticipates graduating an average of four new physics teachers per year from 2013–2016, compared to only 0.8 per year from 2004–2012, before the program began.

I. INTRODUCTION

The nationwide demand for physics teachers is especially pronounced in Missouri, where physics has been listed as a critical shortage area for most of the past 10 years, and where local school districts have struggled to fill open positions [1]. Exacerbating the demand, the number of Missouri high school students taking physics has more than doubled over the past decade.

In recent years, science education faculty members at the University of Missouri (MU) have secured federal funding aimed at attracting students to degree programs in secondary science and math education. A National Science Foundation (NSF)-sponsored project, Tomorrow's Teachers with Dual Degrees (T2D2), aims to increase the number of secondary science education students at MU by offering Robert Noyce Teacher Scholarships and education-related summer internships (e.g., at the Saint Louis Science Center) [2]. The university's M.S. degree program also offers significant support for participants, including scholarships and/or tuition waivers.

Despite the incentives, these programs were initially underenrolled. From 2007 to 2012, MU awarded an average of 13 physics degrees per year (combining B.A. and B.S. degrees in physics), yet physics education degrees averaged only 0.67 per year (combining B.S. and M.S. degrees) during the same time period. Statewide, Missouri institutions of higher education graduated an annual average of only 5.6 new physics teachers from B.S programs in physics education, including the University of Missouri's undergraduate physics teacher preparation program. Clearly, more effort was needed to meet the needs of Missouri's growing number of high school physics classes.

With support from the Physics Teacher Education Coalition (PhysTEC), MU launched a new effort in the fall of 2012 to increase the quantity and quality of future physics teachers, with the goals of increasing the number of MU secondary physics teaching degrees to a three-year total of 10 (or an average of 3.3 graduates per year) by 2015, and of better preparing future teachers with physics-specific content and pedagogy.

To meet these goals, MU incorporated all of the PhysTEC key components [3], including having the principal investigator (PI) become a physics education "champion" who advocates for the recruitment and education of future physics teachers; collaborating with science education faculty and advisors to streamline degree requirements and help keep track of students; adding a pedagogical content knowledge course in physics; actively recruiting to the B.S. and M.S. degree programs; hiring a Teacher-in-Residence (TIR); and starting a Learning Assistant (LA) program [3]. Rather than describe in detail *all* of MU's PhysTEC activities, which are similar to those at other PhysTEC Supported Sites, this article focuses on MU's unique high school LA program.

II. CERTIFICATION PROGRAMS

The University of Missouri, the state's flagship institution of higher learning, has two degree programs leading to secondary physics teaching certification: a Bachelor of Science degree in secondary science education with physics emphasis (B.S. in physics education), and a Master of Science degree in science education with physics certification (M.S. in physics education). Students in the B.S. program complete coursework in the Department of Physics and Astronomy that is the equivalent of a physics major (31 credit hours) in addition to coursework in the College of Education (44 credit hours). Five of these 44 credit hours are for field experiences, which require that students spend between 16 and 24 hours in the secondary school classroom per semester per credit. For clarity, these five credit hours are referred to here as College of Education (COE) field experiences, to distinguish them from other early teaching experiences such as the Learning Assistant program.

For the M.S. degree in physics education, applicants must have earned 30 credit hours in science or related major, with

TABLE I. Elements of the MU high school Learning Assistant program, and how each element was informed by research.

LA program	Supporting research
Follows the Colorado LA model of fostering collaborative learning in a physics classroom using research-based methods.	Physics courses that have been transformed from a traditional lecture or recitation to a collaborative learning environment using research-based curricula and assisted by LAs have shown improvement in student learning [6,12].
Provides almost daily teaching interactions in high school classrooms. Participants get to know students by helping in the same high school classroom every time it meets.	Connecting to students and the sense of helping society are intrinsic rewards that motivate people to pursue teaching careers [15,16].
Participants are placed in local ninth grade "Physics First" classrooms that use a modeling-based pedagogy.	Modeling is an effective instructional practice [18,20]. Ninth grade physics with mathematical rigor and conceptual emphasis provide an excellent foundation for further science training [19,32].
Improves on the one-credit COE field experience courses required for secondary science education majors by connecting the majors' experiences in physics classrooms to mini-lessons about the theory and methods of modeling physics.	Teacher preparation should include extended *physics* teaching opportunities [22]. School-based programs should be supported by "practical theorizing" [24]. "Third spaces" that bridge the gap between campus-based academic training and school-based practitioner knowledge (e.g., Teachers-in-Residence, K-16 partnerships) address weaknesses in field-based experiences that are the norm in teacher preparation programs [26].
Fosters the growth of a new community of future teachers, practicing teachers, and faculty. Participants have regular contact with the physics education community at MU. We plan at least two social gatherings per semester, and invite past, present, and future LAs and others interested in physics education.	Student interactions with peers and with faculty have been demonstrated to improve retention [27–31].

most of the coursework in physics.[1] The M.S. students complete an additional 35 credit hours in the College of Education as a requirement for the degree. To earn physics certification, participants in both the B.S. and M.S. programs must pass the Praxis II Physics test, in addition to completing the degree requirements. The College of Education hosts both degree programs, although B.S. physics education students take all of the 31 credit hours in physics content in the physics department. Physics education majors (B.S. students) are encouraged to meet with both education and physics advisors each semester. The M.S. physics education students have historically had no interaction with the physics department, although physics faculty and staff have recently been trying to find ways to engage these students.

[1]As of 2012, when MU physics faculty began keeping track of physics education students, all three participants in the MS physics education program have previously earned a B.S. or M.S. in physics.

III. DESIGNING A HIGH SCHOOL LEARNING ASSISTANT PROGRAM

The next few subsections describe the rationale behind building this new recruiting, retention, and educational effort that: (1) is centered around a Learning Assistant program; (2) provides almost daily teaching interactions in high school classrooms; (3) uses modeling pedagogy in a "Physics First" classroom; (4) is an improvement over the one-credit hour COE field experience courses required by secondary science education majors; and (5) fosters the growth of a new community of future teachers, practicing teachers, and faculty. Table I summarizes these elements with references to the literature that supports these design choices.

A. Colorado "Learning Assistant" model as a starting point

Under the model developed by faculty at the University of Colorado Boulder (CU Boulder) and replicated by over 30 institutions [4–6], LAs help implement research-based collaborative activities, such as *Tutorials in Introductory*

Physics [7–9], *Workshop Physics* [10], or *Open Source Tutorials in Physics Sensemaking* [11]. LA programs have been shown to contribute to improved student learning [5,6,12]. Importantly, LA programs can also be used as a tool to recruit physics and engineering students into physics education. The number of physics majors seeking teacher certification at CU Boulder has grown dramatically, from new enrollments of 0.3 per year before the LA program commenced (2001–2004), to 2.9 per year after program implementation (2004–2011) [13]. However, in our experience of talking with colleagues at other institutions, it is still typically only a small percentage of LAs who turn to teaching. At CU Boulder, for example, approximately 13% of LAs go on to enroll in certification programs [6,14]. With a limited budget for creating an LA program, we hoped to create a smaller program with a greater return on investment—that is, that a higher percentage of participants would pursue physics teaching.

B. Connecting with students as a recruiting tool

We wondered if students might be attracted to teaching if they experienced some of the intrinsic rewards that motivate teachers, including the connection to students [15] and the sense that one is serving society [16]. With these factors in mind, we designed a Learning Assistant program that places undergraduates in a *ninth grade* classroom during the *same class period almost every day*[2] with *an outstanding physics teacher*. We wanted to offer teaching as an attractive alternative to paid research positions; therefore, we surmised that we should *compensate participants* for their time.

Since the aim of our PhysTEC program is to recruit new *secondary* teachers, we felt that placing LAs in high school courses (rather than in university courses, as is typical in most LA programs) would allow them to explore this career path more directly. Furthermore, by placing them in ninth grade classrooms, we enabled the LAs to take on leadership roles among much younger students, as opposed to working in university courses with students who are often the same age or only slightly younger. Working with the same students and seeing them every day fosters mentoring relationships between the LAs and ninth grade students. One physics and math dual major's optional comments in his end-of-semester evaluation captured the response we were aiming for:

> [The] best part of being an LA was connecting with students. This happened almost daily. Inspiring students was one of the best feelings in the world. My LA experience was something I would never [want to] give up and I definitely hope I can continue.

This aspect of daily contact and connection to students is one of the crucial differences between our LA program and the COE field credits required by the physics education coursework, where undergraduates work with students only about one hour per week. One LA who was also a physics education major expressed his enthusiasm for this opportunity to work with students more frequently than in his education courses:

> I'm in the classroom much more often, which lets me feel like I'm supposed to be there and not just some visitor. This also makes the students much more comfortable with me, and look at me as an actual teacher figure. In my [COE] field experience students always want the teacher to answer their questions, and sometimes wouldn't even ask me the question when I went to see why they raised their hand. I have never had that problem with the LA program.

C. Modeling in a "Physics First" classroom

In addition to the leadership opportunities offered by a ninth grade teaching placement, these freshman courses offer something that most of the MU physics courses lack: a modeling pedagogy. When new teachers feel overwhelmed, they often revert back to teaching techniques that they have experienced as students, rather than the methods they learned in their education courses [17]. If they have practiced using research-based physics teaching methods during their early teaching experiences, they are more likely to employ these skills themselves when they lead the classroom. All of our LAs are placed with physics teachers who use modeling, an approach to teaching and learning physics that has been shown to be effective [18,19]. In this two-stage process of learning, teachers serve as facilitators to help students build robust models of physics concepts while doing science investigations; students later deploy those models to solve complex problems. The models are "conceptual representations of physical systems and processes," and learning involves using multiple tools, including graphs, equations, and diagrams, for representing these systems [20]. Importantly, full development of the model includes understanding *how* the model relates to the physical system it represents as well as understanding the model's limits. Instructional practices in developing these models include designing and conducting investigations to test ideas, using whiteboards to make student thinking "visible," and defending those ideas in oral presentations to the rest of the class.

A short description of a few activities in the uniform motion unit from MU's program gives an idea of what these modeling lessons look like in practice. In an initial activity,

---

[2] In this first year of the program, this was a daily time commitment and all placements were in ninth grade. In the second year, it changed to alternate days, based on a new block schedule in the high schools. A small handful of students have been allowed exceptions, but must still be able to attend at least three days per week. Additionally, one LA in the second year was placed in a twelfth grade classroom, as this was the only placement that would work for her schedule.

students are given three toy cars (a spring-based "pull-back" car, a non-spring "hot wheels" car, and a battery-operated car that moves at a constant velocity—a point that they have not been told). Students describe what they observe as they play with the cars, and begin to develop a vocabulary for motion, as well as a sense that the cars *do* move in significantly different ways. Students then design an experiment to investigate the motion of a bubble through a vertical tube (see Chandrasekhar for a full version of this activity [20]). Through data analysis and discussion, the students arrive at a definition of uniform motion, recognizing that the bubble's position changes by the same amount for equal intervals of time. Students present their findings to each other on whiteboards in whole class discussions. They develop multiple representations of uniform motion, including motion diagrams, position vs. time graphs, and vocabulary to describe the motion. Further questioning by the teacher leads students to interpret the slope of their position vs. time graphs as the velocity of the bubble. In later investigations that include motion detectors, students develop velocity vs. time graphs and algebraic equations to represent an object's motion. By the end of the unit, when the model has been fully developed, students test their understanding by predicting where two cars set to move toward each other at different speeds will meet. All ninth grade physics teachers in the Columbia Public Schools (CPS) district use their own adapted versions of these investigations.

We have been able to place 19 of our 20 LAs with ninth grade physics teachers who use a modeling curriculum developed by the University of Missouri and adopted by CPS. Due to scheduling constraints, one LA was placed with a teacher who is striving to incorporate modeling lessons into what was initially a more traditional twelfth grade classroom. All of the CPS cooperating teachers have had extensive professional development provided by the MU physics faculty. The LAs' enthusiasm for the modeling style of teaching appears to be an important factor in the success of the LA program.

D. Transforming field experiences into physics-specific "third spaces"

The last three subsections describe the major design considerations of the LA program, including its similarities to the Colorado LA model, its emphasis on interacting with high school students, and the modeling pedagogy. The program shares some similarities with typical education field experiences. In this subsection, we describe the MU College of Education (COE) field course, highlighting common features shared by the LA program and COE field credits, but also pointing out two important ways in which the LA program attempts to transcend the COE field experiences. These two improvements include offering physics-specific teaching experiences and placing these opportunities in an intermediate "third space" that bridges the typical university education courses and high school field experiences.

1. Common features of COE field credits and LA program

Despite the widely recognized importance of offering school-based field experiences to preservice teachers, a review of 300 reports on teacher preparation by the University of Washington's Center for the Study of Teaching and Policy concluded that, "field experiences too often are disconnected from, or not well coordinated with, the university-based components of teacher education" [21]. In this regard, MU's COE field experiences are not the norm. The COE field credits required for the secondary physics education majors *do* occur in conjunction with courses in general science teaching methods, and these field experiences occur early and often in the secondary education course sequence. (See Table II.) Likewise, the LA program requires a weekly meeting during which participants reflect on what they have seen in the classroom and what they can expect in the coming week. Therefore, both the COE field students and the LAs make connections between what they are experiencing in the high school class and what they discuss at MU. These connections allow the future teachers to develop an understanding of the design and rationale of the high school lesson; in the COE secondary science courses, emphasis is on the "5E Learning Cycle" [22], whereas Learning Assistants focus on the physics modeling approach [20,23].

In this regard, both of these early teaching experiences also allow undergraduates to participate in what Hagger and McIntyre describe as "practical theorizing," the core activity of a school-based teacher education program. As applied to teacher preparation, practical theorizing means that when the future teachers are in the high school classroom, they are "looking for attractive ideas for practice," while during the secondary education methods course (for COE students) or weekly meetings (for LAs), they are "subjecting these ideas to critical examination" [24].

2. Physics-specific teaching experiences

Yet, in important ways, the LA and COE field experiences differ. Only the LA program strictly follows the National Task Force on Teacher Education in Physics recommendation that "physics teacher preparation programs should include extended physics-specific teaching experiences along with physics-specific field placements for their certification candidates" [25]. Due to the challenge of having one field coordinator for the entire College of Education, physics education majors are frequently placed in non-physics classrooms for their COE field experience. Furthermore, since the COE field experience is connected to a methods course for all secondary science education majors, in-depth discussions about best practices in teaching specific physics content are not possible, even when the physics education majors have been placed in physics courses.

TABLE II. Summary of for-credit College of Education (COE) field experiences required for secondary science students.

Year	COE Field Experiences
First-year	Orientation course culminates in designing and teaching three to four activities for a group of home-schooled children
Sophomore	One semester (16 hours) in K-12 classroom (one credit)
	One semester (20 hours) in service learning
Junior	Two semesters (24 hours each) in learning labs or content-specific classrooms (two credit hours total)
Senior	One semester (24 hours) in learning labs or content-specific classrooms
	16-week internship

In contrast, the LA program was designed to offer participants opportunities to engage in physics teaching experiences *and* to reflect on best teaching practices in physics. At weekly meetings, the LAs discuss challenges specific to teaching physics, such as what questions a teacher can ask to probe student understanding of Newton's laws, how to help a ninth grader build an abstract mental model of circuits, or how to use lab equipment to facilitate understanding of accelerated motion. The LA experience allows our undergraduates to explore critical issues of this nature, as well as the practical concerns of how to implement these ideas in terms of "time, space, and resources available," a key element of Hagger and McIntyre's "practical theorizing" [24].

Another difference is that the COE field experience places heavy emphasis on classroom management, safety, and accessibility. These issues are discussed in the LA program as they come up, but are not put forth as primary goals. The LA program places greater weight on how to foster student understanding and recognize common misconceptions in introductory physics.

Although MU science education faculty would like to see a physics-specific methods course with an associated physics field credit, they have well-founded concerns that it would not meet the university's necessary minimum enrollment. The Learning Assistantships offer such an experience without being constrained by the enrollment requirements of a COE course.

3. LA program as a "third space" to bridge university courses and high school teaching

Another difference between the LA program and the COE field courses is that the weekly LA meetings led by the Teacher-in-Residence help bridge the gap between the practical knowledge gained in high school teaching experiences and the academic knowledge gained in university courses. It is a common problem in field experiences that the cooperating teachers are often uninformed of the specific teaching methods being taught in the university courses. When there has been an effort to make better connections, it has historically been an "outside-inside" model, wherein the higher education institutions try to more effectively transfer their "expert" knowledge to the K-12 teachers [26]. Zeichner recommends building what he calls a "third space" that brings "practitioner and academic knowledge together in less hierarchical ways to create new learning opportunities for prospective teachers" [26]. Examples of these bridging opportunities include inviting K-12 teachers into science education courses (e.g., with a Teacher-in-Residence program), including teacher-produced work in those courses (e.g., websites or articles written from a practitioner perspective), and having science education students regularly visit a single K-12 science class together on weekly basis and discuss the observed best practices as part of their coursework.

The LA program seems to fit Zeichner's definition of shared "third space," in large part due the NSF-sponsored "A TIME for Physics First" project, which is a partnership between physics teachers at Columbia Public Schools and the Department of Physics and Astronomy at MU. Almost all of the CPS physics teachers have participated in one of MU's 10-week Physics First professional development programs, which has trained over 140 ninth grade physics teachers[3] from across all of Missouri in the content and pedagogy of implementing a modeling-based curriculum [23]. The synergy between the PhysTEC project (including the LA program) and the Physics First project is essential. The PhysTEC PI and TIR have both served as instructors and curriculum committee members of A TIME for Physics First. As the LAs visit classrooms, we know that they will observe best teaching practices, and that the teachers will be using the specific modeling pedagogies that we discuss in our weekly meetings. Furthermore, in leading the LAs, the TIR and PI draw on their own practitioner knowledge from teaching high school physics for 29 years (mostly ninth grade) and five years (ninth and twelfth grades), respectively.

Another crucial factor is that the connection to Physics First has greatly facilitated our effort to find excellent, enthusiastic *physics* teachers to host our Learning Assistants. The logistics of placing physics education students is not trivial.

[3]About half were trained during the NSF-sponsored project, and half from an earlier state-sponsored project.

The MU College of Education has a single field coordinator to place all education students in classrooms for their COE field credits. The challenge of managing the huge volume of students has resulted in unsuccessful matches, with education students often placed out of their content areas. Our TIR has a close relationship with local teachers, as well as the time and commitment to make sure that each student is matched with an excellent physics teacher. In comparing the LA program to his COE field experiences, one physics education major said he liked that "I'm in a classroom I want to be in instead of wherever they had room."

There is much overlap between the COE field experiences and the LA program; indeed, some of our LAs are physics education students who are completing COE field experiences simultaneously. (They only get paid for the LA position after they have completed their COE field requirements.) The LA program takes the field experience a step further, however, by offering more physics-specific teaching strategies and more intentionally blurring the boundaries between school-based practitioner knowledge and higher education academic knowledge.

E. Fostering a community of peers and faculty

There is one aspect of our program design that could be overlooked, but that we feel is quite important: the community of peers that the LA program has helped create. Prior to our reforms, only 2.5% of all physics majors at MU were physics education majors. Interviews with the two most recent graduates of the program revealed that both felt isolated when they were undergraduates. One notes, "I felt very alone.... I was mixed in with engineering and physics majors, who sort of teased me for wanting to be a teacher. I always felt that they did not take me very seriously." The student added that in the education classes, "I enjoyed being surrounded by other science teachers, although when we would do projects and create lesson plans, I never had anyone to conference or collaborate with."

We prioritized building a learning community with high peer and faculty contact as a research-based intervention for improving retention of science students [27–31]. In a strong statement about the importance of these interactions, Pascarella and Terenzini note, "student contact with faculty members outside the classroom appears consistently to promote student persistence, educational aspirations, and degree completion, even when other factors are taken into account" [29]. Our new community was branded "Tomorrow's Outstanding Physics Teachers" (TOP Teachers). This name brings under one umbrella the otherwise disjointed collection of labels—physics education majors, physics majors with teaching interest, M.S. education students, etc. It gives a cohesive approach to communications, including our flyers, presentations, website (http://topteacher.missouri.edu), and Facebook page (www.facebook.com/MUTOPteacher).

The past and present LAs form the core of this group, which as of July 2014 included 20 students. Physics education majors who have not yet been LAs (five fit this description) also frequently attend TOP Teacher gatherings, as do future LAs or physics students who are curious about teaching. The sheer number of LAs gives life to our community, creating another reason to meet. Several times per year, the PI hosts a community-building social mixer at her house. These are scheduled during the regular LA weekly meeting, letting others participate in our discussions about what was observed in the classroom and how to prepare for what is coming next. However, we also make sure to make the event social, with time spent getting to know each other. Most importantly, there is always food, including homemade desserts or pizza.

IV. LEARNING ASSISTANT PROGRAM DETAILS

Unlike other LA programs, which typically place LAs in college and university courses, the MU program places LAs in ninth grade classrooms; and unlike the MU College of Education (COE) field experiences (which also place students in local high schools), the LAs attend the high school classes every time they meet for the full semester, and participate in a weekly meeting with the TIR to discuss physics-specific content and pedagogy.

A. Recruitment and hiring

Each semester brings a new cycle of Learning Assistant logistics to be managed: recruiting, processing applications, interviewing and selecting candidates, obtaining background checks and payroll setup, coordinating supporting teacher and LA schedules, weekly meetings with LAs, and meetings with supporting teachers (approximately one per week). Our TIR currently manages these tasks.

For recruiting to the LA program, we have found it helpful to have as many means of advertising as possible, including paper flyers, posters, HDTV slides, presentations in classes, the physics major email list, and word of mouth from other students. Of these strategies, the two most successful recruiting efforts appear to be visiting university courses (all introductory physics sections, as well as selected upper-level physics courses and introductory science education courses) and placing posters around the physics, education, and engineering buildings. Also, we have found it easier to recruit for the spring semester than for the fall (we think due to the shorter break between semesters). For fall LAs, we plan to continue to take applications into the first week or two of the semester, while for spring LAs, we can manage to complete the hiring process before the end of the fall semester. We have also allowed a greater budget for the spring.

To apply, students must complete a two-page document that asks for information about physics background, interest in teaching, previous teaching-related experiences, and past employment history. They must also submit a reference from a former employer and a letter of recommendation (which can be brief and provided by email). Candidates are selected to interview if they meet the minimum requirement for grades (GPA ≥ 2.5) and level of physics content (completion of the introductory physics sequence). During the interview, we evaluate how well we think the applicant would work in a classroom, and whether or not he or she is genuinely interested in teaching. This interest can be very exploratory; we make it clear that they can use this as an opportunity to see if they like teaching, and are in no way obligated to commit to a teaching path. The interview also allows us to ask applicants with a GPA below 3.0 (particularly in physics) to discuss their weaknesses.

Each semester, the program has hired between five and seven LAs and rejected at least one applicant. Twenty LAs have been hired to date; fewer than five applicants have been rejected based on being unqualified for the program; and approximately 10 applicants were told to reapply at a future date, when their schedules could better accommodate the requirement of being in the high school classroom almost daily. Of these 20 LAs, four have been hired for a second semester to work in high school classrooms. Additionally, seven of these high school LAs have been rehired for new pilot program to assist in one of MU's introductory physics courses. This university-based LA program is not the focus of this manuscript, as it has been operating for only one semester.

Administering background checks is the next implementation detail. We point out this minor task, since it is not always obvious to colleagues at other institutions who wish to start similar school-based programs. In Columbia Public Schools, background checks are required of anyone working regularly in the classrooms, as our LAs do. At other institutions, we have heard of program coordinators avoiding the hassle of background checks by having undergraduates work after regular school hours or for a limited amount of time. In our experience, the background check has not been too cumbersome to obtain, and it helps maintain a trusting relationship with our school district administrators. The TIR works with administrative staff in the Columbia Public Schools district office to process the paperwork. Another option would be to have LAs work with the COE field credits placement officer, who handles the paperwork for background checks for all education students.

Payroll is also set up at this point. While many LA programs award a stipend for the whole semester, and a few offer course credit in lieu of pay, we have found it beneficial to pay hourly. The LAs keep track of time cards that are signed by cooperating teachers; these serve as a record to correct any misreporting of hours (which has happened only occasionally). Also, because they are paid hourly, LAs seem more motivated to work an additional after-school help session or other nonstandard responsibility. We feel this allows us greater flexibility and saves us money if an LA ends up attending less frequently than expected (due to scheduling difficulties).

B. High school placement

In Columbia Public Schools, all ninth graders are required to take physics and all of the ninth grade physics teachers have undergone extensive Physics First training. These teachers are identified by their department chairs as being strong in their field and confident in their teaching abilities, and have shown a willingness to work with university students. Typically we find around three such physics teachers per high school. This provides ample opportunities to place LAs in a freshman physics class that uses a modeling curriculum.

Once the LAs have been officially hired, they are matched up with a physics teacher in one of three local high schools. We match the LA to the classroom teacher based on how well their schedules align. Together, they review a list of basic items that includes dress code, parking, obtaining a visitor pass, interaction with students, the LA's role in the classroom, etc.

The students are paid $10 per hour, starting with their first meeting with the supporting teacher. We budget for LAs to work 91 hours per semester (6.5 hours per week over a 14-week semester), of which 77 hours are expected to be in the classroom (one paid hour per week is for the weekly meeting described in the next section). If an LA is also enrolled in a COE field credit course, the first 16 to 20 hours of the semester are unpaid, and instead count toward their credit hours. This strategy has saved us funds in our budget.

The LAs are responsible for their own transportation to and from the high school. This typically requires a vehicle, although LAs who do not have a mode of transportation are assigned to a high school that is within biking or walking distance. We place only one LA per teacher per class; however, if there are multiple LAs at the same school at the same time (with different teachers), carpooling becomes an option. We ask our LAs to be present every time the high school class meets throughout the entire semester. In the first year of the program, this meant attending five days per week. In the second year, after block schedules were implemented, this became a commitment of attending on alternating days (for longer periods).

In the high school classrooms, all LAs start with an initial observation period of two weeks. During this time, they reflect on teacher practices, they learn the students' names (with the help of a seating chart), and they assist with simple tasks such as collecting homework and handing back graded papers (a great strategy for learning names). During the remaining weeks of the semester, the cooperating teachers use the LA in a supporting role in class. This can include helping slower or absentee students, leading small-group discussions, setting up demonstrations or lab activities, introducing these activities to students, and even explaining a portion of a lesson or leading a whole-class discussion. The extent of LA involvement is based on the teacher's confidence in the LA. Some have relied on the LAs to take over a class when the teacher was absent and the substitute did not know (or understand) what was going on. In one physics education major's words, "Whenever we would have a sub, the students would always go to me rather than going to the sub when they needed something."

C. University-based weekly meetings with TIR

LAs attend a paid weekly training meeting, run by the Teacher-in-Residence (TIR), that is similar to the one-credit pedagogy seminar at CU Boulder, in which participants "reflect on their teaching practices, evaluate the transformations of courses, share experiences… and investigate relevant literature" [6]. It also supports the ideas of "practical theorizing" [24] and "third spaces" [26] as described in Section II.D. Once per week, the LAs meet as a group in the MU physics department to share experiences, discuss the Physics First content and pedagogy, identify student misconceptions and strategies to overcome them, and work through homework (see Table III). Thus, the weekly meeting encompasses features of CU Boulder's pedagogy course *and* of the weekly preparation meeting that instructors at CU Boulder are required to have with LAs. These two meetings were combined in the MU high school LA program to avoid overburdening the already overtaxed cooperating teachers from CPS with an additional time commitment. Nonetheless, teachers are expected to communicate with the LAs about upcoming lessons. Some will work with an LA before school to prepare him or her to introduce a new concept; others use email; and one teacher communicates with her LA using a notebook, which the LA reads when he comes into class to prepare for what he will see that day.

At the weekly university-based LA training meetings, the TIR also has LAs engage in specific activities that they will likely see in class that week (e.g., data collection, data analysis and presentation, and practice problems). Since the LAs are placed among several teachers with varying pace and style of teaching, preparing LAs to assist with common classroom activities could be challenging; however, this issue is mitigated by the unanimous use of the MU-developed Physics First curriculum in ninth grade courses at Columbia Public Schools.

D. Timeline

Table IV provides an example timeline for all of the implementation details described above. The Teacher-in-Residence coordinates all of these activities, with oversight from the PI. Since the LA program has been in operation only four semesters at the time of writing, the exact timeline of these details is still being refined.

TABLE III. Content of weekly LA meetings

Focus on experiences:	• LAs share good and bad experiences from the week. • We ask what surprised them about the class setting, teaching style, curriculum, student behavior, abilities, etc. Did the lesson work for all kids, or just a few?
Focus on methods and concepts:	• We identify how to promote discovery, support the modeling process for this particular unit, and consider questioning strategies to foster a greater depth of knowledge. • We identify strategies to overcome common misconceptions that students have with each concept. Some LAs also need help with these same concepts.
Focus on activities:	• We collect data using the same equipment that high school students will use. We discuss technical and conceptual issues with the data collection. • We ask the LAs to do the homework the high school students will be required to do and look at different strategies to help students who struggle with it. This also helps us check to make sure the LAs understand the material.

V. LESSONS LEARNED

Implementing a high school-based Learning Assistant program is not trivial. The coordination effort is great and would not be possible without a dedicated person shouldering that responsibility. It would also be infeasible without solid ties between the physics and education faculty as well as ties between these two groups and the local secondary schools.

A. Building a foundation with local schools

The single most important aspect of implementing our program was that it rested on a solid foundation of long-standing collaborations among Columbia Public Schools, the MU College of Education, and the MU Department of

TABLE IV. Timeline for coordinating the LA program. MU's semesters include 15 weeks of classes followed by a week of exams. The dates indicate the plan for future fall semesters, based on the experiences of coordinating the program during the two academic years from 2012 to 2014.

Timeframe	Action item
Spring semester	Recruit fall LAs. (See recruiting activities described below for spring LAs.) Recruit cooperating teachers from local high schools.
Summer	Visit CPS administration building to clear details of the program with the superintendent and fill out a background check for each of the LAs. Get schedules from cooperating teachers at local high schools.
Two weeks prior to Fall start	Send email to fall LAs to set weekly meeting time and request updated course and work schedules. Make revisions to placements, as updated schedules require.
Fall, week 1	Meet with fall LAs. Review expectations, have LAs contact cooperating high school teachers to set up a time to meet. Ensure that all necessary paperwork for new LAs is completed, including CPS background check and MU HR requirements. *Make spring application available online and provide hard copies outside TIR's office.*
Weeks 2–3	Fall LAs make first visits to high school classrooms, including face-to-face meeting with cooperating teacher and classroom observations. LAs are asked to memorize high school student names. *Visit select upper-level physics and engineering courses to recruit applicants for spring. This is done early in the semester so that if there is an unexpected opening for a fall position, these upper-level students would have an opportunity to apply for the fall.*
Week 2	Host back-to-school gathering at PI's house for all in TOP Teacher community (past and present LAs, physics education majors, and any other interested students).
Weeks 3–15	Fall LAs are fully integrated into the program. They attend weekly meetings led by the TIR and assist in high school classrooms, as described in Section III. TIR checks in weekly with each cooperating teacher to see how LA is doing in the high school class.
Week 10	*Post recruiting flyers for spring around MU physics, engineering, and education buildings.*
Week 11	*Visit all four introductory physics courses to recruit applicants for the spring. This is the same week that MU students register for Spring courses.*
Week 15	*Deadline for spring applications (but application remains open until positions are full).*
Week 15	Host end-of-semester event at PI's house for TOP Teacher community (including fall LAs and new spring LAs). Have fall LAs complete evaluations of their LA experience. *Interview candidates and obtain copies of their schedules for next semester.*
Week 16	*Select new LAs from applicant pool.* *Notify spring applicants of acceptance or rejection.*
Winter break	*Place spring LAs with cooperating teachers (making schedule revisions as needed).*

Note: Activities not in italics are aimed at LAs working in the high schools in the fall. Activities in italics are aimed at recruiting new LAs for the spring semester.

Physics and Astronomy. Additionally, our TIR has strong ties to both local teachers and physics faculty. To others who would seek to create a high school-based Learning Assistant program, we would highly recommend starting by building these foundational relationships. Offer professional development workshops for local teachers (even if they are just for a day or two, at first). Meet for coffee with local physics teachers—or better yet, offer to bring coffee to them in their classroom when they are not teaching. Be prepared to work around teachers' busy schedules, to offer to come to their planning periods or before or after school, and to use weekends for extended interactions like workshops. Ask the district science department chair at local schools if you could come to their science faculty meetings. Explain why physics faculty should be concerned about the shortage of physics teachers, using statistics to demonstrate critical need. (The T-TEP report is a great source of statistics and charts [25].) Get administrators into this conversation, too, by asking to meet with the local school superintendent or with principals. Perhaps most importantly, hire a TIR to help foster these relationships. Our TIR's long-standing and extensive network of contacts across the district has proven to be an invaluable resource.

B. Building a foundation within the university

By the time this project started, the MU physics department had a long history of interest in K-12 science education. In addition to A TIME for Physics First, which has included six physics faculty members and one mathematics professor as instructors, a new NSF-sponsored project, Quality Elementary Science Teaching (QuEST), also offers professional development in physics teaching. The PI on the QuEST grant (not an author of this paper) is jointly appointed in physics and science education; she has been instrumental in helping connect education and physics colleagues across the university. Our successful PhysTEC initiatives rest on the foundation of this supportive environment, wherein almost one quarter of the faculty have devoted significant effort to K-12 educators. Other institutions can create similarly nurturing environments by collaborating with colleagues on science education grants and by seeking a faculty line for a joint appointment in physics and education.

The importance of having a good rapport with colleagues in the College of Education should not be underestimated. We have tried to make it clear in this manuscript that our LA program builds on the great work done by our colleagues in science education, and that weaknesses in the physics education program have mainly been due to the scarcity of students, which makes it challenging to devote extra resources to them. We had to rebuild our relationships with science education faculty after a visit from an external speaker led an education colleague to comment that the physics education community can be antagonistic toward the general education community. Physics faculty have further strengthened connections between these two groups by serving on thesis committees of science education doctoral students, by meeting with education advisors to collaborate on a four-year physics education academic plan, and by serving as key personnel on the NSF-sponsored program for elementary science education (QuEST).

This program potentially overlaps with a lot of other entities, and we wanted to make sure to maintain smooth coordination with all of them. We learned to discuss logistics with all administrators, staff, and faculty who could be involved in physics teacher education, including people in the physics department (faculty involved in research in the high school physics classrooms, financial staff), College of Education (COE field placement coordinator, COE science methods faculty, COE advisors), and Columbia Public Schools (superintendent and his staff, principals, human resources, and teachers).

VI. EVALUATION METHODS

Table V lists specific objectives as they relate to the goals in Section II, and the measures used to evaluate progress toward each objective. Two primary methods of evaluation are described here, listed in order that they appear in Table V.

To track the number of students entering and persisting in physics education degree programs (objectives 1A and 1B), the local PhysTEC management team collects enrollment data each semester. For the B.S. program, the enrollment number is defined as the number of MU students who have officially declared as their major "secondary science education with physics emphasis" (shortened to "physics education" in communications). The M.S. program count is defined as the number of students who have been accepted to and begun coursework in the Science and Mathematics Academy for Recruitment and Retention of Teachers (SMAR^{2}T program) *and* have told the SMAR^{2}T coordinator that they will be seeking physics certification. These data are presented in Figure 1. We track these enrollment data over time to determine the persistence of physics education students.

In addition to the summative data of enrollment numbers (which are a direct measure of Goal 1), the PhysTEC project collects formative data to try to determine if activities such as the LA program are contributing to an increase in enrollment. While we have not explored a causal link between the LA program and physics education enrollment, we have measured how the LA program impacts students' interest in high school physics teaching. We surmised that an increase in interest would suggest a greater likelihood of enrolling in a physics education degree plan, and thus give us an indicator of the effectiveness of the LA program in contributing to our first goal (of increased enrollment). Anonymous online surveys are administered to the LAs after each semester of being enrolled in the program. The questions asked are listed

TABLE V. Goals and objectives of the PhysTEC project, linked to evaluation. The formative assessments described are those that relate specifically to the high school Learning Assistant program.

Objective	Evaluation measure	Outcome
Goal 1. Increase the number of MU secondary physics teaching degrees to a three-year total of 10 (or average of 3.3 students per year) by 2015.		
Objective 1A. More students will enroll in physics certification degree programs.	Summative: Enrollment data obtained from College of Education annually.	Enrollment is increasing (Figure 1).
	Formative: Survey asks LAs about interest in high school teaching.	16 of 20 LAs report being "interested" or "very interested" in high school teaching (Figure 2).
Objective 1B. Students enrolled in physics certification degree programs will persist through graduation.	Summative: Tracking individual students from enrollment through graduation.	Since 2012, all but one have graduated or persisted.
	Formative: Survey asks LAs about interest in high school teaching.	18 of 20 LAs report an unchanged high interest or an increased interest in teaching (Figure 3).
Goal 2. Better prepare teachers with physics-specific content and pedagogy.		
Objective 2A. Physics education students will have more opportunities to teach physics content.	Summative: Track hours that graduates spend teaching physics content. Use COE requirements as "baseline," and compare to hours reported for PhysTEC LA program.	Data are still being collected.
Objective 2B. Physics education students will have greater exposure to pedagogical content knowledge in physics.	Summative: Track how many physics education graduates have participated in LA program and how many have taken the "Teaching Physics" course.	Data on LA program presented in Figure 1.
Objective 2C. Physics teachers will utilize best practices in physics teaching.	Summative: Use Reformed Teaching Observation Protocol to measure graduates' uses of best practices.	Data are still being collected.

in Figure 2 (Q1–Q4) and Figure 3 (Q5). Individual LAs were also given the opportunity to write down additional comments they had about the program. These additional comments were not anonymous, and each LA gave permission to quote his or her comments in publications. Survey data presented in Section V include only the first semester that each LA participated in the high school program.

VII. OUTCOMES

We have made substantial progress on goal 1, increasing the number of future physics teachers in Missouri. Objective 1A, to increase the number of students enrolling in physics education, is being met (Figure 1) and MU is on target to reach or exceed the average goal of 3.3 new degrees per year. For most of the years before reforms, MU graduated zero (the most common number) or one student per year from each program (Figure 1). The data shown are based on actual graduations (2004–2013) and projections based on the number of students who have officially enrolled in degree program (2014–2016). The annual numbers of M.S. degrees in science education with physics emphasis are displayed in red (dark red is actual and light red is projected); the annual numbers of B.S. degrees in secondary science education with physics emphasis are shown in blue (dark blue is actual and light blue is projected). The lightly shaded bars with diagonal blue lines indicate the number of students in the B.S. program who have been LAs at least once. In 2005, the M.S. program had a record high number of graduates, but the reasons for this are unclear. Data from tracking individual students indicate that progress is also being made on objective 1B. Physics education majors are persisting: Only one student has dropped the major. He has indicated that he plans to enroll in the M.S. certification program after earning dual B.S. degrees in physics and math, and he is not counted among the individuals totaled in Figure 1.

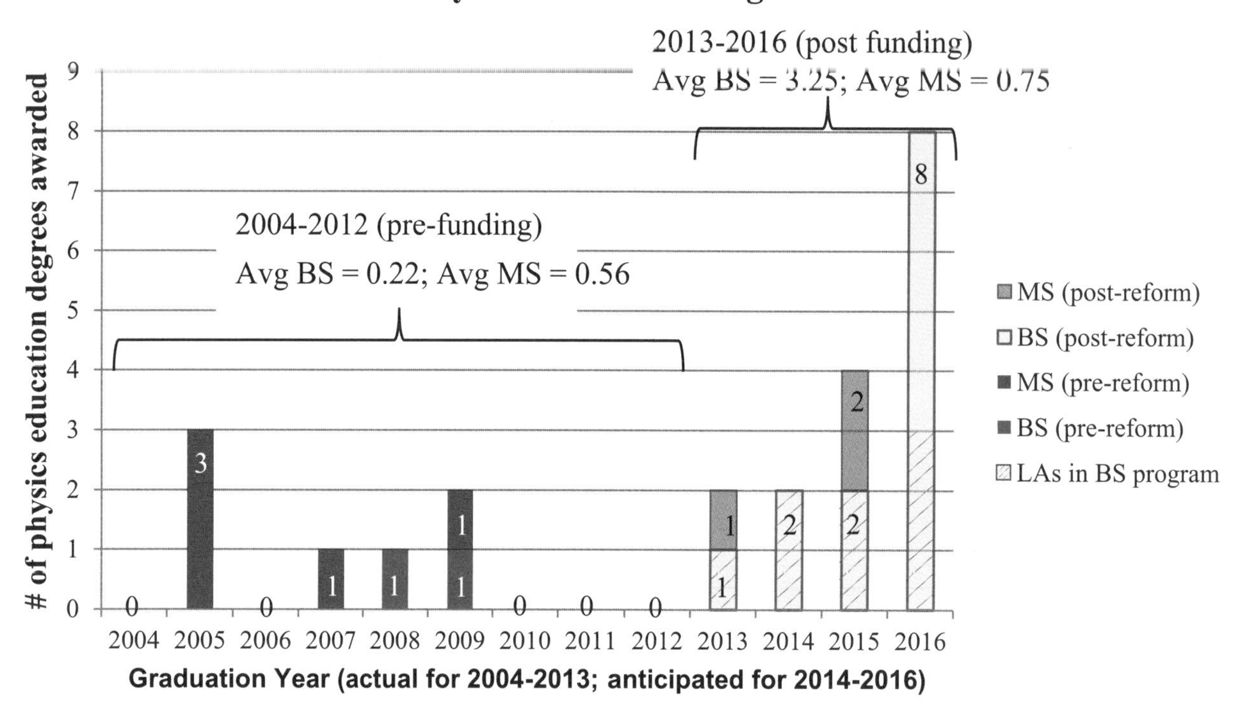

FIG. 1. Total number of MU students earning degrees leading to certification in secondary physics each year, including historical data from the time period before PhysTEC-funded reforms (2004–2012), and expected graduations for the four years since funding and reforms commenced (2013–2016).

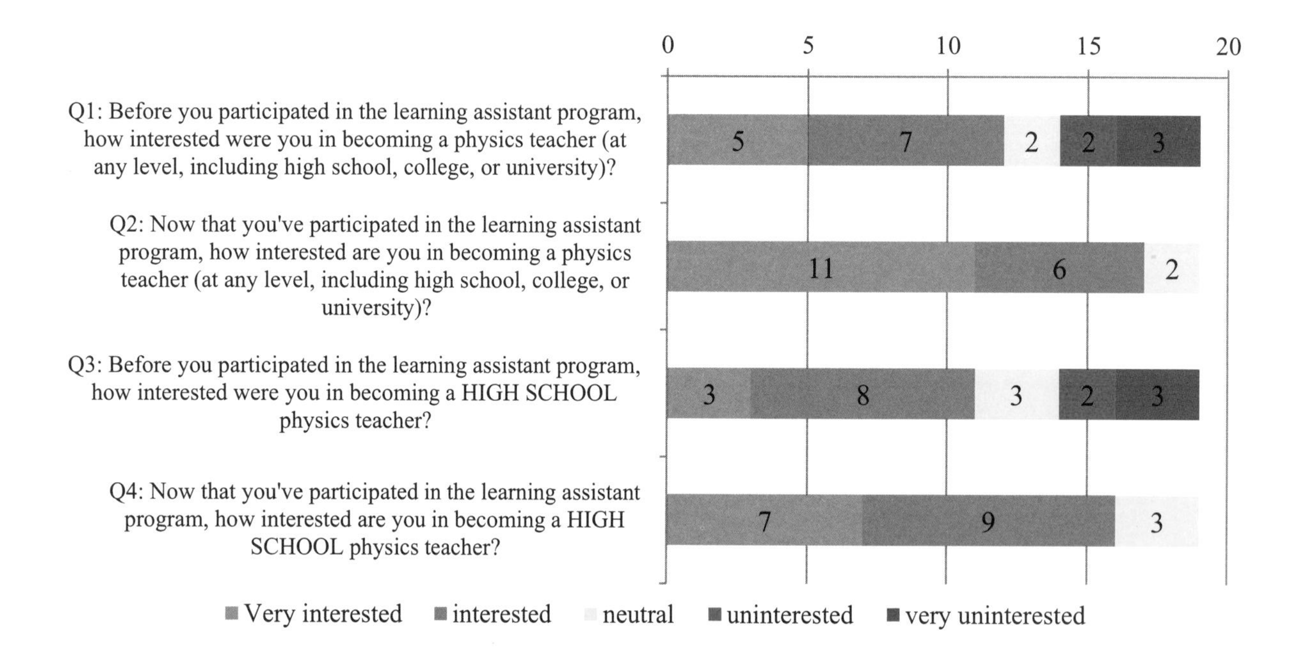

FIG. 2. Results from an anonymous online survey, taken after LAs finished the program. The survey asked LAs about their interest level in teaching physics prior to and after participating in the program. Data labels indicate the number of LAs (out of 19 respondents) choosing each category of interest. One of the 20 LAs did not complete the survey.

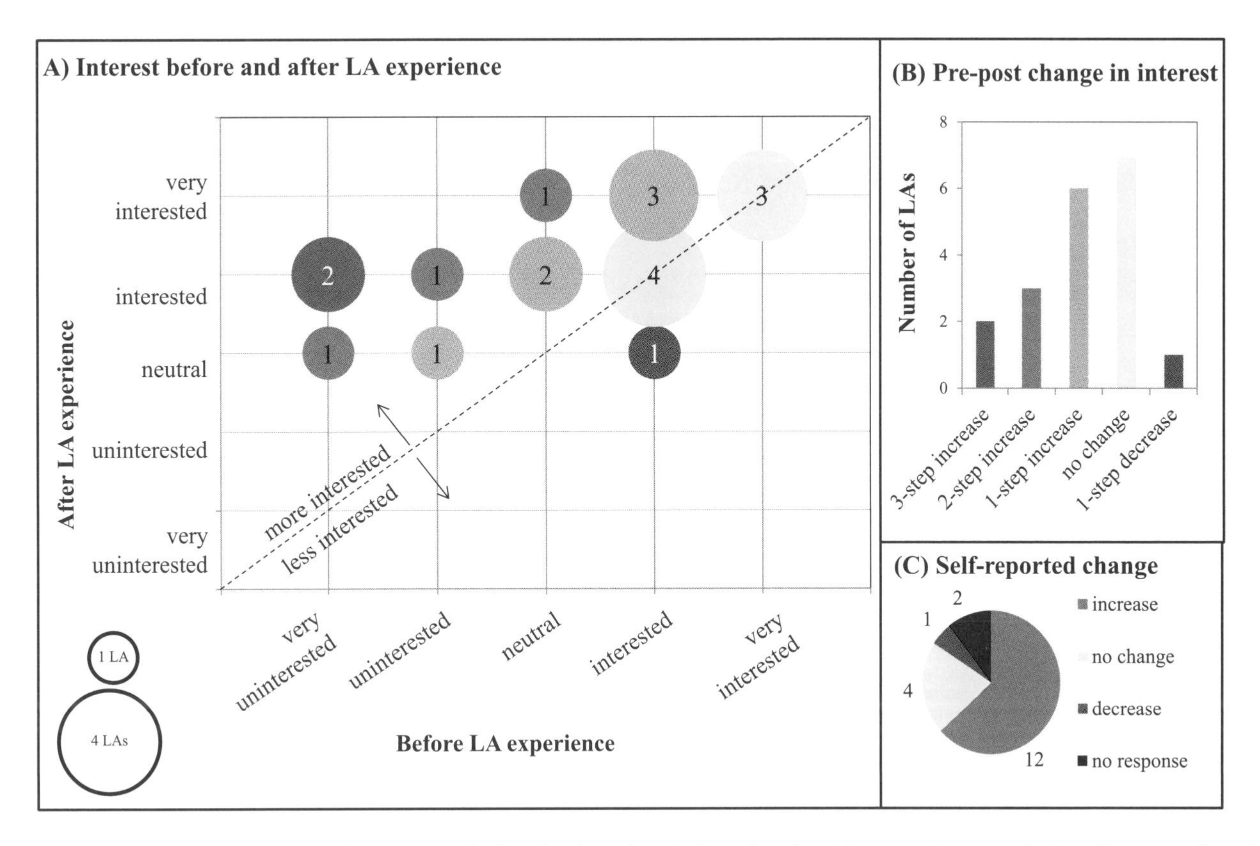

FIG. 3. Change in LA interest in becoming a high school teacher. (A) Each LA's self-reported interest before (horizontal) and after (vertical). The area of each bubble corresponds to the number of LAs. (B) Change in interest, as measured by comparing before and after responses. (C) Survey responses to a direct question about changing interest (see Outcomes).

As described in Section VI, surveys from the LA program were used to determine how the LA program might be contributing toward progress toward enrollment and persistence in physics education (objectives 1A and 1B). The surveys show that 16 of the 20 LAs are either "interested" or "very interested" in teaching high school physics after participating in the program (Figure 2). Only one participant was "neutral," and the other remaining student did not complete the survey despite repeated attempts to reach him.

The data in Figure 3 shed light on LAs' changing interests in high school teaching. The scatterplot in Figure 3A provides the most detail, showing each LA's reported interest in being a secondary physics teacher before (horizontal axis) and after (vertical axis) the LA program. Interestingly, no students report being "uninterested" in teaching after the experience, and the seven who report no change had already ranked their interest high (as "interested" or "very interested"). Figure 3B aggregates the data in Figure 3A; for each LA, we compared his or her reported interests "before" and "after" the LA experience, and calculated the degree of change. For example, the change from "very uninterested" to "interested" would be reported as a "three-step increase" in Figure 3B. Eleven participants were found to have increased interest in high school teaching, to varying degrees (Figures 3A and 3B). Interestingly, when the survey asked LAs more directly, "Did your interest in becoming a high school physics teacher increase or decrease?", 12 participants indicated an increased interest (Figure 3C). One student with decreased interest nonetheless expressed appreciation for the program: "[I]t showed me part of the teacher 'experience' without tying me down to it. It was nice to have a trial run and not have spent years on education only to find out I was not comfortable with my teaching performance."

Objective 2A states that physics education students will have more opportunities to practice teaching physics content. To date, all 5 students in the B.S. program who have either graduated (one student) or are expected to graduate in 2014 (two students) or in 2015 (two students) have participated in the LA program. Therefore, in addition to the 108 hours spent in science (ideally, physics) classrooms as part of their COE field credits, and the semester-long student teaching internships (in physics) that COE requires, these students will have also spent an additional 60+ hours assisting in physics classrooms through the LA experience.

Since the LA program includes a weekly meeting with the TIR to discuss physics content and pedagogy, participation in the LA program is also an indirect measure of objective 2B, that physics education students will have greater exposure to pedagogical content knowledge in physics. It is our aim to have *all* physics education graduates participate in the LA program, in addition to their College of Education field experiences.

One student with experiences in both the LA program and the COE field experience noted that being an LA "helped me learn how to be a teacher much more than any field study course, or course in general for that matter, [I] have ever taken in the College of Education." This comment captured a common sentiment expressed by physics education majors to the TIR and PI on multiple occasions at LA meetings or TOP teacher socials.

Certainly, there is added value (compared to the traditional science education curriculum) in future teachers having the opportunity to see the modeling physics curriculum and methods in action, and to work with the cooperating teacher, the TIR, and the PI to develop practical ideas about how to incorporate these techniques into their own teaching.

Future data collection will determine if we are meeting objective 2C, which states that physics teachers graduating from MU certification programs will utilize best practices in physics teaching. It will be interesting to see if future graduates report any influence of the LA program on their teaching practices.

VIII. SUSTAINABILITY AND NEXT STEPS

As the program enters the second year of a three-year grant, refining our plan for sustainability is a top priority. The PhysTEC funding has made it possible to hire a TIR to coordinate the LA program and to pay the LAs on an hourly basis. Matching funds from MU (specifically from the Department of Physics and Astronomy, the Department of Learning, Teaching and Curriculum, and the deans of the College of Education and the College of Arts and Science) will be applied in the three years after funding to sustain the LA program and to hire a half-time-equivalent teaching faculty member in physics to fulfill the roles of the TIR, including coordinating the LA program and mentoring new graduates. We would also like to expand the LA program to include introductory physics courses at the university, as we turn our attention to course reform.

While the sustained matching funds are helpful, we are already looking beyond 2018 (when these matching funds expire) to find creative solutions for funding the project. Certainly, we hope that continued success of the project will be a selling point as we seek additional external and internal funding for the program. We are also exploring the idea of offering LA positions for course credit, following the example of the University of Arkansas. This credit might be linked to a physics pedagogy course or science methods course, or perhaps we could offer a "teaching credit" in the same manner that undergraduate physics majors may currently earn research credit toward their major for working in a faculty member's lab (as an alternative to a paid position).

IX. CONCLUSION

Data from the Task Force on Teacher Education in Physics show that of the physics departments that have an undergraduate physics teacher preparation program and responded to the survey, fewer than 7% have an "active" program that graduates two or more new physics teachers per years, and only six departments graduate more than four annually [25]. We are optimistic that the MU program will soon join the ranks of this small group of programs graduating high numbers of new physics teachers. The high school LA program, with its extensive early teaching opportunities in a physics classroom, combined with the opportunity to discuss best teaching practices specific to physics, appears to be contributing to this success. Physics and engineering majors have a new attractive undergraduate job opportunity that allows them to explore teaching as a possible career. Physics education students who choose to be LAs have additional opportunities, beyond their degree requirements, to practice and reflect on best practice teaching methods in physics. In separate, private conversations with us and with the American Physical Society and American Association of Physics Teachers representatives who direct PhysTEC, LAs have repeatedly given enthusiastic, positive feedback on the LA program and have indicated that this is the most important of all of the TOP Teacher programs that PhysTEC has funded. "If you change anything," they tell us, "don't change the LA program!"

[1] University of Missouri Department of Elementary and Secondary Education, *Missouri Teacher Shortage Areas* (2013). http://dese.mo.gov/sites/default/files/pictures/Shortage_Areas.pdf

[2] See http://t2d2.missouri.edu, retrieved Sep. 9, 2013.

[3] Physics Teacher Education Coalition, *PhysTEC: Key Components*. http://www.phystec.org/keycomponents/, retrieved Sep. 9, 2013.

[4] Learning Assistant Program, University of Colorado Boulder, *Transforming STEM Education*. http://laprogram.colorado.edu, retrieved Sep. 9, 2013.

[5] S. Pollock and N. Finkelstein, Sustaining educational reforms in introductory physics, Phys. Rev. ST Phys. Educ. Res. **4** (2008). doi:10.1103/PhysRevSTPER.4.010110

[6] V. Otero, S. Pollock, and N. Finkelstein, A physics department's role in preparing physics teachers: The Colorado Learning Assistant model, Am. J. Phys. **78**, 1218 (2010). doi:10.1119/1.3471291

[7] L. C. McDermott and P. S. Shaffer, Research as a guide for curriculum development: An example from introductory electricity. Part I: Investigation of student understanding, Am. J. Phys. **60** (11), 994 (1992). doi:10.1119/1.17003

[8] P. S. Shaffer and L. C. McDermott, Research as a guide for curriculum development: An example from introductory electricity. Part II: Design of instructional strategies, Am. J. Phys. **60** (11), 1003 (1992). doi:10.1119/1.16979

[9] L. C. McDermott, P. S. Shaffer, and M. D. Somers, Research as a guide for teaching introductory mechanics: An illustration in the context of the Atwood's machine. Am. J. Phys. **62**, 46 (1994). doi:10.1119/1.17740

[10] P. W. Laws, Calculus-based physics without lectures, Phys. Today **44** (12), (1991). doi:10.1063/1.881276

[11] A. Elby *et al., Open Source Tutorials in Physics Sensemaking, Suite 1 and 2.* http://www.spu.edu/depts/physics/tcp/tadevelopment.asp, retrieved Sep. 9, 2013.

[12] R. M. Goertzen, E. Brewe, L. H. Kramer, L. Wells, and D. Jones. Moving toward change: Institutionalizing reform through implementation of the Learning Assistant model and Open Source Tutorials, Phys. Rev. ST Phys. Educ. Res. **7**, 2 (2011). doi:10.1103/PhysRevSTPER.7.020105

[13] Learning Assistant Program, University of Colorado Boulder, *Learning Assistant Model for Teacher Preparation in Science and Technology (LA TEST): Project Annual Report to the National Science Foundation* (2013). http://laprogram.colorado.edu/sites/default/files/la-program-11-12.pdf

[14] L. Langdon (private communication).

[15] S. M. Johnson, J. H. Berg, and M. L. Donaldson, *Who Stays in Teaching and Why: A Review of the Literature on Teacher Retention* (The Project on the Next Generation of Teachers, Harvard Graduate School of Education, Cambridge, MA, 2005). http://assets.aarp.org/www.aarp.org_/articles/NRTA/Harvard_report.pdf

[16] C. M. Guarino, L. Santibañez, and G. A. Daley, Teacher recruitment and retention: A review of the recent empirical literature, Rev. Educ. Res. **76**, 173 (2006). doi:10.3102/00346543076002173

[17] M. Eisenhart, L. Behm, and L. Romagnano, Learning to teach: Developing expertise or rite of passage?, J. Educ. Teaching **17**, 51 (1991). doi:10.1080/0260747910170106

[18] J. Jackson, L. Dukerich, and D. Hestenes. Modeling Instruction: An effective model for science education, Sci. Educator **17** (1), 10 (2008). http://modeling.asu.edu/modeling/ModInstrArticle_NSELAspr08.pdf

[19] M. J. O'Brien and J. R. Thompson, Effectiveness of ninth-grade physics in Maine: Conceptual understanding, Phys. Teach. **47**, 234 (2009). doi:10.1119/1.3098211

[20] M. Wells, D. Hestenes, and G. Swackhamer, A modeling method for high school physics instruction, Am. J. Phys. **63** (7), 606 (1995). doi:10.1119/1.17849

[21] S. M. Wilson, R. E. Floden, and J. Ferrini-Mundy, *Teacher Preparation Research: Current Knowledge, Gaps, and Recommendations: A Research Report Prepared for the US Department of Education and the Office for Educational Research and Improvement* (Center for the Study of Teaching and Policy, University of Washington, 2001).

[22] R. W. Bybee, *Achieving Scientific Literacy: From Purposes to Practices* (Heinemann, Portsmouth, NH, 1997).

[23] M. Chandrasekhar, D. Hanuscin, C. Rebello, D. Kosztin, S. Sinha, Teacher professional development must come first for 'Physics First' to succeed, J. Educ. Chronicle **1** (2), 1 (2011).

[24] H. Hagger and D. McIntyre, *Learning Teaching from Teachers* (Open University Press, New York, NY, 2006).

[25] D. E. Meltzer, M. Plisch, and S. Vokos, eds., *Transforming the Preparation of Physics Teachers: A Call to Action. A Report by the Task Force on Teacher Education in Physics* (T-TEP) (American Physical Society, College Park, MD, 2012). http://www.phystec.org/webdocs/2013TTEP.pdf

[26] K. Zeichner, Rethinking the connections between campus courses and field experiences in college- and university-based teacher education, J. Teach. Educ. **61** (1-2), 89 (2010). doi:10.1177/0022487109347671

[27] C. M. Zhao and G. D. Kuh, Adding value: Learning communities and student engagement, Res. High. Educ. **45** (2), 115 (2004). doi:10.1023/B:RIHE.0000015692.88534.de

[28] M. A. Jaasma and R. J. Koper, The relationship of student-faculty out-of-class communication to instructor immediacy and trust and to student motivation, Commun. Educ. **48**, 41 (1999) doi:10.1080/03634529909379151

[29] E. T. Pascarella and P. T. Terenzini, *How College Affects Students* (John Wiley & Sons, San Francisco, CA, 2005), pp. 417-420.

[30] M. L. A. Stassen, Student outcomes: The impact of varying living-learning community models, Res. High. Educ. **44** (5), 581 (2003). doi:10.1023/A:1025495309569

[31] Alliance for Excellent Education, *Tapping the Potential: Retaining and Developing High-Quality New Teachers*, 2007.

[32] R. Goodman and E. Etkina, Squaring the circle: A mathematically rigorous Physics First, Phys. Teach. **46** (4), 222 (2008). doi:10.1119/1.2895672

Preparation in the Knowledge and Practices of Physics and Physics Teaching

Preparing teachers to teach physics and physical science effectively through a process of inquiry

Lillian C. McDermott, Peter S. Shaffer, and Paula R. L. Heron
Department of Physics, University of Washington, Seattle, WA 98195

MacKenzie R. Stetzer
Department of Physics and Astronomy, University of Maine, Orono, ME 04469

Donna L. Messina
Department of Physics, University of Washington, Seattle, WA 98195

The Physics Education Group at the University of Washington (UW) has been teaching university courses, workshops, and institutes for preservice and inservice K-12 teachers of physics and physical science since the early 1970s. Throughout most of this period, the group has also been examining the conceptual and reasoning difficulties that students and teachers encounter while learning physics concepts. Results from this body of research, as well as from the group's collective experience in teaching this material to many different populations, have guided the design of *Physics by Inquiry*, a set of instructional materials specially designed to prepare teachers to teach physics and physical science effectively and confidently through a process of inquiry. These research-based materials have been used at a wide variety of institutions, and the materials' impacts on teacher learning, teacher attitudes toward science, and teacher ability to teach science have been assessed by faculty both at UW and elsewhere. This article includes a brief description of the development of *Physics by Inquiry*, a discussion of some critical features of implementation, and a summary of the impacts on teacher and student understanding of the content and process of science.

I. INTRODUCTION

Since the early 1970s, the Physics Education Group at the University of Washington (UW) has been offering laboratory-based, inquiry-oriented courses, workshops, and institutes for preservice and inservice K-12 teachers. Within this context, we have been developing and refining *Physics by Inquiry* (*PbI*) [1], a coherent set of instructional materials expressly designed to help teachers develop the background needed to teach physics and physical science effectively and confidently.

Well-prepared teachers, it is generally accepted, should have a firm grasp of the content they are expected to teach. However, research and teaching experience indicate that additional aspects of discipline-specific knowledge are also important for helping teachers promote learning. Teachers should be able to engage students not only in learning science *concepts*, but also in experiencing scientific *practices*, such as developing models and constructing scientific arguments based on evidence [2–4]. Teachers should be familiar with common conceptual and reasoning difficulties that students encounter when learning a specific body of material. They should be able to ask questions that allow them to probe student thinking, to assess the degree to which students do or do not understand what is being taught, and to help students deepen their understanding [5–9]. The design and assessment of *Physics by Inquiry* is based on the recognition that all of these skills are important for K-12 teachers.

The need to increase the number of teachers who are well prepared to teach science (and physics in particular) has been widely acknowledged [10–13]. Given the broad range of needs of practicing and prospective teachers, it is clear that not all of these needs can be adequately met in typical introductory courses for physics, engineering, or life science majors. Even when these courses have been reformed on the basis of discipline-based and/or science education research, they cover too much material too quickly to allow students to develop the depth of understanding required of someone who is expected to teach the material. It is also unrealistic to expect that methods courses in a teacher certification program will develop deep understanding of physics content. There is thus a need for special courses in physics for K-12 teachers and, hence, a need for a larger commitment on the part of physics departments to the preparation and ongoing professional development of teachers.

In this article, we discuss the motivation behind the development of *Physics by Inquiry* [14]. Although we and others have used *PbI* in courses for preservice and inservice teachers at all levels, in this article the primary focus is on preservice high school teachers. We describe how various aspects of *PbI* address specific needs of prospective high school teachers and outline practical guidelines followed in its design. We also describe important aspects of implementation and give examples of the impact of the materials on K-12 teachers and students. Although the discussion is in the context of courses for preservice high school teachers, most of the generalizations apply equally well to preservice and inservice teachers at all levels. There are also implications for the preparation of physics graduate students as future faculty.

II. MOTIVATION

Results from physics education research reveal a serious gap between what is taught and what is learned in many physics courses. Students often leave introductory physics courses without a basic understanding of important concepts and principles. Many also have ideas about science that discourage critical thinking and promote rote problem solving. Advanced study often does not solve the problem; common difficulties often persist to the graduate level in physics [15–17].

These observations and others suggest that most introductory physics courses are not well matched to the needs of precollege teachers. Part of the problem is that these courses cover a large body of material too quickly to give students sufficient experience with the phenomena being studied. The emphasis is often on mathematical formalism, with little attention given to the process through which concepts are developed. The approach is often "top-down," based on deductive, rather than inductive, reasoning. Moreover, there is a strong tendency of people to teach as they have been taught. Standard introductory physics courses do not provide a model that teachers, who will be working with younger students in smaller classes, should emulate.

Physics courses for teachers should foster the development of a solid and flexible understanding of basic physics content and the process of science. Teachers need a depth of understanding that allows them to be responsive to the needs and abilities of their students, especially if teaching through a process of inquiry. They should have a functional understanding of the material, by which we mean the ability to interpret a concept properly, distinguish it from related concepts, and do the reasoning required to apply it to real objects and events in a way that a physicist would consider correct for a given stage of learning. Methods courses taught in colleges of education can strengthen some of these skills, but they usually include prospective teachers of biology, chemistry, and earth science, as well as physics. They do not and cannot allot the time to content that is required for the development of a robust and coherent conceptual understanding. Instead, such courses focus on the design, delivery, and assessment of science instruction, often for a range of topics (e.g., all sciences or all physical sciences). *Physics by Inquiry*, by contrast, does not teach methods of instruction explicitly, but rather illustrates by example an approach to teaching through a process of inquiry.

III. DESIGN OF *PBI*

To reduce confusion, in the remainder of this article, we use the term *preservice teacher*, or the more general term *learner,* to refer to undergraduate students in the *PbI* courses that are described. The term student mostly refers to *K-12 students.*

Physics by Inquiry consists of a set of modules, each of which is devoted to the development of a few related concepts and one or more scientific reasoning or representational skills. The first edition of *Physics by Inquiry* consists of the following modules: Properties of Matter, Heat and Temperature, Light and Color, Magnets, Electromagnets, Electric Circuits, Light and Optics, Kinematics, and Astronomy by Sight I and II [1]. The second edition also includes modules on Dynamics, Electrostatics, Balancing, Mechanical Waves, and Physical Optics [18]. Some of the most important reasoning skills developed in *Physics by Inquiry* include proportional reasoning, uncertainty, reasoning by analogy, the control of variables, inductive and deductive reasoning, model-based reasoning, distinguishing observations from inferences, and interpreting graphs. (See Table I for a more complete list.)

Physics by Inquiry is intended for use in a course that is entirely laboratory based, with no lectures. Later in this chapter we discuss the use of *Tutorials in Introductory Physics* in teacher preparation [19]. The *Tutorials* are intended to supplement instruction in courses that feature lectures and labs, that use a traditional textbook, and that cover many topics in a short period of time. However, the depth of learning achieved in such courses, even with a research-based component like the *Tutorials*, falls short of what is typical in a *PbI* course. A course based on *PbI* may be more difficult to develop, given the resources required. However, at UW, enrollments and instructor-student ratios are consistent with those in other senior-level laboratory courses, and, we believe, consistent with enrollments in such courses as other institutions.

Each module has multiple sections, each of which contains a carefully sequenced set of exercises and experiments through which teachers work in groups of two or three. Although some descriptive text is used to frame the activities, answers to questions posed throughout the module are not provided. Teachers are expected to maintain notebooks in which they keep running records of their experimental observations, their responses to the questions, and their evolving ideas about the phenomena they are studying. Supplemental problems provide additional practice in applying and refining relevant concepts and reasoning skills.

Instruction in *PbI* is through a form of *guided inquiry* [20]. Carefully designed sequences of questions guide the learners' investigations of the targeted phenomena and concepts. The framework of questions embedded in the exercises and experiments helps ensure that learners make key observations, reflect on important ideas, and go through the reasoning required for the development and application of the concepts that are being constructed. The explicit guidance helps lay out the logical development of ideas and promotes intellectual rigor, which is particularly valuable for prospective teachers. It enables learners to devote their time to thinking deeply and productively about the material, while helping ensure that they arrive at a scientifically correct

TABLE I. Some scientific reasoning skills and practices that are the focus of various modules in *Physics by Inquiry.*

	Properties of Matter	Heat and Temperature	Light & Color and Light & Optics	Magnets	Electric Circuits	Kinematics	Astronomy by Sight
Designing and interpreting controlled experiments	✔	✔					
Developing and using operational definitions	✔	✔	✔	✔	✔	✔	✔
Calculating uncertainties and using them in making comparisons	✔	✔					
Using patterns in data to derive equations	✔	✔	✔		✔	✔	✔
Constructing and interpreting graphs	✔	✔			✔	✔	✔
Constructing and interpreting diagrammatic representations			✔	✔	✔		✔
Reasoning with proportions and ratios	✔	✔	✔			✔	
Interpreting algebraic expressions and equations	✔	✔			✔	✔	
Developing models from observational evidence	✔	✔	✔	✔	✔		✔
Reasoning by analogy	✔	✔		✔		✔	

TABLE II. Correspondence between practices mentioned in Table I and the scientific practices identified in the *Framework* by the National Research Council [3].

General scientific practices from the *Framework*	Corresponding specific scientific practices emphasized in *Physics by Inquiry*
Developing and using models	• Making "foothold" assumptions • Reasoning by analogy • Testing the ability of models to make predictions • Describing the model-building process in the context of specific models
Planning and carrying out investigations	• Designing and interpreting controlled experiments • Developing and using operational definitions
Analyzing and interpreting data	• Representing data in charts, graphs, tables, and diagrams • Calculating uncertainties and using them in comparing quantities • Using patterns in data to derive equations
Using mathematical and computational reasoning	• Constructing and interpreting graphs (interpreting slopes and derivatives, intercepts, and areas) • Reasoning with proportions and ratios • Interpreting algebraic expressions and equations

understanding that is appropriate for the corresponding stage of learning. Moreover, some of the questions raise conceptual and reasoning difficulties that have been shown to be common and persistent among learners. It is important that prospective and practicing teachers become aware of these problems, even if they themselves do not have these difficulties. They must also have sufficient depth of understanding that, when in the classroom, they can spontaneously come up with questions or alternative experiments to pose to students who are struggling.

Although it may seem that providing learners with a set of specified exercises and experiments is inconsistent with inquiry instruction (especially in contrast to *open-ended* inquiry, in which learners are free to identify their own questions and/or develop their own investigations), we have found that this structure provides necessary guidance, especially at the beginning of such a course. Most teachers gradually develop confidence in their ability to conduct investigations and learn to appreciate the value of dialogues with others as a means of deepening their understanding. As students begin to understand the scientific process, the instructor can take a less directive role. The Appendix contains an example of the kind of exercises and experiments that are used in *PbI* to guide learners toward making appropriate observations and developing and applying a scientific model.

For each module in *Physics by Inquiry* there is also an instructor's guide that includes a set of suggested pretests and examination questions, as well as a writing assignment for a five-to-seven-page summary paper. These are provided to the instructor to illustrate the kinds of questions that we have found useful for assessing teacher progress in the course, as well as for guiding the teachers to reflect on their own learning and to learn how to assess K-12 student progress in their own courses.

The design and development of *Physics by Inquiry* has been driven by ongoing research conducted by us and by others on the learning and teaching of physics from kindergarten through the graduate level in physics. *PbI* has benefited from the concurrent development of *Tutorials in Introductory Physics*, the materials we are developing to supplement instruction in large introductory calculus-based physics courses [19]. *PbI* also incorporates results from informal observations of learners in the instructional environments in which they are embedded. The findings have led to some practical guidelines that underlie the development of *Physics by Inquiry.* Most apply equally well in developing instructional materials for all students, but some are especially important for teacher preparation. Examples include:

- Instruction should be learner centered, not instructor dominated.
- Content and process should be taught together in a coherent body of subject matter.
- Instruction should help learners make explicit connections between concepts, representations, and real-world phenomena.
- The need for, and idea behind, a concept should precede the naming of that concept.
- Concepts should be defined operationally.
- Common conceptual and reasoning difficulties should be addressed explicitly.
- Teachers need multiple opportunities to reflect on their learning.
- Instruction should take place in the laboratory, not through lecture. Moreover, equipment used in courses for teachers should be simple and readily available. The use of "black box" measuring devices should be kept to a minimum, especially at the beginning of instruction.
- Teachers should have the opportunity to learn for themselves through a process of inquiry.
- Most people teach as they have been taught, in terms of both content covered and instructional approaches employed.

This article illustrates how these guidelines are implemented in *PbI.* The following overview describes how they are reflected in the general structure of a *PbI* course and in the expectations of the instructors in a *PbI* course.

At the start of each module, teachers are asked to put aside what they already know, essentially putting themselves in the position of a bright student who has not studied the topic before. In this way, they can go through the process of developing concepts from the ground up (not from the top down) and can reflect on how they have come to know what they know.

Teachers work through carefully constructed experiments and exercises that have limited explanatory text. Intellectual progress is based on observations made in class, inferences drawn from those observations, and the reasoning used in the development and application of concepts. Both inductive and deductive reasoning are emphasized. Discussions with instructors ensure that the conclusions reached by the preservice teachers are appropriate, given the observations and experiments that have been conducted up to that point. Concepts and terms are used only after they have been introduced in class and defined operationally [21].

Throughout each module, preservice teachers are encouraged to consider how various models might be used to account for their observations and to reflect on experiments they might perform that could distinguish between the models [22]. At the end of each module, they are typically asked to write a paper that requires them to articulate, in their own words, the key experiments, observations, and reasoning that led them to develop their model for the phenomenon they have just studied. An important goal is to provide teachers with the experience of constructing for themselves a coherent set of scientific concepts and reasoning skills.

During the course, teachers are asked to make predictions about the outcomes of the experiments they perform and to reflect on how the outcome is or is not consistent with their predictions. They also consider the correct and incorrect ideas that someone unfamiliar with physics might have regarding the phenomena they are studying. In the process, they strengthen their ability to listen to and to interpret student explanations. They also learn how to design and sequence experiments so that the results can be used to guide K-12 students toward constructing scientific models. By the end of the course, each teacher's notebook, together with *Physics by Inquiry*, forms a textbook written by the teacher in his or her own words that can serve as a resource for subsequent teaching.

IV. IMPLEMENTATION IN COURSES FOR TEACHERS

There are some critical elements for successful implementation of *Physics by Inquiry* in a course for prospective teachers. These include issues related to course structure, instructor preparation, and an emphasis on reasoning throughout the course. There are also optional components that focus on helping teachers develop skill in teaching.

At the University of Washington, the *Physics by Inquiry* course for prospective high school teachers is a 400-level (senior-level) laboratory course; introductory calculus-based physics is a prerequisite. (The corresponding courses for elementary and middle school teachers are at the 100- or 200-level and have no prerequisites.) Although introductory physics material is covered, the depth of understanding and the quality of academic work expected is at the level of other senior-level courses. The courses count as electives for physics majors and thus contribute to their upper-division coursework; they are required for majors pursuing the physics education track [23]. The courses do not count toward the teacher certification program in the UW College of Education, which is offered at the graduate level only (a master's in teaching degree is awarded) and which requires a previous undergraduate degree. Completion of the sequence can help candidates for certification obtain an endorsement in physics.

A. Course structure

It is important in a course based on *Physics by Inquiry* that each preservice teacher is actively engaged in his or her own learning. To help promote engagement, the course takes place entirely in the laboratory. There are no lectures, and collaboration is required. Each preservice teacher is expected to go through the requisite inductive and deductive reasoning. Group size is critical. The materials have been constructed to work best for groups of two or three. (We have also found that it can be effective to have two groups of two working across a table from one another. This arrangement allows the groups to discuss their ideas after each has completed a given exercise or section.)

Another important component relates to instructor-teacher interactions. Because of the inquiry nature of *PbI*, answers are not provided in the text. It is therefore important to check progress on a regular basis. Embedded in *PbI* are exercises and experiments that are labeled as "checkouts." At each checkout, preservice teachers review their ideas and reasoning with an instructor. These interactions help instructors ensure that the preservice teachers have made the necessary and appropriate observations and drawn the appropriate inferences based on the results.

It is critical that instructors recognize that the person giving an explanation is usually the one who learns the most. The most critical things instructors can do during checkouts and in their other interactions with preservice teachers are to listen and ask questions. Since a major focus is helping teachers develop skill in reasoning, instructors should encourage students to give complete explanations. Instructors should rarely give answers. Both correct and incorrect responses should be followed by questions that probe more deeply. Preservice teachers should be encouraged to reflect carefully on what they do and do not believe, so that they come to rely on their own reasoning ability and do not simply accept answers given to them. Instruction structured in this way can promote the goal of helping learners recognize when they do or do not understand a concept, and can help them learn how to formulate the types of questions they can ask of themselves in order to arrive at an answer.

Instructor checkouts also provide an excellent opportunity for tailoring instruction to the needs of individual learners (differentiated instruction). In our experience, the prospective high school teachers enrolled in our *Physics by Inquiry* course often differ substantively in terms of subject-matter background (even among the physics majors enrolled in the course). During checkouts, it is possible to spend additional time building and solidifying foundational concepts when working with groups or individuals who are struggling. When working with stronger groups, an instructor can use some time during checkouts to focus on nuanced aspects of the relevant concepts or to explore how one might best teach these same concepts to younger students. In this manner, *PbI* instruction can be optimized based on the needs of the participants. As a result, a variety of different student populations can coexist and thrive in a typical *PbI* course, as discussed later. Readers curious about typical checkouts are encouraged to watch "*Physics by Inquiry*: A Video Resource," which contains several examples of interactions between participants and instructional staff [24].

B. Instructor-student ratio

The interactions described above require a relatively high instructor-to-learner ratio. Such interactions cannot be rushed; sufficient time must be allotted. We have found, in general, that approximately one instructor for eight or 10 preservice teachers works well [25]. This ratio is high, but an argument can be made that courses for K-12 teachers are among the most important that a department teaches. The students will go on to become teachers and will eventually impact many younger learners. A high instructor-to-learner ratio can be achieved in various ways, some of which are relatively inexpensive. For example, individuals who have excelled in the course in a previous year can be brought back as "peer instructors" [26]. In this role, the peer instructors serve as instructors in the course, listening to responses and asking questions that guide their peers to come to their own understanding of the material. (This, in turn, helps the peer instructors deepen their own understanding and provides them with invaluable firsthand experience in teaching physics via guided inquiry.) We compensate peer instructors through either course credit or pay as departmental teaching assistants (TAs).

C. Instructor preparation

It cannot be assumed that instructors (e.g., university faculty, graduate and undergraduate TAs, and undergraduate peer instructors) can teach a course based on *PbI* without having in-depth preparation. Even the best instructors need to reflect on the logical development of concepts in the instructional materials. They must be able to identify, *at each stage* in the module, the observations that preservice teachers will have made and the reasoning that is appropriate *at that stage.* They need to be able to challenge learners when they use terms that have not yet been operationally defined and be able to refer back to relevant experiments and observations. The instructors must be able to put themselves in the position of learners in the course and force themselves to go through the requisite reasoning. Few prospective instructors, even among physics majors and physics graduate students, have thought carefully about the logical development of ideas in a given domain of physics. This is especially true for topics that are taught at the K-12 levels.

D. Promoting the development of conceptual understanding

There is a great deal of evidence that students in introductory physics courses often fail to develop a functional understanding of the basic physics concepts that were covered. It is not surprising then that many prospective physics teachers graduate from university programs with some of the same conceptual and reasoning difficulties as high school students. Often the same difficulties can be found among physics graduate students [15–17] and practicing K-12 teachers [27]. Some difficulties are extremely persistent. Even the act of teaching a topic does not ensure the development of a robust understanding.

Various elements of a *PbI* course are directed at promoting significant and lasting conceptual change. One example is having learners make and record their predictions and explanations before they make experimental observations. This occurs in pretests that are administered before a given module or before individual sections of a module. Without this step of committing to an answer, it is easy for learners to change their minds about what they believe they would have predicted—or to dismiss the extent to which they may have held incorrect ideas. Even if the prospective teachers themselves do not have common incorrect ideas, it is likely that their students will. To ensure that teachers become aware of common conceptual and reasoning difficulties, *PbI* contains fictional dialogues in which students articulate both correct and *incorrect* ideas and lines of reasoning. The comments by students in the dialogues are drawn both from research on student learning and from observations of students in the classroom.

Other aspects of the course also play a role in strengthening correct conceptions. For example, many homework exercises present situations that we have found to be effective at eliciting incorrect ideas. In addition, as part of paper assignments, preservice teachers are often asked to review their predictions for various experiments and to reflect on how their thinking has changed as a result of instruction.

E. Emphasis on reasoning

Prospective (and practicing) teachers can reap significant benefits from learning (or relearning) introductory physics through a process of inquiry. In particular, the process helps make explicit the reasoning skills needed for the development and application of scientific concepts. It is not sufficient to require reasoning from the teachers solely while they are in the laboratory; they need to be given repeated practice in formulating and articulating explanations. This can be accomplished by asking them to keep detailed laboratory notebooks and by requiring complete explanations of reasoning on homework and papers.

F. Time

Significant time is required to promote the depth of understanding that has been described in this paper. The process cannot be rushed. The *Physics by Inquiry* courses at the University of Washington meet three times each week for two hours, for a total of six hours every week. Preservice teachers are expected to take the course all three quarters in which it is offered during a given academic year. During the three-quarter sequence, preservice teachers complete modules on up to six different topics—typically kinematics, dynamics, electric circuits, waves, optics, and astronomy by sight. These subjects appear in many high school physics courses; however, we do not attempt to cover all the topics a teacher of physics might teach. Instead, we choose to go into depth on only a few topics that provide especially good opportunities for developing scientific reasoning skills. For example, the Electric Circuits module covers much more than current, resistance, etc. A major focus is the development of scientific models. Kinematics and Dynamics each emphasize graphical and algebraic representations, and address many well-known conceptual difficulties with force and motion. Table I contains more examples. After a single quarter, gains in conceptual understanding are evident. However, we believe that progress toward other instructional goals takes a sustained effort over a longer time period.

G. Grading

Grading is important and must be done consistently. High standards should be set both in class and in written work. It is important to present learners with opportunities to articulate their reasoning in untimed assignments (e.g., on homework and papers) before they are asked to do so on timed exams. We have found that few undergraduates have had courses in which they have turned in assignments and then been asked to revise their assignments on the basis of general comments by the instructor.

Grading of written work follows the same pattern as comments by instructors during oral responses during class. Correct answers are not given, but rather guiding questions are used to indicate incorrect ideas and incomplete explanations. For example, comments like, "Good, but you should try to identify any steps that are missing in your explanation," or, "Does your explanation hold in all cases?" can require a greater intellectual investment of learners than do comments pointing out the missing steps in reasoning.

When determining final course grades, we have found it productive to weigh exams and paper assignments similarly. Some learners who excel on the exams struggle with the paper assignments and vice versa. (Indeed, for preservice teachers in the course, this is typically the first time they have been asked to write a paper about the development of physics concepts.) By ascribing equal weight to both types of summative assessments, the instructor can make it clear that both activities are equally important and valuable for future teachers.

H. Enrollment

At most universities, the number of undergraduate students planning to become precollege teachers of physics or physical science is relatively small. Thus it becomes important to find other populations for whom the course is relevant and interesting. Many of the students in our courses for high school teachers are prospective teachers of mathematics or other sciences (some of whom are pursuing masters in teaching degrees through the College of Education). These students can be motivated to take the course when they recognize that in the U.S. many physics classrooms are taught by teachers without a degree in physics [13]. We also conduct a 100-level version of the course intended for preservice elementary teachers. We advertise the 100-level *PbI* course to students in the liberal arts. These students are often interested in understanding the process of science and find that a *PbI* course can give them firsthand experience in how science is done. We have also found that the course is particularly valuable for graduate students in physics who are interested in teaching at liberal arts colleges or community colleges. Graduate students working with our group as they conduct research in physics education also benefit from taking the course.

I. Additional (optional) components

There is good evidence of the impact of *PbI* on teacher understanding of basic physics concepts. (See Section V.) To address the increased emphasis on teaching scientific practices as expressed in the Next Generation Science Standards, teachers need to have a firm grasp of these practices themselves [28]. Moreover, effective teachers also need *pedagogical content knowledge,* including knowledge of how students learn specific concepts [6–9]. A number of components can be added to a *Physics by Inquiry* course to enhance these other aspects of teacher preparation and help prospective teachers develop: (1) knowledge of common student difficulties, (2) the ability to recognize these difficulties, and (3) the ability to design written and verbal questions to probe student thinking. Below we discuss several of these optional components, including a teaching practicum in a local high school.

1. Enhancing teacher knowledge about common student difficulties

A major emphasis in *Physics by Inquiry* is on eliciting and addressing some of the most common conceptual and reasoning difficulties that have been identified through research. Specific activities in each module are intended to elicit particular difficulties, help learners recognize a contradiction between their ideas and experimental results, and guide them in resolving the conflict. In this way, teachers become aware of the most common difficulties that students encounter in their study of a particular topic.

Within this context, it is relatively easy to include readings that document findings from physics education research on K-20+ students. These readings should be given after preservice teachers have completed the relevant module or portions of the module. For example, after the Light and Color or Light and Optics modules, teachers can be assigned readings from articles on K-12 student understanding of geometrical optics [29–35].

2. Improving teacher ability to recognize student difficulties

Having abstract knowledge about student difficulties does not necessarily translate into the ability to recognize these and other difficulties while interacting with students during class and when examining student work. The student dialogues in *Physics by Inquiry*, in which incorrect ideas are articulated by fictional students, provide some practice in helping teachers identify incorrect conceptions held by students. We have found, however, that skill in interpreting student thinking can be further enhanced by occasionally providing preservice teachers with authentic student work, typically in the form of anonymous student responses to questions that have been posed in precollege classrooms or in introductory university physics courses. The teachers can be asked to categorize and/or analyze the student responses as part of written homework or paper assignments or to examine the responses and discuss them in small groups.

3. Promoting teacher ability to design assessments of student thinking

It is difficult to design good assessments (formative or summative) that probe student thinking clearly and concisely. Even when an assessment seems to be well-constructed from the perspective of an expert, novices may not interpret questions or context in the intended way. Knowledge of common modes of student thinking can help, but is not sufficient. In our courses for preservice teachers, we try to incorporate opportunities for the teachers to design assessment questions and to administer them to real students in precollege physics and physical science courses. The teachers then have the opportunity to analyze the responses, both to assess student thinking and to reflect on how the questions could be modified to probe student understanding more effectively. An example is discussed in the next section.

4. Teaching practicum

For the past 10 or so years, we have incorporated a *teaching practicum* into the final quarter of our course sequence for preservice teachers. The practicum provides the teacher with an opportunity to design a pretest, instructional materials, and a posttest on a topic they have studied in the *Physics by Inquiry* course, and then to teach a lesson (or a series of lessons) in the classroom with middle or high school students. Typically teachers work in groups, designing their course materials together, with input from the course instructors as well as the K–12 teacher in charge of the precollege classroom. The practicum gives preservice teachers an opportunity to implement what they have learned in the *PbI* course in a controlled environment. Often it is their first chance to reflect deeply and carefully on the complete, iterative process of design, implementation, and assessment of instruction.

Another way in which we have found it possible to provide a similar (although less in-depth) teaching experience for preservice teachers is through participation as undergraduate or graduate teaching assistants in introductory physics courses that use *Tutorials in Introductory Physics.* (At some institutions, undergraduate Learning Assistants, or LAs, also have the opportunity to teach in tutorial-supported introductory physics courses [36].) The *Tutorials* are the instructional materials that our group has been developing and refining for more than 20 years for use in small-group sessions designed to supplement standard instruction in algebra- and calculus-based physics courses. Like *PbI,* the tutorials consist of worksheets that pose sequences of questions designed to help guide learners toward developing a functional understanding of basic physics concepts for themselves. The tutorials also focus on the concepts and reasoning skills that research on the learning and teaching of physics has shown to be hard for most students. The TAs thus have an opportunity to obtain practice in listening to students and in asking probing questions that elicit student conceptions and guide students toward strengthening their understanding of the course material.

V. ASSESSMENT

Physics by Inquiry is the product of an iterative process of research, curriculum development, and instruction. Ongoing assessment has guided the design of both the materials and the courses in which they are used. The impact on various aspects of teacher and student thinking has been examined by the Physics Education Group at the University of Washington as well as by faculty at a variety of other colleges and universities. These assessments include examining the materials' and course's impact on (1) teacher understanding of physics concepts (including long-term retention), (2) teacher understanding of scientific reasoning skills, (3) teacher attitudes toward science and teaching, and (4) K-12 student learning of science and attitudes toward science [37,38]. The following discussion highlights some of this research and the relevant aspects of *Physics by Inquiry* and/or course structure.

A. Impact on teacher understanding of physics concepts

A crucial component of the preparation of K-12 teachers is helping them develop a sufficiently deep and flexible understanding of the material they must teach. Especially when teachers are expected to teach through a process of inquiry, they need to have expertise with the subject matter that goes beyond memorized facts and formulas. In order to guide students in their learning, teachers must understand the experimental basis for the relevant theories, principles, and models, and be familiar with the inductive and deductive reasoning that links real-world observations and physics formalism. The impact of *PbI* on teacher understanding of physics concepts is extensively addressed in a variety of papers, both by our group [37] and by faculty at other institutions [38].

Research conducted at UW has taken place in the classes for preservice high school teachers that are the focus of this article, as well as in classes for preservice elementary teachers and for inservice teachers [27,34,39–46]. Pretesting and posttesting show that in some cases, these elementary teachers outperform undergraduate science students with significantly more experience. For example, in a study related to the principles of buoyancy, 85% of elementary teachers who had taken *PbI*-based courses were able to answer a question requiring a prediction about sinking and floating, whereas only 10% of students in an introductory algebra-based physics course did so after traditional instruction [46]. (Introductory students who received targeted research-based instruction in a lab and a tutorial-based interactive lecture performed at a similar level as the elementary teachers.)

At other institutions, faculty have implemented a variety of assessment strategies, including the use of multiple-choice tests [47,48]. For example, in a study at the University of Cyprus, preservice elementary teachers were given the Determining and Interpreting Resistive Electric Circuit Concepts Test [49] on electric circuits. Those who had taken a course based on *Physics by Inquiry* outperformed those whose course had used constructivist pedagogy, but had not used research-based materials [50]. Moreover, the difference persisted a year later. Similar evidence of long-term retention was reported at the University of Kentucky [48]. At The Ohio State University, preservice elementary teachers who had studied from the Electric Circuits module in *PbI* were given an assessment on electric circuits. They performed at a higher level than students in introductory physics [51].

B. Impact on teacher understanding of scientific reasoning skills

At UW, we conducted an in-depth study of the ability of university students and precollege teachers to apply the control-of-variables reasoning strategy [52]. We found that most of the participants in both populations recognized the need to control variables, but many had difficulty with the underlying reasoning. Often they could describe the experiments that might be conducted under ideal conditions to test for whether or not a given variable influences the outcome of an experiment. However, most struggled when confronted with experimental results from which they needed to draw inferences about the influence of one or more variables. We have used sinking and floating as the context for teaching control-of-variables reasoning to K-8 teachers and have found that the experience can have a dramatic impact on their reasoning skills [48].

Faculty implementing *PbI* at other institutions have also examined the growth of facility with important scientific reasoning skills [53]. For example, at Ohio State, large gains on the Lawson Test of Scientific Reasoning have been reported [54,55].

C. Impact on teacher attitudes toward and understanding of the nature of science

The impact of *PbI* on teacher attitudes and beliefs about the nature of science has been studied at several institutions [56–60]. In one example, gains on the Colorado Learning Attitudes about Science Survey (CLASS) were reported at several institutions that had slightly different implementations of *PbI* [61,62]. In contrast, in typical courses, responses to the CLASS tend to regress toward non-expert views.

D. Impact on teaching practice

At the UW we have examined the impact of an inservice program based on *PbI* on teachers' classroom practice. The Reformed Teaching Observation Protocol (RTOP) [63] has been used in observations of local teachers participating in our summer institute for inservice teachers, both prior to the institute and approximately one year later. Significant increases in scores were documented [64,65]. We found positive impact in how the teachers demonstrated their own understanding of the topics, embedded fundamental concepts in their lessons, and engaged students in probing dialogues for teaching and for assessing student understanding.

E. Impact on K-12 student learning and attitudes toward science

Faculty at several institutions outside of UW have conducted studies to examine the impact on K-12 students of *PbI* courses and workshops for teachers. Improvements in both student learning [41,66] and attitudes toward science have been documented [67,68]. There is also indirect evidence that the use of *PbI* with K-12 teachers may have a positive impact on K-12 student learning. This evidence comes from studies that suggest that changes in teaching practice (as measured by the RTOP) may be correlated with improved student achievement [69,70].

VI. USE OF *TUTORIALS IN INTRODUCTORY PHYSICS* IN COURSES FOR PROSPECTIVE TEACHERS

In order to be able to teach effectively through a process of inquiry, teachers should have the opportunity to learn a body of material for themselves in the same way. They should have a chance to develop concepts from the ground up and to come to understand *how* they know *what* they know. However, the time available for professional development of preservice and inservice teachers is often limited, and the use of *Physics by Inquiry* may not be possible.

Our group drew extensively on our experience in developing *Physics by Inquiry* when we began preparing *Tutorials in Introductory Physics*, our supplemental materials for small-group sections of introductory algebra- and calculus-based physics [19]. The *Tutorials* have a strong research base that demonstrates their impact on student learning at UW and at a wide variety of educational institutions [16,34,44,71–73]. Unlike *PbI*, the *Tutorials* are not meant to stand alone. They rely on students having been introduced to basic concepts in lecture, homework, and laboratory. *PbI* and *Tutorials* share a major focus on developing a functional understanding of introductory physics and on addressing common conceptual and reasoning difficulties. However, *PbI* has a much greater emphasis on the step-by-step development of concepts. Nonetheless, there is evidence that the use of *Tutorials in Introductory Physics* in programs for preservice and inservice teachers can contribute to significant gains in teacher understanding of physics concepts.

One example of the use of the *Tutorials* for preservice teacher preparation is from the University of Colorado Boulder's (CU's) Learning Assistant program, which is designed to support science, technology, engineering, and mathematics course transformations and to raise interest in teaching careers through comprehensive efforts to prepare and support undergraduates as novice instructors [36,74]. In the introductory calculus-based physics courses at CU, the LAs help teach the small-group tutorial sections. In the course-specific part of their preparation, which is modeled after the TA preparation seminar developed at UW, the LAs work through the tutorials they will be teaching and examine student responses to the tutorial pretests in order to gain insight into common conceptual difficulties. The impact of the LA program on the physics content knowledge of the LAs has been assessed in multiple ways. Results from both the Force and Motion Concept Evaluation and the Brief Electricity and Magnetism Assessment indicate strong gains for the LAs. Moreover, results from the Colorado Learning Attitudes about Science Survey indicate a dramatic favorable shift in student beliefs about science, interest in science, and mastery of science, relative to other students taking the survey [36].

The tutorials were also used in a short course for the development of inservice teachers at the Tecnológico de Monterrey in Monterrey, Mexico [75]. In addition to other activities, the teachers took the tutorial pretests, worked through the tutorials and homework, and then reflected on the difficulties that their students have with related materials. The authors found good gains on the Force Concept Inventory, one of the means by which the impact of the course was assessed.

As is the case for *PbI*, resources for adopters of *Tutorials* can be found on our website [76]. We also offer workshops at national meetings so that prospective adopters can learn more about either set of materials.

VII. CONCLUSION

Courses for teachers of physics have multiple purposes. In addition to helping teachers develop a functional understanding of physics concepts and principles, such courses should help teachers understand the nature of science (the methods, the models, and the role of explanations) and promote critical thinking. They should also help teachers develop skill in reflective thinking (e.g., learning to ask the types of questions necessary to recognize when one does or does not understand, as well as learning to ask the types of questions necessary to come to an understanding).

These outcomes would benefit all students taking physics, and *PbI* has proved useful in working with a wide variety of undergraduates, including those in liberal arts programs and those who are underprepared in science but intend to pursue careers in science and engineering [77–79]. However, teachers of physics also need to understand the challenges that students encounter in learning physics. They need to become familiar with common conceptual and reasoning difficulties and with instructional strategies that have proved effective with students. Moreover, if teachers are expected to teach through a process of inquiry, they need to be able to construct and implement coherent instructional materials that serve to guide students in developing an understanding of scientific concepts and practices [80]. These are in addition to the critical issues of class management, course structure, depth of coverage, etc.

The preparation of preservice K-12 teachers of physics is thus a complicated problem. It requires the active participation of faculty from both colleges of education and physics departments. *PbI* represents an attempt to produce instructional materials that physics faculty can readily use as part of their department's contribution to physics teacher preparation. The focus is on helping teachers learn the content of physics and the process of science in ways that are consistent both with the practice and culture of physics and with the pedagogical and practical issues that teachers encounter in education courses.

We have given some examples of how this approach to teacher preparation has proved effective at helping preservice high school teachers develop some of the skills they need in ways that they can draw upon in their own classrooms. A growing body of research is demonstrating the impact of *Physics by Inquiry*, both on the teachers themselves and on their students. Critical to the development of these materials has been ongoing research on learning and teaching, observations of teachers and students in a variety of classrooms, and input from a wide range of faculty, postdocs, graduate and undergraduate students, and practicing K-12 teachers. As noted in the section on Assessment, faculty using *PbI* at other institutions have described their experiences and shared the results of their assessments in publications. While their courses share several essential features with those at UW, the evidence indicates that *PbI* can be used effectively in other formats and environments [75,81].

The development of *PbI* and the design of the courses in which they are embedded have taken place over many years. Both continue to change as the needs of teachers and the structure of their classrooms evolve. However, certain components will remain the same. It takes time for learners to go through the reasoning required to construct concepts for themselves. Learners also need repeated practice in constructing scientific arguments and models. Moreover, it is unreasonable to assume that programs for preservice teachers are sufficient to prepare teachers. Teachers need help after they enter the classroom. It is therefore also necessary that faculty in departments of physics conduct courses, workshops, or institutes to provide ongoing support to inservice teachers as they begin and continue to teach physics in precollege classrooms.

---

[1] L. C. McDermott and the Physics Education Group at the University of Washington, *Physics by Inquiry*, 1st ed. (Wiley and Sons, Inc., New York, NY, 1996).

[2] See http://www.nextgenscience.org

[3] National Research Council, *A Framework for K-12 Science Education: Practices, Crosscutting Concepts, and Core Ideas* (The National Academies Press, Washington, DC, 2011). http://www.nap.edu/openbook.php?record_id=13165

[4] R. A. Duschl, H. A. Schwingruber, A. W. Shouse, and the National Research Council, *Taking Science to School: Learning and Teaching Science in Grades K–8* (The National Academies Press, Washington, DC, 2007). http://www.nap.edu/openbook.php?record_id=11625

[5] These and other skills are often called pedagogical content knowledge (PCK).

[6] L. S. Shulman, Those who understand: Knowledge growth in teaching, Educ. Res. **15** (2), 4 (1986). doi:10.3102/0013189x015002004

[7] D. L. Ball, M. H. Thames, and G. Phelps, Content knowledge for teaching: What makes it special?, J. Teach. Educ. **59** (5), 389 (2008). doi:10.1177/0022487108324554

[8] H. C. Hill, D. L. Ball, and S. G. Schilling, Unpacking pedagogical content knowledge: Conceptualizing and measuring teachers' topic-specific knowledge of students, J. Res. Math. Educ. **39** (4), 372 (2008). http://www.jstor.org/stable/40539304

[9] S. Magnusson, J. Krajcik, and H. Borko, Nature, sources and development of pedagogical content knowledge, in *Examining Pedagogical Content Knowledge,* edited by J. Gess-Newsome and N. G. Lederman (Kluwer Academic Publishers, the Netherlands, 1999), pp. 95-132. doi:10.1007/0-306-47217-1_4

[10] U.S. Department of Education, *Our Future, Our Teachers: The Obama Administration's Plan for Teacher Education Reform and Improvement* (Washington, DC, 2011). http://www.ed.gov/sites/default/files/our-future-our-teachers.pdf

[11] Committee on Prospering in the Global Economy of the 21st Century: An Agenda for American Science and Technology, National Academy of Sciences, National Academy of Engineering and Institute of Medicine, *Rising Above The Gathering Storm: Energizing and Employing America for a Brighter Economic Future* (National Academies Press, Washington, DC, 2007). http://www.nap.edu/catalog.php?record_id=11463

[12] National Science Board, *America's Pressing Challenge – Building a Stronger Foundation* (Arlington, VA, 2006). http://www.nsf.gov/statistics/nsb0602

[13] D. E. Meltzer, M. Plisch, and S. Vokos, eds., *Transforming the Preparation of Physics Teachers: A Call to Action. A Report by the Task Force on Teacher Education in Physics* (T-TEP) (American Physical Society, College Park, MD, 2012). http://www.ptec.org/webdocs/2013TTEP.pdf

[14] For an overview of the process of research, curriculum development, and instruction by the group, see L. C. McDermott, Oersted Medal Lecture 2001: Physics education research–The key to student learning, Am. J. Phys. **69** (11), 1127 (2001). doi:10.1119/1.1389280

[15] Research has shown that specific difficulties with introductory material can extend to the upper-division and/or graduate levels in physics. See, for example, references [16] and [17].

[16] P. S. Shaffer and L. C. McDermott, A research-based approach to improving student understanding of the vector nature of kinematical concepts, Am. J. Phys. **73** (10), 921 (2005). doi:10.1119/1.2000976

[17] M. R. Stetzer, P. van Kampen, P. S. Shaffer, and L. C. McDermott, New insights into student understanding of complete circuits and the conservation of current, Am. J. Phys. **81** (2), 134 (2013). doi:10.1119/1.4773293

[18] The 2nd edition of *Physics by Inquiry* is in preparation.

[19] L. C. McDermott, P. S. Shaffer, and the Physics Education Group at the University of Washington, *Tutorials in Introductory Physics* (Prentice Hall, Upper Saddle River, NJ, 2002).

[20] The term inquiry is often applied to a wide variety of teaching methods. The approach taken in *Physics by Inquiry* is based on years of research and teaching experience. A primary goal is to enable learners, after instruction, to be able to describe in their own words how they came to develop and apply scientific models based on observations they themselves have made. An additional goal for K-12 teachers is to illustrate an instructional approach that has proved effective at helping students learn. Teachers must have a flexible understanding of the subject matter that they are expected to teach, be familiar with the difficulties that the material presents to young students, and be aware of alternate paths that might prove effective at helping students learn. For these reasons, *Physics by Inquiry* includes exercises that range from being somewhat directive to being open-ended.

[21] The term *operational definition* was introduced to many physicists by P. W. Bridgman in *The Logic of Modern Physics* (MacMillan, New York, NY, 1927). There is more than one interpretation of the term. In this paper and in curricula by the Physics Education Group, an operational definition is a series of steps that, if followed by different individuals, would lead unambiguously to the same result (e.g., a specific number or event). For example, in the Properties of Matter module in *Physics by Inquiry*, students construct an operational definition for mass. Their definition describes a series of steps that one could perform with an equal arm balance to find the mass of an object. The steps include balancing the balance arm, finding a set of standard objects that balance one another, and counting the number of standard objects required to balance the unknown object.

[22] Students are encouraged to consider various models that can account for the observations they have made during class. However, students who are struggling with the material or who are new to the idea of modeling in science often find it difficult to think about more than one model as they are learning the material for the first time. This is particularly true of students whose backgrounds in physics are especially weak. The judgment of the instructor is critical in assessing how far to take each individual student in exploring alternative models and in designing experiments to help distinguish between them.

[23] Students at the University of Washington can pursue one of four different tracks as a physics major: the applied track, the biological track, the comprehensive track, and the teacher preparation track.

[24] L. C. McDermott and the Physics Education Group in partnership with WGBH-Boston, *Physics by Inquiry: A Video Resource*, 2000. http://main.wgbh.org/wgbh/learn/instructional/physics.html, retrieved Sep. 16, 2013.

[25] Smaller instructor-to-student ratios can also be used effectively, but there is a tradeoff in terms of the quality of the discussions between the instructor and students. In a course for teachers, the ratio should remain at about this level. For a discussion of a course for non-science (liberal arts) majors in which there was one experienced instructor and about 80 students, see R. E. Scherr, An implementation of *Physics by Inquiry* in a large-enrollment class, Phys. Teach. **41** (2), 113 (2003). doi:10.1119/1.1542051

[26] The use of the term *peer instructor* in this way differs from that of Eric Mazur in his book, E. Mazur, *Peer Instruction: A User's Manual* (Prentice Hall Inc., Upper Saddle River, NJ, 1997). Peer instructors in our courses serve as novice instructors. They listen to students and ask questions that guide students in coming to their own understanding of the material. They thus require a much broader view of the material than the students themselves.

[27] L. C. McDermott, Millikan Lecture 1990: What we teach and what is learned—Closing the gap, Am. J. Phys. **59** (4), 301 (1991). doi:10.1119/1.16539

[28] NGSS Lead States, *Next Generation Science Standards: For States, By States* (The National Academies Press, Washington, DC, 2013).

[29] Relevant articles for students to read after working through the Light and Color module include references [30-34].

[30] E. Feher, Interactive museum exhibits as tools for learning: Explorations with light, Int. J. Sci. Educ. **12** (1), 35 (1990). doi:10.1080/0950069900120104

[31] K. Rice and E. Feher, Pinholes and images: Children's conceptions of light and vision. I, Sci. Educ. **71** (4), 629 (1987). doi:10.1002/sce.3730710413

[32] E. Feher and K. Rice, Shadows and anti-images: Children's conceptions of light and vision. II, Sci. Educ. **72** (5), 637 (1988). doi:10.1002/sce.3730720509

[33] S. Bendall, F. Goldberg, and I. Galili, Prospective elementary teachers' prior knowledge about light, J. Res. Sci. Teach. **30** (9), 1169 (1993). doi:10.1002/tea.3660300912

[34] K. Wosilait, P. R. L. Heron, P. S. Shaffer, and L. C. McDermott, Development and assessment of a research-based tutorial on light and shadow, Am. J. Phys. **66** (10), 906 (1999). doi:10.1119/1.18988

[35] F. M. Goldberg and L. C. McDermott, An investigation of student understanding of the real image formed by a converging lens or concave mirror, Am. J. Phys. **55** (2), 108 (1987). doi:10.1119/1.15254

[36] V. Otero, S. Pollock, and N. Finkelstein, A physics department's role in preparing physics teachers: The Colorado Learning Assistant model, Am. J. Phys. **78** (11), 1218 (2010). doi:10.1119/1.3471291

[37] UW Physics Education Group articles that illustrate assessments of K-12 teacher and university student learning of physics can be found at http://depts.washington.edu/uwpeg/research/publications

[38] There has been a great deal of research by faculty and students not at UW regarding the impact of *Physics by Inquiry* on K-12 teachers and K-12 students. An extensive list of these articles, graduate dissertations and theses, and research talks and posters can be found at http://depts.washington.edu/uwpeg/impact-pbi

[39] L. C. McDermott, Preparing K-12 teachers in physics: Insights from history, experience, and research, Am. J. Phys. **74** (9), 758 (2006). doi:10.1119/1.2209243

[40] L. C. McDermott, P. R. L. Heron, P. S. Shaffer, and M. R. Stetzer, Improving the preparation of K-12 teachers through physics education research, Am. J. Phys. **74** (9), 763 (2006). doi:10.1119/1.2209244

[41] L. G. Ortiz, P. R. L. Heron, and P. S. Shaffer, Student understanding of static equilibrium: Predicting and accounting for balancing, Am. J. Phys. **73** (6), 545 (2005). doi:10.1119/1.1862640

[42] P. S. Shaffer and L. C. McDermott, A research-based approach to improving student understanding of the vector nature of kinematical concepts, Am. J. Phys. **73** (10), 921 (2005). doi:10.1119/1.2000976

[43] L. C. McDermott and P. S. Shaffer, Research as a guide for curriculum development: An example from introductory electricity. Part I: Investigation of student understanding, Am. J. Phys. **60** (11), 994 (1992). doi:10.1119/1.17003; Printer's Erratum, Am. J. Phys. **61** (1), 81 (1993). doi:10.1119/1.17448

[44] P. S. Shaffer and L. C. McDermott, Research as a guide for curriculum development: An example from introductory electricity. Part II: Design of instructional strategies, Am. J. Phys. **60** (11), 1003 (1992). doi:10.1119/1.16979

[45] L. C. McDermott, P. S. Shaffer, and M. D. Somers, Research as a guide for teaching introductory mechanics: An illustration in the context of the Atwood's machine, Am. J. Phys. **62** (1), 46 (1994). doi:10.1119/1.17740

[46] P. R. L. Heron, M. E. Loverude, P. S. Shaffer, and L. C. McDermott, Helping students develop an understanding of Archimedes' principle. II. Development of research-based instructional materials, Am. J. Phys. **71** (11), 1188 (2003). doi:10.1119/1.1607337

[47] R. K. Atwood, J. E. Christopher, R. K. Combs, and E. E. Roland, Inservice elementary teachers' understanding of magnetism concepts before and after non-traditional instruction, Sci. Educator **19** (1), 64 (2010). http://eric.ed.gov/?id=EJ874155

[48] K. C. Trundle, R. K. Atwood, and J. E. Christopher, Preservice elementary teachers' knowledge of observable moon phases and pattern of change in phases, J. Sci. Teach. Educ. **17** (2), 87 (2006). doi:10.1007/s10972-006-9006-7

[49] See http://www.compadre.org/PER/items/detail.cfm?ID=12388&Attached=1

[50] L. C. McDermott, P. S. Shaffer, and C. P. Constantinou, Preparing teachers to teach physics and physical science by inquiry, Phys. Educ. **35** (6), 411 (2000). doi:10.1088/0031-9120/35/6/306

[51] B. Thacker, E. Kim, K. Trefz, and S. M. Lea, Comparing problem solving performance of physics students in inquiry-based and traditional introductory physics courses, Am. J. Phys. **62** (7), 627 (1994). doi:10.1119/1.17480

[52] A. B. Boudreaux, P. S. Shaffer, P. R. L. Heron, and L. C. McDermott, Student understanding of control of variables: Deciding whether or not a variable influences the behavior of a system, Am. J. Phys. **76** (2), 163 (2008). doi:10.1119/1.2805235

[53] C. L. Bao, T. Cai, K. Koenig, K. Fang, and J. Han, J. Wang, Q. Liu, L. Ding, L. Cui, Y. Luo, Y. Wang, L. Li, and N. Wu, Learning and scientific reasoning, Science **323** (5914), 586 (2009). doi:10.1126/science.1167740

[54] A. E. Lawson, The development and validation of a classroom test of formal reasoning, J. Res. Sci. Teach. **15** (1), 11 (1978). doi:10.1002/tea.3660150103

[55] B. Patton and J. Esswein, The development of conceptual thinking in inquiry-based physics, in *Proceedings of the National Association for Research in Science Teaching (NARST) Annual Meeting* (Baltimore, MD, 2008).

[56] V. L. Akerson, D. L. Hanson, and T. A. Cullen, The influence of guided inquiry and explicit instruction on K–6 teachers' views of nature of science, J. Sci. Teach. Educ. **18** (5), 751 (2007). doi:10.1007/s10972-007-9065-4

[57] Y. Kim, L. Bao, and O. Acar, Students' cognitive conflict and conceptual change in a Physics by Inquiry class, AIP Conf. Proc. **818**, 117 (2005). doi:10.1063/1.2177037

[58] R. S. Lindell, D. Franke, E. Peak, T. Withee, and T. Foster, Meeting the needs of our future and inservice teachers: The development and implementation of a PER-based course to teach instructional strategies in astronomy, AIP Conf. Proc. **818**, 23 (2006). doi:10.1063/1.2177014

[59] H. R. Sadaghiani, Physics By Inquiry: Addressing student learning and attitude, AIP Conf. Proc. **1064**, 191 (2008). doi:10.1063/1.3021251

[60] S. Ucar and T. Demircioglu, Changes in preservice teacher attitudes toward astronomy within a semester-long astronomy instruction and four-year-long teacher training programme, J. Sci. Educ. Tech. **20** (1), 65 (2010). doi:10.1007/s10956-010-9234-7

[61] W. K. Adams, K. K. Perkins, N. S. Podolefsky, M. Dubson, N. D. Finkelstein, and C. E. Wieman, New instrument for measuring student beliefs about physics and learning physics: The Colorado Learning Attitudes about Science Survey, Phys. Rev ST: Phys. Educ. Res. **2**, 010101 (2006). doi:10.1103/PhysRevSTPER.2.010101

[62] B. A. Lindsey, L. Hsu, H. Sadaghiani, J. W. Taylor, and K. Cummings, Positive attitudinal shifts with the Physics by Inquiry curriculum across multiple implementations, Phys. Rev. ST Phys. Educ. Res. **8**, 010102 (2012). doi:10.1103/PhysRevSTPER.8.010102

[63] M. Piburn, D. Sawada, K. Falconer, J. Turley, R. Benford, and I. Bloom, *Reformed Teaching Observation Protocol (RTOP) Reference Manual* (2000). http://physicsed.buffalostate.edu/AZTEC/RTOP/RTOP_full/about_RTOP.html, retrieved Sep. 16, 2013.

[64] D. L. Messina, *Bringing Reform Teaching to the K12 Science Classroom: The Role of Research-Based Professional Development*, Ph.D. thesis, University of Washington (2008).

[65] D. L. Messina, Learning and teaching through inquiry: Bringing change to the science classroom, in *Science and Mathematics Education Conference Proceedings* (Dublin, Ireland, 2010).

[66] K. C. Trundle, R. K. Atwood, and J. E. Christopher, Fourth-grade elementary students' conceptions of standards-based lunar concepts, Int. J. Sci. Educ. **29** (5), 595 (2007). doi:10.1002/tea.10039

[67] L. Kuppan, S. K. Munirah, S. K. Foong, and A. S. Yeung, On the attitude of secondary 1 students towards science, AIP Conf. Proc. **1263**, 118 (2009). doi:10.1063/1.3479846

[68] R. N. Steinberg, S. Cormier, and A. Fernandez, Probing student understanding of scientific thinking in the context of introductory astrophysics, Phys. Rev. ST Phys. Educ. Res. **5**, 020104 (2009). doi:10.1103/PhysRevSTPER.5.020104

[69] D. Sawada, M. D. Piburn, E. Judson, J. Turley, K. Falconer, R. Benford, and I. Bloom, Measuring reform practices in science and mathematics classrooms: The Reformed Teaching Observation Protocol, School Sci. Math. **102** (6), 245 (2002). doi:10.1111/j.1949-8594.2002.tb17883.x

[70] M. R. Blanchard, S. A. Southerland, J. W. Osborne, V. D. Sampson, L. A. Annetta, and E. M. Granger, Is inquiry possible in light of accountability?: A quantitative comparison of the relative effectiveness of guided inquiry and verification laboratory instruction, Sci. Educ. **94** (4), 577 (2010). doi:10.1002/sce.20390

[71] The impact of tutorials on student learning has been documented in many publications. See, for example, references [72] and [73].

[72] T. O'Brien Pride, S. Vokos, and L. C. McDermott, The challenge of matching learning assessments to teaching goals: An example from the work-energy and impulse-momentum theorems, Am. J. Phys. **66** (2), 147 (1998). doi:10.1119/1.18836

[73] K. Wosilait, P. R. L. Heron, P. S. Shaffer, and L. C. McDermott, Addressing student difficulties in applying a wave model to the interference and diffraction of light, Am J. Phys. **67** (S1), S5 (1999). doi:10.1119/1.19083

[74] V. K. Otero, S. J. Pollock, R. McCray, and N. D. Finkelstein, Who is responsible for preparing science teachers?, Science **313** (5786), 445 (2006). doi:10.1126/science.1129648

[75] G. Zavala, H. Alarcon, and J. Benegas, Innovative training of in-service teachers for active learning: A short teacher development course based on physics education research, J. Sci. Teach. Educ. **18** (4), 559 (2007). doi:10.1007/s10972-007-9054-7

[76] See http://depts.washington.edu/uwpeg

[77] L. C. McDermott, L. Piternick, and M. Rosenquist, Helping minority students succeed in science: I. Development of a curriculum in physics and biology, J. Coll. Sci. Teach. **9** (3), 135 (1980). http://eric.ed.gov/?id=EJ215068

[78] L. C. McDermott, L. Piternick, and M. Rosenquist, Helping minority students succeed in science: II. Implementation of a curriculum in physics and biology, J. Coll. Sci. Teach. **9** (4), 201 (1980). http://eric.ed.gov/?id=EJ220109

[79] L. C. McDermott, L. Piternick, and M. Rosenquist, Helping minority students succeed in science: III. Requirements for the operation of an academic program in physics and biology, J. Coll. Sci. Teach. **9** (5), 261 (1980). http://eric.ed.gov/?id=EJ226310

[80] Teachers also need to be able to participate in the selection process for the curriculum used in their own classrooms, in their schools, and in their school districts. They must be able to choose instructional materials that have the science correct and that represent a coherent approach to a given topic. The process of developing their own conceptual framework for a given topic helps teachers identify gaps in the treatment of concepts in texts and can guide their recommendations for textbook adoption.

[81] A more complete list of references can be found on http://depts.washington.edu/uwpeg

APPENDIX

This Appendix illustrates the instructional approach used in *Physics by Inquiry*. The exercises and experiments are drawn from Section 2 of the Electric Circuits module (A model for electric current). The excerpted materials are preceded by an overview of the entire module, the background for Section 2 that is provided by Section 1, and a brief description of Section 2.

Overview of the *Electric Circuits* module:

In the *Electric Circuits* module, students (who may be preservice and inservice teachers) perform experiments with batteries, bulbs, and wires. On the basis of their observations, they develop a model (*i.e.*, a set of assumptions and rules) that they can use to predict and account for the behavior of a variety of simple dc circuits. Many students do not have a good understanding of the process of scientific model building. This material represents a good opportunity for them to come to recognize how observations can guide informed inferences that can be tested and modified on the basis of experimental results. As they work through the module, students repeatedly encounter circuits that help them recognize limitations of their model. They are guided in extending the model to include new concepts that broaden the range of circuits for which it is applicable.

The model that students develop is based on macroscopic observations they can make. Initially they reason on the basis of bulb brightness. Eventually, ammeter and voltmeter readings are used to quantify their results. Since a primary goal is to help students recognize not just what they know but also how they know it, the issue of what is flowing (*i.e.*, electrons) is not addressed.

Background for *Section 2:* In Section 1 of the *Electric Circuits* module *(Single-bulb circuits)*, teachers examine the brightness of a bulb when it is connected to a battery in various ways. Based on their observations, they construct operational definitions for *complete circuits, conductors, and insulators.* They use these ideas to understand the construction of light bulbs, switches, and bulb holders.

Schematic symbols for wires, bulbs, batteries, and switches are introduced and students are given practice in drawing and interpreting circuit diagrams. The idea that circuit diagrams show electrical connections, *not* physical layout is stressed. It should be noted that the term *electric current* and the idea that there exists a flow through a circuit are not introduced until Section 2. In Section 1, no inferences are drawn regarding what is happening inside the circuit.

Description of *Section 2 (A model for electric current)*: Section 2 introduces two assumptions that form the basis for the model for electric current. These are motivated, in part, by observations that a complete circuit is required to light a bulb and that a wire becomes uniformly warm along its length (Experiment 2.1). These observations are consistent with the assumptions that (1) something is flowing in an electric circuit and (2) the brightness of a light bulb can be used as an indicator of the flow. The name given to the flow (after Exercise 2.3) is electric current. Discussion as to what constitutes the flow in the circuit is avoided because students have no direct evidence on which to base an answer.

Students then experiment with simple two-bulb circuits and, drawing on their assumptions, determine that the current through a battery depends on the number and arrangement of circuit elements. This result is in conflict with a belief that many students have about a battery providing the same current to all circuits. Many students are unable to differentiate among the ideas of current, energy, and power. Often they have a naive conception that something is "used up" in an electric circuit. By focusing on a quantity that is conserved (current), the student can begin to distinguish among these ideas.

Section 2. A model for electric current

In working with electric circuits, you may have noticed some regularities in the way they behave. Perhaps you have begun to form a mental picture, or model, that helps you think about what is happening in a circuit. In this section, we will begin the process of developing a scientific model for an electric circuit.

A scientific model is a set of rules that applies to a particular system that makes it possible to explain and predict the behavior of that system. We would like to build such a model for electric circuits that will enable us to predict the behavior of any circuit of batteries and bulbs. If we connect several bulbs and batteries together in a circuit, we would like to be able to predict which bulbs will light, which will be brightest, dimmest, and so forth.

The first step in building our model will be to incorporate within it features that can account for the behavior we have already observed. As we learn more about electric circuits, we shall add to our model or change it if we need to do so. At any time, we should be ready to discard any part of our model that is in conflict with our observations and that cannot be modified to be consistent with them.

Experiment 2.1

Briefly connect the terminals of a battery with a single wire until the wire feels warm. Does the wire seem to be the same temperature along its entire length or are some sections warmer than others? What might this observation suggest about what is happening in the wire at one place compared to another?

✓ Check your results with a staff member.

When a wire or a light bulb is connected across a battery, we have evidence that something is happening in the circuit. The wire becomes warm to the touch; the bulb glows. In constructing a model to account for what we observe, it is helpful to think in terms of a flow around a circuit. We can envision the flow in a continuous loop from one terminal of the battery, through the rest of the circuit, back to the other terminal of the battery, through the battery, and back around the circuit. We have found that a light bulb included in this circuit will light. We shall assume that the brightness of the bulb is an indicator of the amount of flow through the bulb.

The *assumptions* that something is flowing through the entire circuit (including the bulb) and that a light bulb can be used as an indicator of the flow are both consistent with our observations. We cannot claim, however, that we have direct evidence for either of these assumptions.

Exercise 2.2

A. In Section 1, you found that a complete circuit was necessary for a bulb to light.

Does this observation suggest that the flow in an electric circuit is one way (e.g., from battery to bulb) or round trip (e.g., from battery to bulb and back again through the battery)? Explain.

What does your answer above suggest is a major difference between the flow in an electric circuit and the flow of water in a river?

Can you tell from your observations thus far the direction of the flow through the circuit?

B. Base your answers to the following questions on the assumptions about the flow in an electric circuit.

If two identical bulbs are equally bright, what does this indicate about the electric flow through them?

If one bulb is brighter than another identical bulb, what does this indicate about the flow through the brighter bulb?

Exercise 2.3

Consider the following dispute between two students:

Student 1: *"When the bulb is lit, there is a flow from the battery to the bulb. There is also an equal flow from the bulb back to the battery."*

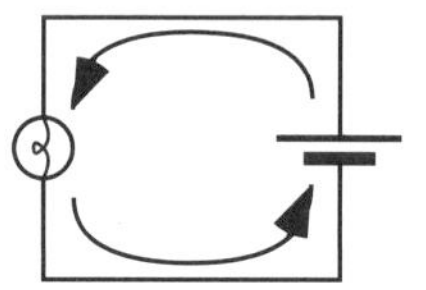

Student 2: *"The flow is only from the battery to the bulb. We know this is so, because a battery can light a bulb, but a bulb can't do anything without a battery."*

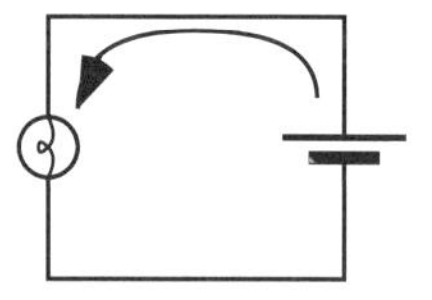

Do you agree with student 1, student 2, or neither?

✔ Explain your reasoning to a staff member.

Since we cannot see anything flowing in an electric circuit, we cannot be sure what kind of flowing process to envision. In keeping with custom, we will use the term *electric current* to refer to the flow. It is important, however, to remember that assigning the name "current" tells us NOTHING about the nature of what flows.

As we build our model step by step, we will draw only on what we can observe in the laboratory and on what we can infer from our observations. A good scientific model is as simple as possible and includes the fewest features necessary for making correct predictions. Thus, we will not include additional features for which we have no direct evidence, such as "electrons." We will find that it is not necessary to identify what flows in an electric circuit in order to predict its behavior.

Thus far we have worked with circuits that contain only one bulb. We will continue developing our model for electric current by investigating what happens in circuits with more than one bulb. Unless stated otherwise, assume all bulbs used in this module are identical. Similarly, assume that all batteries are identical.

Experiment 2.4

A. Set up a two-bulb circuit with the bulbs connected one after the other as shown.

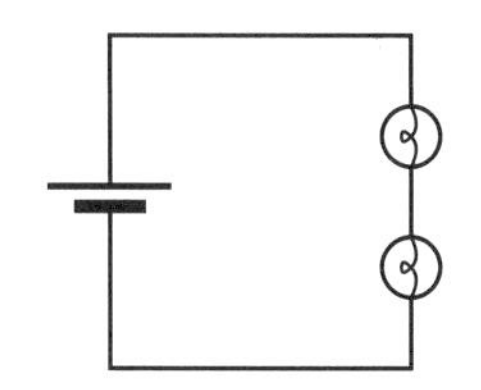

Two bulbs connected one after the other are said to be connected in *series*.

Compare the brightness of each of the bulbs with the brightness of an identical bulb in a single-bulb circuit.

Recall the assumptions we have made in developing our model for electric current. How does the current through a bulb in a single-bulb circuit compare with the current through the same bulb when it is connected in series with a second bulb? What does this imply about the current through the *battery?*

B. Compare the brightness of the two bulbs in the two-bulb series circuit with each other. What can you conclude from this observation about the amount of current through each bulb?

Pay attention to large differences you may observe, rather than minor differences that may occur if two "identical" bulbs are, in fact, not quite identical. (How can you test whether minor differences are due to manufacturing irregularities?)

C. On the basis of your observations and the reasoning you used above, respond to the following questions:

Is current "used up" in the first bulb, or is the amount of the flow the same through both bulbs?

Do you think the order of the bulbs in this circuit might make a difference? Verify your answer by switching the two bulbs.

EC 12 *§2 A model for electric current*

Can you tell the direction of the flow through the circuit?

How does the *amount* of the flow through the battery in a single-bulb circuit compare with the flow through the battery in a circuit with two bulbs connected in series?

Exercise 2.5

Consider the following dispute between two students.

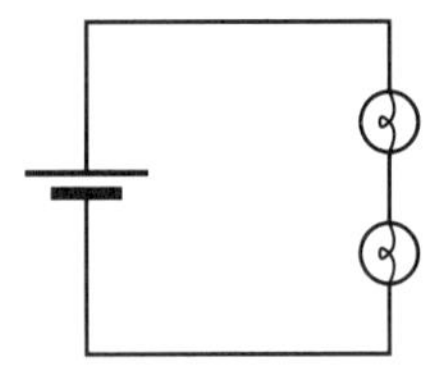

Student 1: *"In this circuit, the flow is from the battery to the first bulb, where some of the current gets used up. Then the rest flows to the second bulb, where all the remaining current gets used up."*

Student 2: *"We know that the current flows back through the battery since we know that we need a complete circuit in order for bulbs to light. If current were used up, there wouldn't need to be a path back to the battery. Furthermore, the bulbs are equally bright so both must have the same amount of current through them."*

Characterize the model of electric current each student is using. Do you agree with student 1, student 2, or neither?

✓ Explain your reasoning for Experiment 2.4 and Exercise 2.5 to a staff member.

In Experiment 2.4, we investigated the behavior of circuits in which two bulbs are connected one after the other in series. We now consider circuits in which the bulbs are connected in a different way.

 McDermott & P.E.G., U.Wash./*Physics by Inquiry*

Experiment 2.6

A. Set up a two-bulb circuit with two identical bulbs so that their terminals are attached together as shown.

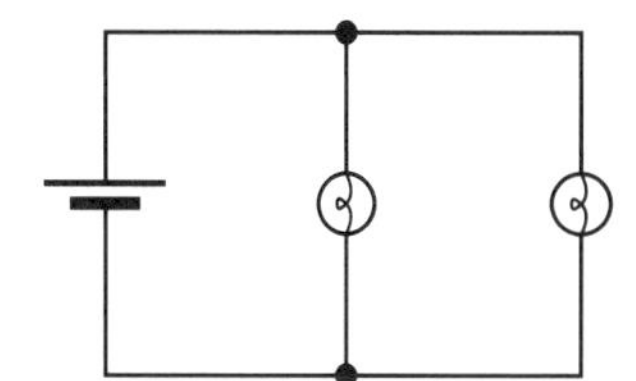

Two bulbs with their terminals attached together in this way are said to be connected in *parallel.*

Compare the brightness of each of the bulbs with the brightness of an identical bulb in a single-bulb circuit.

> Recall the assumptions we have made in developing our model for electric current. How does the current through a bulb in a single-bulb circuit compare with the current through the same bulb when it is connected in parallel with a second bulb?

B. Compare the brightness of the two bulbs in the two-bulb parallel circuit with each other. What can you conclude from this observation about the amount of current through each bulb?

Concentrate only on any large differences you may observe, rather than the minor differences that may occur if two "identical" bulbs are, in fact, not quite identical.

C. On the basis of your observations and the reasoning you used above, respond to the following question:

> Do you think it is the physical layout of the circuit that makes a difference or the electrical connections? You can investigate this question by comparing what happens:
>
> (1) when the two bulbs are both on the same side of the battery and when they are on different sides.
>
> (2) when each bulb has separate leads to the battery and when the terminals of the bulbs are connected together and then connected to the battery.

D. Describe the flow around the entire circuit for the two-bulb parallel circuit. What do your observations of bulb brightness suggest about the way the current through the battery divides and recombines at the junctions where the circuit splits into the two parallel branches?

E. What can you infer about the relative amounts of current through the battery in a single-bulb circuit and in a circuit in which two identical bulbs are connected in parallel across the battery?

F. Does the amount of current through a battery appear to remain constant or to depend on the number of bulbs in a circuit and how they are connected?

Exercise 2.7

Consider the following dispute between two students.

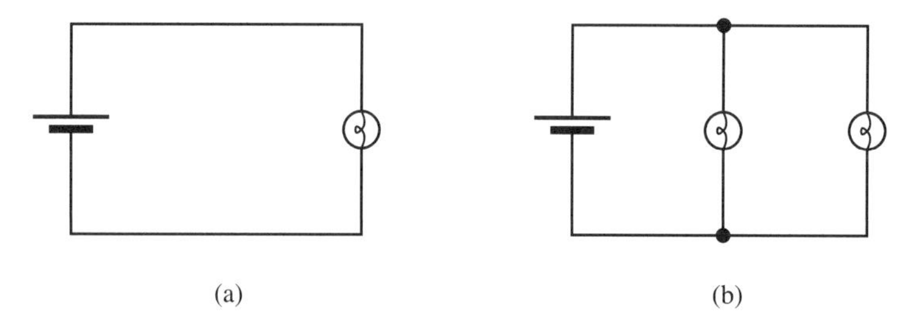

(a) (b)

Student 1: *"The current through the battery in each circuit is the same. In circuit b the current from the battery is divided between the two bulbs—so each bulb has half the current through it that the bulb in circuit a has through it."*

Student 2: *"We know the current through each of the bulbs in circuit b is the same as through the bulb in circuit a. That's because the bulbs are all about the same brightness —and bulbs that are equally bright have the same current through them. So the flow through the battery in circuit b is more than that through the battery in circuit a."*

Do you agree with student 1 or student 2?

✓ For Experiment 2.6 and Exercise 2.7, explain your reasoning to a staff member.

Strengthening a physics teacher preparation program using key findings from the SPIN-UP report

Bruce Palmquist and Michael Jackson
Department of Physics, Central Washington University, Ellensburg, Washington 98926

For the past decade, the Department of Physics at Central Washington University has implemented numerous recommendations from the Strategic Programs for Innovations in Undergraduate Physics (SPIN-UP) report to strengthen its physics program along with its physics teacher education program. The four changes that have had the most positive impacts on our teacher education program are the following: (1) The physics department partnered with the mathematics department to develop a dual-degree program in physics and mathematics education, and with the College of Education and Professional Studies to streamline the college's Professional Education Program. This improves time to degree and gives students teaching certification in two high-need areas. (2) All first-year courses were converted from separate large lecture sections having loosely coupled small labs to 40-to-50-student lecture/lab classes for which mini-lectures, activities, and collaborative problem-solving sessions are utilized in most class meetings. This format models effective physics teaching practices for preservice teachers and provides a venue for meaningful Learning Assistant experiences. (3) All majors, including preservice teachers, are mentored in a research project in which they present the results at the campus research symposium or the state physics teaching conference. These projects give preservice teachers authentic research experiences so they can model the nature of science and the discovery process to their students. (4) Finally, the Learning Assistant courses were developed to facilitate student learning in first-year physics courses, provide preservice teachers and other physics majors with an early teaching experience, and provide training in two important methods for monitoring student learning: basic action research and formative assessment. These changes have been essential features contributing to the more than doubling of the number of physics degrees awarded and the number of physics teacher candidates.

I. INTRODUCTION

During the 2001–2002 academic year, the National Task Force on Undergraduate Physics visited thriving physics departments to answer the question: Why, in the 1990s, did some departments increase the number of bachelor's degrees awarded in physics during a time of declining enrollment in physics across the country? [1,2] This question was answered in a report called Strategic Programs for Innovations in Undergraduate Physics (SPIN-UP). While the goal of the SPIN-UP report was to increase the number of physics degrees across the board, it offered a lot of information for improving physics teacher education programs as well. The main benefit to physics teacher education programs is that more physics majors means a larger pool of potential preservice teachers! But along with advice for simply increasing the number of majors, the SPIN-UP report offers specific advice that is directly applicable to improving teacher education programs. Based on site visits to 21 physics departments housed at small liberal arts colleges, comprehensive state universities, and flagship research universities, the authors of the SPIN-UP report developed a list of characteristics that describes why these departments were successful. In general, these characteristics fall into three main areas: leadership; a supportive, encouraging, and challenging environment; and experimentation and evaluation.

Institutionalizing the findings of the SPIN-UP report was a major step in strengthening the physics department and the physics teaching program at Central Washington University (CWU). Many of the report's recommendations were adopted by the department. The four that we believe have had the greatest impact in strengthening our physics teaching program are:

(1) flexible degree programs;

(2) an emphasis on undergraduate research;

(3) ensuring that most faculty are aware of the findings of physics education research and encouraging physics education research within the department; and

(4) ensuring that the department pays special attention to the introductory physics courses, and that any course revisions be based on findings from physics education research.

After describing our university context, we will explain how each of these findings improved our physics teacher education program and list some of the lessons we learned from implementing the findings.

II. BACKGROUND

Central Washington University is a four-year, public university located in rural Washington, with a student population of about 11,000. The university largely serves state residents and graduates about 2,500 students per year. More than 75% of our students have either a subsidized or unsubsidized federal loan, while state funding makes up just 16% of the university's total budget. The physics department has 5.5 full-time-equivalent tenured and tenure-track faculty members. Courses are taught on the quarter system, with three academic-year quarters and a summer session. The summer session is supported solely by tuition.

A. Physics teaching program in the campus context

One of the key ingredients of the success of the physics teacher education program at CWU is the support it receives from numerous campus organizations, most importantly the science education department. The science education program started in the 1960s with biology and chemistry faculty members who spent about half of their teaching load educating preservice and inservice K-12 science teachers. Over the subsequent 40 years, the program added faculty in physics and Earth science, and eventually became a department. Currently, the science education department works collaboratively with other science departments, the mathematics department, and the College of Education and Professional Studies (COE) to deliver teaching programs in biology, chemistry, Earth science, general science, middle level mathematics, middle-level science, and physics. The science education department is in the College of the Sciences, with all but one faculty member holding a dual appointment in the science education department and a science department. Science education faculty members, who are experts in their content areas and in how to best teach the content, serve their colleagues in the science departments by jointly developing curriculum, providing feedback on curricular issues, and leading department assessment initiatives. The first author has a joint appointment in the physics department and the science education department.

The CWU physics department offers two majors (Bachelor of Science and Bachelor of Arts) and two minors (physics and astronomy). This includes the option for students to participate in a dual-degree physics-engineering program or a dual-degree physics-mathematics education program. All physics majors are required to have a research or senior design experience. Students are also required to present at a venue outside the department. Most students present at SOURCE, the university-wide Symposium on University Research and Creative Expression. In addition, students are encouraged to submit their research for presentation at professional meetings.

From the CWU physics department's formation in the 1960s, enrollment has typically followed the national trend, which was increasing enrollments until the early 1970s and a steady decline since then [1]. However, in 2007, a new department chair was hired from a thriving physics program as identified by the National Task Force on Undergraduate Physics in its SPIN-UP report [2]. Over the next six years, the physics program grew from fewer than 20 to greater than 80 majors and from a three-year average of 4.5 graduates to over 10 graduates, about twice the national average of physics graduates from four-year comprehensive institutions [3]. The number of qualified physics teacher candidates has increased from about one every two years to two per year. In 2012, the department received a Physics Teacher Education Coalition (PhysTEC) grant to develop a Learning Assistant (LA) program and a mathematics-physics teaching dual-degree track leading to dual certification [4].

Along with fulfilling the recommendations of the SPIN-UP report, all programmatic changes in the physics department, including those influencing preservice teacher education, are done to enhance students' abilities to meet the department's six student learning outcomes. In brief, these outcomes are proficiency in physics content knowledge, technical skills, critical thinking, communication skills, civic engagement, and lifelong learning. All of the programmatic changes listed in the abstract address increasing student content knowledge. Additionally, reforming the first-year courses also improves critical thinking skills, and undergraduate research also increases a student's technical and communication skills. The dual-degree physics-mathematics education program also helps increase civic engagement, while the LA program increases communication skills and promotes lifelong learning. The department evaluates a subset of these student learning outcomes annually.

B. The role of broad collaborations in improving teacher programs

While it is important for science education faculty to collaborate with faculty in their science content area, it is equally important for them to work with COE faculty. As with most universities, CWU students seeking teacher certification take content courses and professional education courses. In an effort to shorten time to degree and to better meet new Washington state teacher preparation standards, starting in 2010, science education faculty worked with COE faculty to develop a competency-based program. This means that science teacher candidates have to meet specific competencies in order to be certified to teach. Examples of these competencies include "Develop and provide a rationale for a classroom management plan appropriate for grade level and content area," and "Use the National Education Technology Standards for Teachers to inform instruction and other educational decisions" [5]. The science education department has incorporated many of these competencies into the required science teaching methods courses, which subsequently eliminated several courses in the Professional Education Program. Additionally, physics and other science teacher candidates do not have to take generic courses in classroom management or educational technology, because these skills are taught in science teaching methods courses.

The elimination of courses duplicating specific skill development has been crucial toward shortening time to degree.

Aside from the science education department and the COE, there are other organizations at CWU that support physics teacher education. The Center for Excellence in Science and Mathematics Education (CESME) supports teacher education on campus and throughout central Washington. CESME was awarded a National Science Foundation Robert Noyce Teacher Scholarship grant to provide scholarships and internship opportunities for preservice teachers. Faculty who are members of CESME can apply for reassigned time to develop innovative curricula. For example, the first author received a CESME grant to plan PHYS 106, the first class in the department to use an integrated lecture/lab format. Other CWU programs such as the Supplemental Instruction (SI) and Tutoring programs provide physics teaching candidates with an opportunity to explore their interests in teaching and facilitating learning.

Nationwide, approximately four out of 10 teachers take at least some of their math and science courses at a community college [6]. At CWU, nearly 50% of physics majors, including preservice teachers, are transfer students. Therefore, no reform effort will be successful without collaboration with community colleges. The CWU science education department has had a partnership with Green River Community College in Auburn, Washington, since 2002. The partnership started with an elementary education program that evolved to a middle-level science teaching program, and it has blossomed into a partnership between the two physics departments. This collaboration includes the joint review of all major physics education initiatives at CWU, along with assessment of the impact of their implementation. The Washington section of the American Association of Physics Teachers (AAPT) has facilitated additional collaborative efforts with two-year institutions, an organization in which members of both departments have routinely held leadership positions.

III. INFLUENCE OF THE SPIN-UP REPORT ON PHYSICS TEACHER EDUCATION AT CWU

As noted in the introduction, there were four recommendations from the SPIN-UP report that we believe have had a significant impact on improving our physics teacher education program. These are:

(1) flexible degree programs;

(2) an emphasis on undergraduate research;

(3) ensuring that most faculty are aware of the findings of physics education research and encouraging physics education research within the department; and

(4) ensuring that the department pays special attention to the introductory physics courses, and that any course revisions be based on findings from physics education research.

A recognition of the importance of physics education research (PER), and its subsequent findings over the past several decades, played a critical role in our implementation of the SPIN-UP recommendations. The successful revision of the introductory physics sequence could not have been achieved without knowledge of PER findings. As such, the last two recommendations have been combined into a single point of discussion.

A. Flexible degree programs

Very few high schools in central and eastern Washington offer enough physics courses to hire a full-time physics teacher. To be competitive for teaching positions, students must be certified to teach more than one subject. Pedagogically, there is a natural fit between physics and mathematics. Additionally, both mathematics and physics have been on the U.S. Department of Education Office of Postsecondary Education list of teacher shortage areas nearly every year and in nearly every state since 1990 [7].

Until recently at CWU, completing both programs required students to take two majors and complete the professional education sequence with no significant course overlap. This typically required five or more years for students to complete. In 2010, the mathematics department and science education departments started working with the COE to address certain professional teaching competencies in content courses within the major. This reduced the number of professional education courses in these programs by up to five courses. The physics department subsequently developed a dual-degree memorandum of understanding with the mathematics department. Patterned after the dual-degree physics-engineering programs that are successful at CWU and many other schools, each department agreed to a set of courses satisfying the bachelor's degree in mathematics teaching and the bachelor's degree in physics. Due to the collaboration of the COE and the mathematics and physics departments, students in the dual-degree physics-mathematics education program save nine courses compared to what would be required if the students took each program separately; this amounts to a reduction of two quarters. For example, a student prepared to start calculus in his or her first quarter at CWU could complete both degrees in four years and one quarter. An added advantage is that both departments are permitted to count the students as graduates of their respective programs.

Beyond the dual-degree option, there are two other pathways in the College of the Sciences that allow our students to fit the definition of a PhysTEC Secondary Graduate, meaning they have, at least, the equivalent of a physics minor [8]:

(1) Receiving a degree (either Bachelor of Arts or Bachelor of Science) in physics and completing the Professional Education Program.

(2) Minoring in physics, majoring in another science or mathematics degree program, and completing the Professional Education Program.

Finally, the COE also manages an alternative certification program that allows individuals with bachelor's degrees to obtain their teacher certification in 15 months.

B. Emphasis on undergraduate research

The physics department was the first department in the college to incorporate an undergraduate research project into the curriculum for all students, whether they earn the Bachelor of Arts or Bachelor of Science degree. Although potentially costly in terms of faculty time, financial resources, and infrastructure, the physics department determined that the benefits to students and the department outweighed the costs [9–11]. This initiative was integrated into the physics curriculum nearly 15 years ago, when all physics majors were required to take at least one two-credit course in undergraduate research in order to perform a research project. To strengthen the justification for students to have meaningful, faculty-mentored research experiences, the physics department took advantage of an administrative initiative requiring all departments to have, document, and assess student learning outcomes for all of their degree programs. Starting in 2009, all graduating physics majors have submitted to the department a portfolio containing a representative sample of their work as it relates to each of the department's six outcomes for a CWU physics graduate. Undergraduate research can be used to meet up to half of these outcomes. Along with technical and intellectual skill development, undergraduate research can be used to demonstrate effective communication skills. One of the portfolio requirements is for students to present the results of their research projects at venues outside the department. Potential venues include the university research symposium or the Washington section meeting of AAPT [12], the latter being a particularly meaningful venue for preservice teachers.

CWU physics undergraduates can choose to conduct research in a physics content area such as optics, complete an engineering-related design project, or complete a pedagogical project, with the latter proving to be particularly popular among students with an interest in teaching. Preservice teachers benefit from undergraduate research for all of the same reasons that students in any major do: increased content knowledge and experience with the methods of conducting a scientific investigation. There are also additional benefits to preservice teachers and their future students. Teachers need to be familiar with a variety of science teaching resources in their community. Inservice and preservice teachers stress the importance of making connections with working scientists as a primary benefit of participating in scientific research [13,14]. Additionally, if teachers are expected to teach science practices, as they are called on to do in the newly developed Next Generation Science Standards (NGSS), they need experience with such practices. Doing physics pedagogical projects as an undergraduate research project allows preservice teachers to use and perfect NGSS science practices such as asking questions, developing and using models, planning and carrying out investigations, analyzing and interpreting data, constructing explanations, engaging in argument from evidence, and communicating information [15]. Pedagogical projects have the added benefit of being immediately applicable to the high school classroom that the preservice teachers will soon enter. The most common pedagogical projects are either building, testing, and assessing physics demonstrations, similar to the Build-it, Leave-it, Teach-it program at SUNY Geneseo [16], or developing, teaching, and assessing online lab activities such as the PhET activities at the University of Colorado Boulder [17].

Here are summaries of two recent projects. One preservice teacher built an updated version of the "monkey and hunter" demonstration that shows the independence of vertical and horizontal motion. Our old demonstration was dangerous, unsanitary, and unreliable, because it required the user to blow into a metal tube that was hooked up to a current source. The authors refused to use it. The preservice teacher made a spring-loaded projectile launcher with an electric eye switch at the end of the muzzle. He also developed, and administered to a university introductory physics class, a pre- and post-test instrument to evaluate if the demonstration improved student understanding.

Another preservice teacher knew that the first author uses PhET activities in his classroom, and asked whether students learn as much doing these activities as they do from doing similar hands-on lab activities. The preservice teacher subsequently developed a simple projectile motion lab that was similar to the PhET projectile motion lab [18]. He administered the labs to a university introductory physics class and analyzed a basic pre-test and post-test to assess the students' understanding. Both of these preservice teachers presented posters summarizing their work at SOURCE, the campus research symposium.

C. Physics education research and the introductory physics courses

To better address the SPIN-UP recommendation to explicitly improve the introductory-level physics courses, inquiry-based teaching and other progressive pedagogies were introduced as the department transitioned from the traditional format of separate lecture and lab courses to an integrated lecture/lab model. This iterative process began when the first author was asked by the second author to develop an inquiry-based course for preservice K-8 teachers (PHYS 106) similar to those found in other physics departments [2]. The first author was able to secure a small university grant to develop the curriculum for that course. PHYS 106 proved to be an effective test case for evaluating this type of course reform and transformation on a "friendly" audience, namely preservice K-8 teachers who are open to seeing different teaching techniques modeled. The success of this course, as measured by student satisfaction on course evaluations, convinced the department to begin moving the integrated approach into the introductory physics courses, beginning with those offered for scientists and engineers.

The primary concern raised by students in the interactive courses when initially implemented by the first author was how long it took to get help from the instructor. Because of this, the authors sought and acquired PhysTEC [4] funding to develop a Learning Assistant program [19] to train physics majors to facilitate learning in an interactive classroom setting (described in detail in the following section). The timing was ideal for two reasons. First, department faculty could use the Force Concept Inventory (FCI) [20] data they had been gathering for years to support their argument that interactive courses helped students learn physics better, while simultaneously demonstrating that more instructional assistance was needed. Second, the physics department was starting the process of designing its teaching and research spaces for a new building. Such data were important in convincing the administration and architects to design spaces that facilitated learning using interactive pedagogy. The PhysTEC grant provided assistance in the short term. The combination of supporting data and department initiative facilitated a financial commitment from the administration to better support the pedagogical goals of the department in the long term.

1. What does the introductory physics sequence look like?

As the introductory physics course sequence is critical to recruiting students to and retaining students in a physics degree program, the introductory sequence is also the best place to introduce inquiry-based instructional practices to students. We focus on two types of inquiry in our classes. During short class discussions and problem-solving episodes, we use structured inquiry, in which students determine an answer to a question or problem posed by the instructor using methods set forth by the instructor. An example of this would be small groups of students collaboratively solving a problem organized around the principle of energy conservation. For lab activities, we often use guided inquiry, in which the instructor poses the question but students have the freedom to develop their own hypotheses, procedures, and conclusions [21,22]. Guided inquiry is similar to what is used in *Physics by Inquiry* and *Tutorials in Introductory Physics*, two well-known physics curricula [23].

Neither structured nor guided inquiry is lecture based. In fact, lectures of more than 15 minutes have largely been removed from our introductory physics course sequence. From 2010 to 2012, the department transitioned from the traditional format of separate 80-student lecture-based courses and 20-student lab sections to an integrated lecture/lab model with about 40 to 50 students in a section. Because of positive learning gains and student evaluations in the inquiry-based course for preservice K-8 teachers, the first author, a senior faculty member, was encouraged to expand the integrated approach the next time he taught PHYS 181, the first course in the introductory calculus-based physics sequence, which focuses on topics related to mechanics. But before this method would become the department norm for all introductory physics courses, we reviewed the student performance on standardized tests.

The department had gathered assessment data for the PHYS 181 course using the Force Concept Inventory [20]. These data were collected for several years and are summarized in Table I. In the years before adopting the integrated lecture/lab model, physics faculty introduced a limited number of inquiry-based activities into their courses, such as those described in Table II. However, the types of activities were often limited by the way the independent lecture and lab sections were scheduled as well as by the classroom space in which each were taught. The average FCI gain for PHYS 181, as taught in this format, was 0.40, which is in the low-to-medium range on the Hake curve [24]. In the interactive courses, which permitted the full range of activities as described in Table II, the average gain increased to 0.52.

TABLE I. Normalized fractional gains for assessment exams taken by CWU students, including physics teacher candidates, in the calculus-based introductory physics sequence.

	Pre-reform fractional gain (N)	Post-reform fractional gain (N)
PHYS 181 (FCI)	0.40 (118)	0.52 (169)
PHYS 183 (CSEM)	0.32 (31)	0.39 (77)

This distribution is comparable with the meta-analysis by Richard Hake [26,27]. The average gain for interactive courses is in about the 70th percentile of the "interactive-engagement" courses listed in Hake's meta-analysis. In addition to the FCI, the department started administering the Conceptual Survey of Electricity and Magnetism (CSEM) [28] in 2009; the CSEM assesses students' understanding of concepts in electricity and magnetism, topics covered in the third course of the calculus-based introductory physics sequence (PHYS 183). Fractional gains improved slightly during the first year of offering the introductory electricity and magnetism curriculum in the integrated format. Additionally, a variety of instructors have taught the 181 and 183 courses both before and after the reforms. Therefore, the gains are independent of the instructor.

What does an interactive class look like? Classes meet four days a week for 80 minutes each day. The classroom has 11 tables that seat four to five students each. During a typical class period, students engage in a variety of inquiry activities. Most of the "inquiry" in our courses on a typical day is either structured inquiry, wherein the instructor provides the questions and sets the parameters for addressing the questions, or guided inquiry, wherein the instructor provides the questions but the students develop their own methodology in formulating their answers. Table II gives a sample schedule for two PHYS 181 class meetings showing the general range of activities.

2. Benefits to preservice physics teachers from reformed introductory courses

While the interactive teaching method benefits all students [26,27], it has additional benefits for preservice physics teachers. The primary additional benefit is that the way people teach a subject is strongly influenced by the way they were taught that subject [33,34]. According to a landmark study by Lortie, teacher actions in the classroom are influenced by 13,000 hours of the "apprenticeship of observation" based on their own education [35]. If preservice physics teachers see their introductory physics courses taught as lecture courses with separate labs that are only loosely aligned with the lectures, they are strongly influenced to teach that way. After all, this introductory course sequence is the one that most closely models what the preservice teachers will be teaching at the high school level. But if students see introductory physics as a highly participatory experience in which concepts are immediately reinforced by interactive demonstrations, hypothesis-testing labs, and collaborative problem solving, the students are engaging in physics as a physicist would. Thus, the integrated teaching approach provides an effective model for preservice physics teachers to follow. Additionally, this method of teaching provides opportunities for Learning Assistants, nearly half of whom are preservice physics teachers, to participate in teaching the class.

Over the next few years, the science curriculum in nearly every state will be revised to better address the newly developed science standards called the Next Generation Science Standards [15]. Each standard in the NGSS is organized in a way that combines the three dimensions of practices, crosscutting content, and core disciplinary ideas. Practices describe the behaviors that scientists engage in. Crosscutting concepts are the unifying themes across all sciences, such as patterns and similarity or stability and change. Disciplinary

TABLE II. A typical set of activities for the calculus-based introductory physics course, PHYS 181.

Class meeting	Activities
Monday	This day starts with an entry task wherein each base group (the table where students first sit down) must come to a consensus about an interpretation of two position-versus-time graphs. The entry task is a way to immediately have students talk about the material with each other (and the LA, if needed). It also gives students a physics-related task while the instructor attends to any last-minute administrative issues. Next, the instructor goes over how to sketch position-, velocity-, and acceleration-versus-time graphs based on a description of motion. Then, each group is given one of the three graphs or a description of motion and must come up with the other two, similar to those exercises found in [24,25,29]. As small groups of students work on this, the instructor and the LA listen to group conversations and provide input as needed. Ideally the instructor and the LA spend more time listening to conversations than providing feedback. Often, we have let students try something we know to be incorrect before giving feedback. Next, the instructor randomly selects table numbers and each group shares their results and demonstrates the motion in front of a motion sensor, to verify their results. Finally, the instructor assigns a 1-D motion problem for students to work on with their neighbors. After a few minutes of students working together, the instructor elicits how to set up the problem.
Tuesday	This day starts with an entry task in which students determine displacement from three straightforward velocity-versus-time graphs. Next, the instructor, with student input, finishes solving the problem from the previous class. Then, the students get into their three- or four-person problem-solving cooperative groups [30,31] to solve a context-rich problem [32]. The instructor formed these groups, which are heterogeneous by ability, earlier in the quarter. When every group is done with the problem or after 30 minutes, whichever comes first, students go back to their base groups and develop a hypothesis for the open-ended lab about acceleration that they will do the following day.

core ideas are the key concepts in the field. Preservice teachers will be expected to teach lessons with standards that incorporate these three dimensions, and will need effective models for how this is done. The two lessons briefly described in Table II provide an example. Students in a CWU introductory physics course regularly engage in physics practices (collaborative problem solving, formulating and testing hypotheses) and incorporate crosscutting concepts (sketching graphs based on descriptions of motion) while learning core physics concepts.

The interactive teaching method provides a model to help preservice physics teachers meet at least three of the National Science Teachers Association's Preservice Science Standards [36]. Standard 1 states that "effective teachers of science understand and articulate the knowledge and practices of contemporary science." Standard 2 states that "preservice teachers use scientific inquiry to develop this [scientific] knowledge for all students." Standard 3 states that "candidates design and select learning activities…to achieve those [science] goals." Thus introductory physics students at CWU are immersed in an environment in which they use inquiry to develop their science knowledge, while preservice teachers in these classes have the additional benefit of seeing how a variety of activities can be combined to help students learn physics.

3. Challenges of reforming introductory courses

Significant changes in pedagogy necessitate changes in classroom design in order to implement that pedagogy. We could not teach an integrated lecture/lab course in a room designed for lecture. When another department in our building vacated its 25-foot-by-53-foot computer lab, we received funding from the university for a minor remodel that converted this space into a modified SCALE-UP classroom [37]. Getting the optimal table configuration was more difficult than first envisioned. Round tables are used in many SCALE-UP classrooms [38], but our room was too long and skinny to optimize both seating capacity and maneuverability with round tables. The first quarter that the class was taught in the room, existing rectangular tables were used. Finally, the authors experimented with trapezoid-shaped tables. Two small trapezoid-shaped tables placed with their short sides together in a butterfly configuration worked best, and is the current table configuration. Depending on whether mini-lectures or collaborative activities are being performed, four students could all face forward, face each other, or work around the table on a lab activity. In fact, when the architects designing our new science building evaluated our table layout, they were unable to improve upon the design.

Another concern with revising our teaching approach was the need for more faculty. The college dean was concerned that converting to 40-student integrated courses would cost significantly more than the traditional arrangement of 80+ student lectures with separate lab sections. At CWU, this conversion did not increase the need for more faculty and, in some cases there even was a cost savings.

IV. LEARNING ASSISTANT PROGRAM

The recommendation of the SPIN-UP report, along with the findings of physics education research, led us to reform our introductory physics courses. This involved making our courses more inquiry based [21–23]. This method works best when students are given adequate feedback on their ideas as they emerge. In a small class, the instructor can typically provide that feedback. But with enrollments of 40 to 50 students, class time was initially used inefficiently, because groups would often reach a point at which they didn't understand a concept or instruction and simply stop working. We believed the best way to address this issue was to implement a research-based Learning Assistant program following the University of Colorado [19] and Seattle Pacific University [39] models.

A. What our LAs do

LAs in our courses assist students with conceptual in-class questions, dialog with students as they solve content-rich problems in collaborative groups, answer content and procedural questions, and occasionally organize study sessions outside of class time. The number of LAs in a given class period or working with a specific instructor varies from quarter to quarter and depends on the schedules of the available LAs. There would never be more than two LAs in a typical class period, and an individual instructor would never have more than four LAs for a single course with multiple sections. LAs work about four hours per week at minimum wage, with most of our PhysTEC funding going toward paying LAs. In the seven quarters we have had LAs, as of the writing of this paper, the number of LAs per quarter has ranged from four to nine, with an average of seven.

Our LA preparation curriculum consists of two two-credit courses in which LAs work in an introductory physics or astronomy classroom for a few hours a week and discuss teaching-related issues with other LAs for two hours a week. These two courses, along with the undergraduate research project, also provide a progressive introduction to pedagogical research. Each course has a capstone assignment wherein students must identify a classroom learning issue and assist in developing a solution to resolve the issue.

B. The LA courses

PHYS 292, Exploring Physics Teaching I, introduces LAs to the basic skills of facilitating learning and is taken solely by first-quarter LAs. During a typical PHYS 292 class meeting, the LAs will discuss issues such as participating in dialogue with students, active listening, communication skills, and formative assessment. Formative assessment is the process by which teachers gather information about what students are learning and immediately use that information to improve student learning [10].

In most meetings, the LAs will watch and discuss a video from the Video Resource for Learning Assistant Development [41]. These videos provide effective examples of how LAs react to real-world teaching situations. The capstone assignment for the course is a case study written by the LAs about an interaction that LA had with a group of students. The LA summarizes the interaction, including the dialogue, and describes how he or she helped resolve the issue. The case study is essentially a written version of an LA training video using situations from a CWU classroom. It also introduces preservice teachers to action research, which is a way for teachers to gather and interpret data about student learning [42,43]. Action research "seeks to bring together action and reflection, theory and practice, in participation with others, in the pursuit of practical solutions to issues of pressing concern to people" [44]. More and more, teachers are being asked to justify their instructional decisions with data [45]. This assignment provides preservice teachers with basic training and an early opportunity to summarize and reflect on a real classroom situation in a safe and familiar setting. In addition to the case study, course assignments include written reflections on all assigned articles, a one-page "tip sheet" of helpful hints for LAs, and brief reflections of at least six teaching-learning interactions in the class the LAs are assisting in.

PHYS 392, Exploring Physics Teaching II, is taken after completing the first LA course and requires LAs to explore pedagogical issues in greater depth. The two LA courses meet concurrently, with experienced LAs acting as informal mentors to the new LAs. Experienced LAs in the second LA course are expected to take more initiative by writing and teaching an actual lesson plan. They also do a more involved action research project. Similar to the first LA course, LAs in the second course are required to observe and diagnose a learning issue. But instead of just summarizing the issue with short dialogues and reporting on their responses, they need to develop formal means of remediating the learning issue, create assessment tools to determine if the remediation was successful, and evaluate the effects of the interventions. If a student wanted to use it as such, he or she could fully develop and present a PHYS 392 action research project as his or her undergraduate research project.

C. Benefits of the LA program

In planning the PHYS 292 course, the curriculum was aligned with two other courses on campus. First, the mathematics department agreed to allow the physics-mathematics education dual-degree students to substitute PHYS 292 for their entry into the mathematics education major, provided that an assignment similar to their tutoring reflection assignment was included. Second, all students who work in the CWU tutoring center, called the Learning Commons, must pass a tutor training course. The LA training curriculum [19,39] has significant overlap with the International Tutor Training Program Certification guidelines followed by our university [46]. We worked with the tutoring coordinator to align our curriculum with the campus tutoring course. As a result, upon successful completion of PHYS 292, LAs are certified tutors and receive a certificate from the College Reading and Learning Association.

The LA program has greatly benefited the department, department faculty, majors, and preservice teachers by providing assistance in first-year physics courses. The department has used the success of the LA program to more than double the student salary budget from the college dean; this increased funding is set to begin when the PhysTEC grant is completed. Faculty members receive help in the classroom from trained and highly capable peer mentors, many of whom bring new teaching ideas to the course instructor. The LA program also exposes students who may have otherwise never considered a career in teaching physics to the possibility of a teaching career. Some of these individuals may subsequently pursue a career in education later in life or at an institution of higher education. Additionally, the skills students gain from participating in the LA program are transferrable to all careers, not just to teaching. Majors have the opportunity to revisit the content they first learned in the introductory class. Many LAs comment on how they understand the material better after helping others learn it. It also helps them hone their ability to communicate science in both technical and nontechnical ways. Finally, preservice teachers receive an early teaching experience in which they participate in action research projects. Another potential benefit, which has not been fully realized yet at the writing of this paper, is that the LA program provides a means to recruit content majors to become teacher candidates. Of the 32 students who have been LAs during the two years of the program, five were already physics teacher candidates. Two students became interested in high school teaching after their LA experiences, and one went on to be awarded a Noyce Scholarship.

V. LESSONS LEARNED

The recommendations provided in the SPIN-UP report were critical in guiding our department's curricular reform and physics teacher education program. As such, we believe that many of these recommendations can also be appropriately customized and adopted by other physics teacher education programs. The following 10 lessons highlight the actions we believe have been the most beneficial to our program.

A. Administrative lessons learned

(1) Develop strong partnerships with the college of education, and, especially, with other science departments. It may be possible, as we have done, to work with the college of education to address some of the professional education competencies in physics education program courses. This may shorten your program's time to degree for students pursuing teacher certification. Such partnerships with other science departments may be especially important if these programs have relatively small numbers of preservice teachers. Small programs working together will have more influence on the professional education curriculum.

(2) If appropriate to your setting, work closely with community colleges when you revise your programs. Developing programs that do not articulate well with community college curricula may eliminate a significant number of potential students. For example, requiring a first-year physics course that community colleges do not offer will place community college students at a disadvantage if they transfer to your four-year school.

(3) Develop a variety of pathways to physics teacher certification, including a dual-degree program, if appropriate for your institution [47]. In our service area, there are very few jobs for those certified to teach only physics. Both mathematics and chemistry certification make good dual-degree partners with a physics certification. Additionally, there are typically more mathematics education majors than in any of the sciences. If just 10% of them become dual-degree students, this would greatly increase the size of a typical physics education program. While CWU didn't pursue this route initially, a physics-chemistry teaching program may have proven an even more effective dual-degree partner to start with, because of the close alignment between the physics and chemistry curricula.

B. Curricular lessons learned

(1) Before overhauling your entire introductory physics curriculum, make changes in lower-stakes or lower-enrollment courses that meet for a single term. The first author was able to evaluate the interactive lecture/lab arrangement in a one-quarter physics course designed for preservice K-8 teachers. The course was not a prerequisite for any other course, and the students were a "friendly" audience due to their interest in learning effective teaching techniques along with the physics content.

(2) When reforming your courses, assign senior faculty who have experience and an interest in pedagogical reform. Often the professional standing of these individuals would be unaffected by student evaluations. Trying new teaching and learning techniques with science and engineering majors may lead to lower course evaluations because these techniques often don't meet student expectations for what a college science course should look like. Hence, new faculty members may be uncomfortable taking a risk by trying new teaching techniques, and they may lack experience in dealing with some of the complications that may arise.

(3) Offer pedagogical research as an option for undergraduate research projects. As mentioned, CWU physics students are required to perform research for their degrees. Accommodating the recent increase in majors would have been impossible without allowing pedagogical research as an option. Including pedagogical research has several benefits beyond expanding the breadth of topics students can participate in. Pedagogical research allows undergraduate student participation in research within the department to be scaled up with minimal cost, and allows all department faculty having the ability to serve individually or collaboratively as mentors. The results of the research can also be readily implemented in existing courses.

(4) Perform a detailed budget analysis of traditional course arrangements versus an integrated lecture/lab approach or whatever approach you decide to pursue. At CWU, large lecture classes with numerous smaller labs had the same instructional cost as a few mid-sized integrated lecture/lab courses. This was the key factor in gaining administrative support for the department's transition to this pedagogical format.

C. LA program lessons learned

(1) Visit a successful LA program and talk to numerous participants. The first author was lost as to how to organize an LA program until visiting Seattle Pacific University. He visited a class in which LAs were working, attended two LA meetings, and met with the LA program leaders.

(2) Align the LA course curriculum with similar courses on campus. This increases your potential LA pool, helps with time to degree for students by eliminating courses that address the same learning outcomes, and improves your LA course curriculum. For example, we used assignments and outcomes from the campus tutor-training course. The "Teaching Tip Sheet for LAs" assignment in PHYS 292 was suggested by the campus tutoring coordinator and is one of the most popular assignments. We also borrowed the case study assignment from the introductory course in

the mathematics education program. Because of this collaboration with the tutoring and the mathematics education programs, students who successfully complete the first LA course are excused from a course in each of these programs.

(3) Incorporate action research into the LA curriculum. Preservice teachers need practice collecting formative assessment data and using that data to make instructional decisions. This is a critical component of the edTPA, a preservice teacher assessment required for certification by many states [48]. Two of the three sections of the assessment require preservice teachers to make instructional decisions based on information they gather while teaching. The action research activities in the two LA courses introduce preservice physics teachers to some of the skills necessary to succeed on the edTPA.

VI. CONCLUSIONS

Over the past seven years, our department has increased the number of physics teacher candidates by a factor of four. This has gone hand-in-hand with our decision to implement key findings from the SPIN-UP report. For example, our dual-degree physics-mathematics teaching program gives teacher candidates a more efficient pathway into two high-need teaching areas [7]. Transformation of the introductory physics sequence allows the instructor and the curricular design to be effective models for preservice teachers to follow. This is important because the way people teach a subject is strongly influenced by the way they were taught that subject [33–35]. Increased undergraduate research opportunities give physics majors, and particularly preservice teachers, opportunities to connect with working scientists [13,14] and experiences in science practices and research methodologies, an important aspect of the NGSS the majors will be expected to teach [15]. The LA program provides instructional assistance in first-year physics courses. Additionally, it gives potential teacher candidates early experience in action research [42–44] and formative assessment [40,45], two important methods for monitoring student learning.

Our physics department is not highly funded, and we are understaffed compared to most physics departments with similar numbers of majors. Nevertheless, following a few key findings from the SPIN-UP report has allowed us to greatly strengthen our physics teacher preparation program.

ACKNOWLEDGMENTS

This material is based upon work supported by the Physics Teacher Education Coalition. The authors would also like to thank the following members of the Central Washington University physics department for their work in improving physics teacher education: Michael Braunstein, Andrew Piacsek, and Sharon Rosell.

[1] R. C. Hilborn and R. H. Howes, Why many undergraduate physics programs are good but few are great, Phys. Today **56** (9), 38 (2003). doi:10.1063/1.1620833

[2] R. C. Hilborn, R. H. Howes, and K. S. Krane, eds., *Strategic Programs for Innovations in Undergraduate Physics: Project Report* (American Association of Physics Teachers, College Park, MD, 2003). http://www.aapt.org/Programs/projects/spinup/upload/SPIN-UP-Final-Report.pdf

[3] For example, see http://www.aip.org/statistics/trends/highlite/edphysund/figure7e.htm, retrieved Aug. 18, 2013.

[4] More information about PhysTEC can be found at http://www.phystec.org. Our institution page can be found at http://www.phystec.org/institutions/institution.cfm?ID=775

[5] A complete list of the competencies is available online at http://www.cwu.edu/education-foundation

[6] An overview of the role community colleges play in teacher preparation can be found at http://nacctep.riosalado.edu/Drupal/PDF/CR_2012.pdf

[7] U.S. Department of Education Office of Postsecondary Education, *Teacher Shortage Areas, Nationwide Listing, 1990-1991 through 2014-2015* (2014). http://www2.ed.gov/about/offices/list/ope/pol/tsa.pdf

[8] See http://www.phystec.org/webdocs/Definitions.cfm, retrieved Aug. 18, 2013.

[9] S. A. Lei and N. Chuang, Undergraduate research assistantship: A comparison of benefits and costs from faculty and student perspectives, Educ.**130** (2), 232 (2009). http://eric.ed.gov/?id=EJ871657

[10] B. A. Nagda, S. R. Gregerman, J. Jonides, W. von Hippel, and J. S. Lerner, Undergraduate student-faculty research partnerships affect student retention, Rev. High. Educ. **22**, 55 (1998). doi:10.1353/rhe.1998.0016

[11] R. S. Rowlett, L. Blockus, and S. Larson, Characteristics of Excellence in Undergraduate Research (COEUR), in *Characteristics of Excellence in Undergraduate Research*, edited by Nancy Hensel (Council on Undergraduate Research, Washington, DC, 2012), pp. 2-19. http://www.cur.org/assets/1/23/coeur_final.pdf

[12] For the B.S. portfolio cover sheet, see http://www.cwu.edu/physics/forms

[13] J. Westerlund, D. Garcia, J. Koke, T. Taylor, and D. Mason, Summer scientific research for teachers: The experience and its effect, J. Sci. Teach. Educ. **13**, 63 (2002). doi:10.1023/A%3A1015133926799

[14] S. Brown and C. Melear, Preservice teachers' research experiences in scientists' laboratories, J. Sci. Teach. Educ. **18**, 573 (2007). doi:10.1007/s10972-007-9044-9

[15] See http://www.nextgenscience.org

[16] See http://www.geneseo.edu/phystec/blt

[17] See http://phet.colorado.edu

[18] See http://phet.colorado.edu/en/simulation/projectile-motion, retrieved Aug. 18, 2013.

[19] V. Otero, S. Pollock, and N. Finkelstein, A physics department's role in preparing teachers: The Colorado learning assistant model, Am. J. Phys. **78**, 1218 (2010). doi:10.1119/1.3471291

[20] D. Hestenes, M. Wells, and G. Swackhamer, Force concept inventory, Phys. Teach. **30**, 141 (1992). doi:10.1119/1.2343497

[21] J. Schuab, Inquiry, the science teacher, and the educator, School Rev. **68** (2), 176 (1960). http://www.jstor.org/stable/1083585

[22] R. Bell, L. Smetana, and I. Binns, Simplifying inquiry instruction, Sci. Teach. **72**, 30 (2005). https://www.mun.ca/educ/undergrad/scied/files/bell_simplifying-inquiry_2005.pdf

[23] L. C. McDermott, Melba Newell Phillips Medal Lecture 2013: Discipline-based education research – A view from physics, Am. J. Phys. **82**, 729 (2014). doi:10.1119/1.4874856

[24] L. C. McDermott and the Physics Education Group at the University of Washington, *Physics by Inquiry: An Introduction to the Physical Sciences* (John Wiley & Sons, New York, NY, 1996).

[25] L. C. McDermott, P. S. Shaffer, and the Physics Education Group at the University of Washington, *Tutorials in Introductory Physics* (Prentice Hall, Upper Saddle River, NJ, 2002).

[26] R. R. Hake, Interactive-engagement versus traditional methods: A six-thousand-student survey of mechanics test data for introductory physics courses, Am. J. Phys. **66**, 64 (1998). doi:10.1119/1.18809

[27] E. Mazur, *Peer Instruction: A User's Manual* (Prentice Hall, Upper Saddle River, NJ, 1997).

[28] D. Maloney, T. O'Kuma, C. Hieggelke, and A. Van Heuvelen, Surveying students' conceptual knowledge of electricity and magnetism, Am. J. Phys. **69**, S12 (2001). doi:10.1119/1.1371296

[29] See http://www.physicsclassroom.com, retrieved Aug. 18, 2013.

[30] P. Heller, R. Keith, and S. Anderson, Teaching problem solving through cooperative grouping. Part 1: Group versus individual problem solving, Am. J. Phys. **60**, 627 (1992). doi:10.1119/1.17117

[31] P. Heller and M. Hollabaugh, Teaching problem solving through cooperative grouping. Part 2: Designing problems and structuring groups, Am. J. Phys. **60**, 637 (1992). doi:10.1119/1.17118

[32] See http://groups.physics.umn.edu/physed/Research/CRP/crintro.html, retrieved Aug. 18, 2013.

[33] R. Putnam and H. Borko, What do new views of knowledge and thinking have to say about research on teacher learning?, Educ. Res. **29**, 4 (2000). http://www.jstor.org/stable/1176586

[34] A. Demir and S. K. Abell, Views of inquiry: Mismatches between views of science education faculty and students in an alternative certification program, J. Res. Sci. Teach. **47** (6), 716 (2010). doi:10.1002/tea.20365

[35] D. Lortie, *Schoolteacher: A Sociological Study* (University of Chicago Press, Chicago, IL, 1975).

[36] See http://www.nsta.org/preservice/docs/2012NSTAPreserviceScienceStandards.pdf

[37] R. Beichner, J. Saul, D. Abbott, J. Morse, D. Deardorff, R. Allain, S. Bonham, M. Dancy, and J. Risley, The Student-Centered Activities for Large Enrollment Undergraduate Programs (SCALE-UP) project, in *Research-Based Reform of University Physics*, edited by E. F. Redish and P. J. Cooney (American Association of Physics Teachers, College Park, MD, 2007). http://www.per-central.org/items/detail.cfm?ID=4517

[38] For example, see photos online at http://scaleup.ncsu.edu

[39] L. Seely and S. Vokos, Creating and sustaining a teaching and learning professional community at Seattle Pacific University, American Physical Society Forum on Education Newsletter (Summer 2006). http://www.aps.org/units/fed/newsletters/summer2006/seattlepacific.html

[40] R. Dufresne and W. Gerace, Assessing to learn: Formative assessment in physics instruction, Phys. Teach. **42**, 428 (2004). doi:10.1119/1.1804662

[41] See http://www.ptec.org/lavideo

[42] P. Blanton, Action research to evaluate student achievement, Phys. Teach. **41** (4), 252 (2003). doi:10.1119/1.1564515

[43] K. S. Volk, Action research as a sustainable endeavor for teachers: Does initial training lead to further action?, Action Res. **8**, 315 (2010).

[44] P. Reason and H. Bradbury, eds., *Handbook of Action Research: Participative Inquiry and Practice* (Sage Publications, London, England, 2001).

[45] P. Black and D. Wiliam, Inside the black box: Raising standards through classroom assessment, Phi Delta Kappan, **92**, 81 (2010). doi:10.1177/003172171009200119

[46] See http://www.crla.net/ittpc/certification_requirements.htm

[47] See http://goo.gl/Aw05bg for a copy of the CWU MOU between the physics and mathematics departments. For more information about the course numbers in the MOU, go to http://catalog.acalog.cwu.edu

[48] See http://edtpa.aacte.org

Engaging future teachers in having wonderful ideas

Leslie J. Atkins
Department of Science Education and Department of Physics, California State University, Chico, Chico, CA 95929

Irene Y. Salter
Department of Science Education, California State University, Chico, Chico, CA 95929

This chapter describes a course in open scientific inquiry for preservice teachers. The course uses neither a textbook nor a lab manual, but instead engages students in developing models of puzzling phenomena through an iterative process of designing experiments, crafting models, debating, and refining ideas. We outline strategies for planning such a course, including materials and classroom setup. We discuss basic structures in the class, including introducing an initial question or phenomenon, engaging in small-group investigations, leading whole-class conversations, and assessing students' work. Included are a range of examples of student ideas and student work, and results of surveys on students' progress.

I. A PEDAGOGY OF WONDERFUL IDEAS

Writing of her experience as a developmental psychologist, Eleanor Duckworth describes one seven-year-old boy, Kevin, arranging a set of straws she brought to his class [1]. Upon seeing the straws he has an idea of how he wants to arrange them, struggles, succeeds, and is clearly delighted by the wonderful idea he had—the idea to arrange an assortment of straws in order by length. Reflecting on Kevin's joy, Duckworth notes (p. 1):

> The having of wonderful ideas is what I consider the essence of intellectual development. And I consider it the essence of pedagogy to give Kevin the occasion to have his wonderful ideas and to let him feel good about himself for having them.

It's tempting to believe that Duckworth is referring to *childhood* intellectual development and *primary school* pedagogy; in other words, that we cannot expect high school students or undergraduates to be delighted by their ideas about Snell's Law in the same way that a first-grader enjoys arranging straws, or that we cannot expect students to delight and feel good about the misconceptions they bring to physics class. Reading this story, however, one author (Leslie Atkins) was reminded of her first experience working in a physics lab as a summer research experience for undergraduates (REU) student:

> My task was mundane—aligning a laser with a mirror—but at some point I saw an unexpected, though easily explained, pattern of light. I approached the postdoc to show him and was relieved when he shared in my excitement, saying, "You have to show Eric! He will get a kick out of this!"

Though most of this summer research experience is lost to the author (Atkins), this event stands out: Having and sharing a wonderful idea with a member of the physics community, witnessing his enthusiasm for it, and figuring out that this was the *kind of thing* that was to be celebrated and shared with a faculty supervisor. It was striking not only because of the magical quality of that moment, but also because of the absence of such moments from the rest of my undergraduate experience.

Such experiences, however, are routine for practicing scientists; as part of the job description, we have and share ideas with colleagues. It is only during formal science education that we find scant opportunities for students to have and share novel ideas and feel good for having them. Our exams and surveys ask students about scientifically correct ideas; that is, if they have learned and understood *others'* wonderful ideas. In reform-based curricula, we walk students through observations and simulations by which they can re-create wonderful ideas. But opportunities for students to construct their own ideas in response to their own questions and to share these ideas as the cornerstone of a curriculum are rare. Given that having and sharing novel ideas are hallmarks of scientific practice, their absence is troubling for those concerned with students' conceptual development. Given the sense of joy, wonder, and esteem that having and sharing novel ideas can bring, their absence from our curricula is troubling in an ethical sense as well.

In response to these concerns, we have developed a course that prioritizes the "having of wonderful ideas," first and foremost because *it is the central practice of scientific inquiry*—the axle around which all other practices revolve. Scientists do science because we're curious; we pursue investigations because we think we're onto something and we want to figure out if we're right. We attend conferences and publish papers because we believe our ideas to be important and enjoy sharing those ideas with colleagues.

A secondary rationale for prioritizing such a pedagogy, and the reason we believe the course is particularly critical for future teachers, is with an eye toward their future classrooms. We model a teaching practice that values students' ideas, with all the attendant scientific practices that arise in constructing, debating, and refining those ideas, so that they may enact a similar practice in their future teaching careers.

As Catley notes:

> The majority of pre and in-service teachers have never had the opportunity to undertake an open-ended research project. Yet, we expect teachers to feel comfortable using inquiry-based methodology in the classroom. The unique perspective gained from having "done research" cannot be overemphasized and will play a vital role in persuading teachers to commit to an inquiry-based pedagogy. [2]

In a study of scientific inquiry among secondary teachers, the teachers "who eventually used inquiry in their own classrooms were those who had significant research experiences in careers or postsecondary study" [3]. That is, the teachers who used inquiry had extensive prior experience in conducting open-ended inquiry in a scientific community. And yet there are few opportunities for future teachers to participate in such research at universities [4–6]. Windschitl finds that 80% of preservice teachers in a secondary science methods class had never conducted open inquiry [7]. Even more dramatically, 90% of elementary science methods students had never experienced science as an investigation [8]. America's Lab Report notes that current courses for teachers "rarely address laboratory experiences and do not provide teachers with the knowledge and skills needed to lead laboratory experiences" [9]. This becomes all the more problematic with the emphasis the recent Next Generation Science Standards places on crosscutting scientific practices (e.g., "asking questions," "developing and using models," "planning and carrying out investigations") and the nature of science [10].

One solution to this challenge has been the development of programs that bring teachers into practicing laboratories. Such programs, while often powerful experiences for teachers, are not practical as a widespread solution: Numbers alone are prohibitive, and participants self-select and are therefore frequently already interested in and familiar with scientific inquiry. At a comprehensive university like ours, future K-12 teachers outnumber all science majors combined; clearly, placing all of these students in campus laboratories is not feasible.

A second solution has been to develop courses that replicate the work of research labs, engaging students in open inquiry and creating a scientific community in the classroom. Though relatively rare in the literature and varying in structure, such courses have been shown to create significant changes in students' understanding of science and their interest in teaching science as inquiry, while promoting deep conceptual understanding [2,11–13]. This is the approach that we have taken. However, replication and dissemination of these courses poses a unique challenge: The stocks in trade of reform curricula—guided-inquiry worksheets (e.g., [14]), interactive computer programs (e.g., [15]), and laboratory materials (e.g., [16])—are not suitable for the student-generated questions and investigations that characterize open-inquiry courses. Below we provide an attempt at such dissemination, describing the structure, sequence of activities, and the kinds of ideas and outcomes that result. While this chapter leaves many questions unanswered—particularly questions pertaining to the finer-grained, moment-to-moment faculty-student interactions that really define the course and are a departure from traditional instruction—we hope it is sufficient for a curious instructor to get started. Additional materials, designed to provide guidance on the moment-to-moment interactions, are in development.

II. SCIENTIFIC INQUIRY

A. The Scientific Inquiry course at CSU, Chico

In the scientific inquiry course taught at California State University, Chico (CSU, Chico), students examine complex phenomena and construct ideas and representations that account for these phenomena. The course uses neither a textbook nor a lab manual. Instead, students work in groups with everyday items, sharing ideas and findings with other groups as the class moves toward consensus models of phenomena. The syllabus describes the goals of the class as follows:

> This is a class on Inquiry—not a class on Light & Color. (Just like a drawing class might spend time drawing flowers, but the class is not a class about flowers.) You are not assessed on how accurately your ideas mirror the ideas of scientists, but on how accurately your activities mirror the practices that scientists engage in when they study light & color—problematizing phenomena, creating careful definitions, constructing models of phenomena that are consistent with evidence, designing tests to further test those models (particularly competing models), using those models to explain and predict, reading and following the ideas of colleagues (your classmates), critiquing and improving ideas over time, and sharing work via writing.

The course, titled Scientific Inquiry, is taken by two sets of majors: liberal studies majors (primarily future K-8 teachers) and natural science majors (primarily future "foundational science" teachers, which is a California credential for middle and high school science instruction). A similarly structured course, Advanced Inquiry into Physics, is taught to upper-division physics majors and, again, to natural science majors. The examples in this chapter are primarily from Scientific Inquiry, because it is older and has more data to illustrate the approach; however, the pedagogy in Scientific Inquiry can easily be adapted to include content relevant to a future high school physics teacher.

For the K-8 teachers, Scientific Inquiry is part of a sequence of five required courses, three of which explicitly address content learning goals (Physical Science and Everyday Thinking [16], Life Science and Everyday Thinking [17], and a geoscience course) and one of which addresses pedagogy (the Hands-on Science Lab teaching practicum). For the preservice middle and high school teachers, the inquiry course counts as one of their upper-division physics courses. Because calculus is not a prerequisite for the course, it is a popular option for preservice secondary science teachers who need an upper-division physics class.

B. The structure of the course and classroom

The scientific inquiry course is taught to sections of 24 students in five hours of "activity" (lab) per week. Each week, the students attend two two-hour classes and one one-hour class. This meeting structure is driven by institutional constraints around scheduling; we find one-hour sessions too brief, and recommend longer sessions, when possible, in a course that meets four to six hours per week.

Students are arranged into six "research" groups of four students each; these groups are assigned randomly at the start of the term and are reshuffled every five weeks to provide a mix of abilities and talents in each group.

Our current classroom setup, pictured in Figure 1, has rolling tables, chairs, and whiteboards. The room has storage space for classroom materials, and we keep all materials in the room in clearly labeled drawers. This allows students to design their own experiments, as needed, from the available materials. Occasionally we supplement the classroom materials with items from home (e.g., containers of various shapes and sizes, liquids of different viscosity) or with specialized equipment from the science stockrooms (e.g., dissecting tools, bulbs with unusual filaments) when we sense that these additional materials would help groups move forward with their investigations. In addition, students often bring their own materials to class, such as spoons (when discussing what makes an image "flip") and clear "bath beads" that create images similar to lenses.

FIG. 1. Classroom setup, with rolling whiteboards and materials storage in the back.

While this setup is ideal, it is not necessary. In the initial semesters of the course we taught in a classroom in which foreign languages are typically taught. The room had slanted "chair desks" and no tables, lab benches, sinks, storage space, or whiteboards. We rolled in a cart of materials and smaller handheld whiteboards. Though the setup was less convenient, there was no appreciable difference in the quality of students' work.

Ideally, at a minimum, a Science Inquiry classroom contains chairs that are portable (i.e., no fixed lab benches), handheld whiteboards, and room for students to work in groups of four and then reconvene as a whole class. Students are required to purchase a lab notebook (of their choosing). You may also consider asking students to purchase materials, such as a Maglite flashlight or hand lens, depending on the topics you plan to discuss and whether you will want students to use those materials while doing homework.

Conceptually, we divide the 15-week semester into three units that are roughly five weeks long. We often let a unit go on longer than anticipated if engagement remains high and if a natural stopping point seems imminent. Each unit is framed by an opening question or observation, then moves from whole-group discussion to small-group investigations, and concludes with an individual paper and a group lab exam. The units are often structured to build on one another (e.g., light, then the eye, then color), but not always (e.g., light, then the eye, then astronomy).

III. THE INSTRUCTIONAL SEQUENCE

A. First-day activities

We set the tone of the class on the first day using the following activities:

(1) After a brief overview of the syllabus, students introduce themselves and we learn everyone's name. While we are generally averse to "icebreaker" activities, we believe that knowing one another's names is critical for this course. We provide a roster with student photos and emails to each student.

(2) We have students generate a rubric for their lab notebooks using a guided activity, as described in [18]. This activity underscores that students' work will mirror scientific research, and engages students in developing the criteria by which they will be assessed.

During the first week we also schedule time for small groups to develop a list of their rights and responsibilities in this course, along with a list of the instructors' rights and responsibilities. These are shared, discussed, and refined by the whole class, and the final list is discussed and modified throughout the semester as needed. Generally, we have this discussion toward the end of the week, once students have an idea of how the course is run.

B. An initial prompt

At the start of the second day of class, we introduce a question or observation for all students to discuss. These prompts will ideally launch a range of ideas and models that will inspire five weeks of ongoing investigations and modeling. Commonly used prompts are listed below. (SI stands for Scientific Inquiry, which focuses on light and sound; AI stands for Advanced Inquiry into Physics, which focuses on mechanics.)

- Is every color in the rainbow? (SI)
- Observe a pinhole theater [19] and explain what is happening. (SI)
- Dissect a cow eye and speculate about the role of each part. (SI)
- Discuss whether the pitch rises or falls when water is added to a glass and why this might be. (SI)
- What path does the sun take from the moment it rises until the moment it sets, when viewed by someone sitting on the North Pole? (AI)
- Create a realistic stop motion animation of something falling. (AI)
- Observe a Gaussian gun and discuss where the energy comes from. (AI)

We try to find prompts that are intriguing, for which we can imagine a range of reasonable answers, and that lend themselves to ongoing investigations. We particularly like starting with the question, "Is every color in the rainbow?" Students rarely think they should know the answer to this question; they bring a wealth of everyday experience to the topic; and, in the many years of asking this question, no student has had a formal school experience of learning this content before.

Students are given a brief period (10 to 20 minutes) to sketch out ideas and questions on their own or in small groups. This allows them to have carefully considered their own ideas before making sense of others' ideas. We then move into a large circle and launch into a whole-class discussion, filling whiteboards with ideas and questions that arise. This often fills the entire two-hour class.

The roles of the instructor in facilitating this conversation are primarily:

(1) understanding the ideas that students offer by revoicing (e.g., "I heard you say…did I get that right?") or asking for clarity (e.g., "When you say…I'm not sure what you mean. Can you/can someone else put that another way?");

(2) highlighting dissenting ideas (e.g., "I think what you just said, Andy, is actually a little different from what Kait is arguing…"). This is done not to resolve these differences, but simply to highlight them;

(3) promoting a conversation among students in which they listen and respond to one another's ideas (by explicitly noting that this is a goal, offering pauses where students may jump in; or by asking, for example, "Do others agree with what Janeal just said?"); and

(4) prioritizing students' everyday ideas by treating these experiences as important and legitimate data in a scientific argument (e.g., "Ulani, when you say that you think every color makes black, I think you're probably thinking about mixing up a bunch of paints and seeing the water turn a murky color; is that right?," or "I really like Cory's point. If I put my hand in front of a video projector, I can see an image projected on it, but if I put my hand in front of a window I can't. So if you're saying an image comes in through the window, I think you're going to have to address Cory's point that you can't see an image on your hand...").

This initial conversation is not an opportunity to reach a conclusion, nor should the instructor serve as an arbiter of the ideas. The goal of the conversation is to open up the range of ideas that students might pursue in the coming weeks, and the instructor aids in articulating and reinforcing the kind of reasoning and ideas students will bring to bear.

C. Moving forward: Small-group work

By the third day of class or the second day of a new unit, students will have generated a range of ideas and (very) nascent models for a given phenomenon. In their small research groups, students should generate a list of one to three questions they are interested in pursuing further. From that list, students should then choose a direction for their group. As instructors, we often counsel students away from certain questions. For example, when we are discussing color, students often generate questions about colors that hew toward psychological explanations (e.g., Do all people experience red the same way?). We point this out and suggest that, while the question may be an interesting one, it may be better for students to pursue something that will have more in common with other groups. At times the question may seem ill-posed (e.g., how does a lens work?), but as students begin their investigations with concrete materials, these questions become more focused (e.g., what happens when light from a bulb goes through a lens?). For more structured questions, such as the question above about stop motion animation, student groups will immediately launch into creating their representations before comparing and discussing one another's work.

From this point forward, the majority of the class time is spent with students working in these small groups on their

FIG. 2. Students examining the path a light ray takes through gelatin.

FIG. 3. Students sketch rays through a model of the cornea and lens.

investigations. Some need little guidance to generate ideas and observations; other groups struggle to turn their questions into a research question that they can pursue. The instructor should circulate between groups, asking students about their efforts, suggesting materials they might use, listening for productive ideas, coordinating activities, and getting a sense of where groups are headed and how to coordinate those efforts. At times, two groups will be working on related problems, and the instructor will suggest that these groups work together, sharing ideas and observations. Many groups will initially tend towards what we call "science fair" experiments, asking, "what will happen if..." followed with an experiment to answer that question. (For example, they may find that a "square-shaped" pinhole doesn't change the image on the screen, or determine that a black marker's ink, when separated out, includes dyes from a range of colors.) While a reasonable first step, we find that the groups often need to be prompted to consider the "so what" question and use their data to inform and modify nascent models of what's going on.

If you were to look into our classroom during the eye unit, for example, in the days following our eye dissection, you might see the following:

- Several groups are interested in the lens. One is building a mock "eye" consisting of a box with a hole cut in one side, a biconvex lens inside the box, and a plano-convex lens (representing the cornea) covering the hole. They find that some lens shapes make blurry images on the back of the box while others don't, and they begin to speculate about the differences between the lenses, examining the width and the curvature of the lens.
- A second group is curious how the lens works, and, using a Maglite and lens, have found the "fry-an-ant" spot where rays from the Maglite seem brightest. They sketch out what is happening to the light to make it create this spot. Using dry-erase markers, they color half the lens to test their model.
- A third group is modeling the vitreous humor (the gelatinous filling in the bulk of the eye) using gelatin. The gelatin allows the students to see the path that a laser takes, and they find that light usually bends as it enters, as shown in Figure 2.
- The fourth group is curious about distortion, believing that images projected on a curvy surface (the back of the eye) should be distorted (like a funhouse mirror). They wonder if the lens can correct this. They build a model eye with a rectangular box and another one with a curved back to examine distortion before modeling it on their whiteboard, unwittingly explaining why objects at greater distances appear smaller.
- The fifth group is intrigued by the iris and the effect of iris color. They take lenses and cover them with colored "gels" to see whether this will "dye" the image or make it blurrier.
- Finally, one group is interested in the shape of the iris, noting that cats and goats have differently shaped "slits" to their eyes. They too are constructing model eyes and are surprised to find that a shaped aperture only seems to make the image brighter or dimmer, rather than changing the shape of the image.

The instructor, circulating among groups, pushes the first group to sketch out very carefully how light might be traveling to create the blurry image. That is, the students need to move beyond phenomenology toward a model. They carefully diagram possible pathways for light through the combined lenses (as seen in Figure 3).

The second group is proceeding well, with a model and a test of that model. They are encouraged to take photos of the fry-an-ant spot using their cell phones, and post it to the class website. The third group is asked if they can tell which way the light will bend upon entering; in other words, is there some kind of rule it seems to follow? They struggle to find precise ways of describing the angle a line makes with a curved surface, and invent a method for finding the tangent. The fourth group has very clear models and diagrams—the precision of their work could serve as a model for other groups to follow—though their particular topic may be the hardest to connect to others' investigations. The instructor makes a mental note to be sure that this group is following the other groups' conversations and ideas. The fifth group could also work on carefully diagramming their ideas to account for their observations; the instructor imagines that doing so would not only account for the blurriness of a semi-transparent iris (which scatters some light), but would also explain why any image close to the lens is not in focus. And, finally, the sixth group's ideas, originally related to the iris, will also begin to address how lenses work. They are encouraged to photograph their data carefully and query the lens groups about their findings [11].

D. Reaching consensus: Whole-class discussions

The heart of the class, in our view, lies in the whole-class conversations, in which groups share their questions, observations, and models. Here ideas are introduced, challenged, refined, and moved to consensus (or further questions), in much the same way as a research scientist's lab meeting allows colleagues in the same research group to share and critique one another's ideas.

As the instructor circulates among small groups, he or she will notice that some have reached conclusions or have a model to share, while other groups are stuck: They have observations they cannot interpret and may benefit from seeing others' models. When the majority of the class seems at a stopping point (either "done" or stuck), we ask groups to take 15 minutes to sketch their ideas on their whiteboards to present to the class. If relevant, they might also prepare a demonstration to enable the class to share their observations.

Prior to the presentation of the whiteboards, the instructor often can imagine a conversation unfolding among groups. Perhaps one group will present puzzling data that students in another group can interpret using their model, which then is complicated by a third group's data. Or there are two competing models that the instructor has identified while circulating between groups, and the instructor wants to have those models presented early on in the discussion, so that other groups can use their observations to select between competing models. Rather than simply going around the room in a circle and asking groups to present in turn, you may want to ask a specific group of students to begin the conversation by sharing their observations with the class. Alternatively, you might mention to one group, "There's another group that has a really different idea from yours. Please be sure to get your model on your whiteboard and I will start by having each group present its model."

To begin, we have students arrange their chairs in a circle. This simple restructuring of the room has proved absolutely necessary: When students remain in small groups, awkwardly turning to view one another's work, the conversations are much more stilted and instructor directed.

Once chairs are arranged, the conversations unfold fairly naturally, as groups find one another's work surprising or helpful, comment on that fact, and start a conversation. Especially at the beginning of the course, students may need more active facilitation on the part of the instructor to ensure that they are actively engaged in a true discussion with the "presenters" and not just waiting quietly until their turn to present. The instructor should allow time for students to digest ideas, note them in their lab books, and respond to one another's presentations.

Occasionally, we find that students need help in seeing different ideas as critically different: They may present their boards and say, "Basically we're saying the same thing as that group...," when, to the instructor, the ideas are critically different, or one group has been detailed and precise while the other is vague. We try to highlight these differences and ask students to comment on them, perhaps noting, "While you're both showing that the lens redirects the rays, this group, I think, is trying to show that the light bends when it enters and when it exits the lens, and your diagram shows the light ray curving as it travels *through* the lens. Is one more accurate?" Other times we recognize areas of overlap that students haven't recognized: One group, perhaps, has ideas about how light bends, and another group's diagrams don't account for this finding. We might ask the first group to comment on the second group's ideas.

Though the groups are all pursuing different investigations and ideas, these will all inform the initial question (e.g., "Is every color in the rainbow?" or "How does the eye work?"), and students should be able to understand and use ideas from other groups in their own developing understanding of the material.

E. Assessing ideas: Homework, exams, and participation

This course functions in our department as a "writing proficiency" course for the major and, since its inception, writing has figured prominently—so much so that we can no longer imagine teaching the course without asking students to write extensively. We construe writing broadly, considering "scientific writing" to be exemplified not only by writing that mimics a formal paper, but also by a host of practices that scientists engage in to develop, share, critique, and refine ideas. Scientists scribble on chalkboards;

construct representations; annotate graphs and photos; jot ideas in notebooks; send emails; scrawl notes in the margins of papers; build models; plan experiments; write grants; draft conference proceedings; put together presentations; and, ultimately, tidy up all of this prior work into a publishable journal article, which is then critiqued, edited, and revised, all in a kind of living, iterative process of idea development. In this way, to scientists, writing in the sciences is not just a means to *communicate* knowledge, but a way to *build* knowledge within the scientific community; writing generates, persuades, critiques, challenges, and defends ideas in conversation with other scientists.

Our goal has been to help our students use writing in similar ways: to develop new ideas, critique and refine others' ideas, and to communicate ideas to others. We emphasize not only final products but also the ongoing, daily composition practices that students employ: whiteboards, notebooks, and peer evaluation. We avoid the lab report, literature review paper, and other standard school science writing genres, because, in our experience, these are often not taken up by students as a way of making sense of their own ideas, but simply as a way of mimicking a genre. Homework assignments and even exams are crafted with this goal in mind.

Homework assignments are not written in advance, but rather emerge from groups' conversations and ideas, and allow students to further develop and refine their ideas over the weekend. For instance, a typical homework assignment in our course would feature students' own images of their data and ask them to develop a mechanistic explanation for that observation and defend their idea. Another typical assignment emphasizes argumentation. For example, we might ask students to pick a side in an unresolved class debate, defend their points of view, and respond to possible counterarguments in a written homework assignment. Initially it often helps to tell students to assume that the instructors are firmly convinced of an opposing view, and thus their job is to convince us away from our idea and toward their idea. As students become more practiced at generating explanations and providing evidence to support their assertions, we remove some of these scaffolds and simply ask them to explain an observation from class, knowing that they will be graded on how "convincing" their arguments are.

Occasionally, we dedicate a homework assignment exclusively to developing students' abilities to construct mechanistic, theoretical diagrams. If used well, these diagrams can explain many puzzling phenomena and can lead to counterintuitive predictions that may be tested. For example, in our eye unit we often ask students to imagine an eyeball with just an iris and a pupil (no lens and no humors). The assignment asks them to draw a diagram to show how light rays would enter this eye, explain the diagram in words, and predict what a person might see if he or she had an eye like this. Our goal is to encourage students to view diagrams not only as a means of sharing scientific ideas (that is, illustrating an idea they have already contemplated), but also as a way to generate ideas. In this assignment, students often come to recognize that a pupil without a lens would result in a blurry image at the back of the eye and that the lens must either restrict the rays coming through the pupil or redirect them to "unblur" the resulting image.

For many homework assignments, we engage students in providing peer feedback and critique. One way of doing this is to anonymize and post every student's diagram on a class website or free online image hosting website, as described in [20]. Students are asked to consider strengths and weaknesses of the diagrams and then recreate their own. They can then view and evaluate one another's diagrams with an eye toward appreciating the simplicity of Allie's approach, the clarity afforded by Emma's use of color, and the precision of Mikayla's ruler-drawn lines, so that one's own work can be improved with such techniques. Moreover, it positions the students as authorities on a scientific idea. They may not know whether or not an idea is correct, but they can evaluate ideas for their coherence and their consistency with data.

Exams provide an opportunity for groups to extend their theories and ideas to novel scenarios. We typically assign a group in-class exam and a take-home individual exam at the end of each five-week unit. We frequently find that the group exams in particular inspire students to have a final "Eureka!" moment [21]. For instance, on the group exam at the end of the eye unit, we asked students to compare two lenses (a flat versus a bulgy lens), consider how the lens in the eye might work to focus at different distances, and consider the effect of the bulgy cornea on a cat's eye. Tui was part of a group that had been investigating pupils of different shapes. During the group exam, she had an exciting realization: "Oh, but maybe the cornea *would* interact with the peripheral vision. Because it's going to allow more in and bend in and get in the slit...[laughs] Eureka! I learned something new on the last day." On another exam, students were shown the shadows seen during a partial eclipse. One student commented to her classmate, "But if those were just because of pinholes in the trees, we would be able to look outside right now and see circular sun-shaped shadows." They turned to look out the window and, to their surprise, saw circles in the dappled light on the ground.

F. Challenges

In general, students respond enthusiastically to the course, many describing it as a favorite of their undergraduate career. In the first weeks, however, they are often uncertain, and worried that they will be held accountable for scientific ideas and yet getting no answers from faculty. This reaction is most common among our science majors, who pride themselves on having the right answers in science class and in explaining those right ideas to their peers.

For this reason, assessing ideas early, fairly, and in ways that hold students accountable to one another's ideas and experiments (but not necessarily for scientifically correct ideas) is critical. In addition, we assign readings [1,22] that underscore the goals of this course and how they differ from traditional classes. The syllabus also discusses these goals. We may share examples of excellent homework assignments, showing how students with very different ideas may nonetheless score equally well. Occasionally we will ask students who seem concerned or frustrated to talk to us outside of class about their concerns.

Another challenge arises when students look up information online and introduce this into the class. Surprisingly, we have found time and again that this information is rarely useful: Students often don't know how to make sense of this information in the absence of a well-formed model; when they introduce these ideas, it is clear they are simply reporting on information they read and do not deeply understand; other students in the class indicate that they find it distracting and contrary to the intent of the course. That said, in the 13 semesters of teaching this course, we have never had to actively curtail a student's use of external sources. Moreover, if a student can use such sources appropriately (that is, making sense of the ideas and bringing these ideas to bear on the ongoing investigations accurately), it is entirely in keeping with the course's goals.

G. Answer day

Following the exam, we have "answer day" during which students may ask any question relating to the prior unit and get a straight answer from faculty. Student questions from the "eye" unit include:

- How can you change where the eyes focus?
- Why does light bend "in" when both entering and leaving a lens?
- Why do things look life-sized if they are so tiny on the retina?
- How does the tapetum help animals see in the dark?
- Why do differently shaped pupils make no difference in how we see?
- Do the [vitreous and aqueous] humors do anything except keep the eye inflated?

Students enjoy answer day because they can finally get direct answers to the questions they have struggled with all semester. Admittedly, we instructors also enjoy answer day because we can share our scientific expertise with a room full of students who care about these scientific questions: precisely how and why lenses bend light; whether light "scatters" or "shatters" during diffuse reflection; what, exactly, does a scientist think a light ray looks like; or the differences between rods and cones.

IV. STUDENT OUTCOMES FROM THE SCIENTIFIC INQUIRY CLASS

A. Qualitative data: Student ideas

Many science education researchers believe that the book is closed regarding the merits of open inquiry: Such instruction has failed to achieve the same outcomes that guided inquiry and direct instruction achieve. We do not disagree: Open-inquiry courses do not have the same kinds of outcomes as guided inquiry or traditional instruction. The question then remains: What content knowledge *do* students learn in such a course?

In this section we present a range of questions and content ideas that students generate, with examples of student work. We believe that readers will find the examples, which come from a range of students (none of whom are physics majors), to be profound. In addition, we highlight the questions that they ask on "answer day," highlighting how these questions touch on many core ideas in physics, but in unanticipated ways.

1. Student questions

One example of insightful questions arose with a lab group, as they considered how a lens could both focus *and* magnify an image. Kristin argued to her group that magnification must occur when the rays that leave one point on the object occupy an *area* on the screen. Others countered that this was the definition for blurriness. They diagrammed and were stumped: It seemed that magnification without blurriness was impossible. The crux of the argument was the idea that if the image occupies a larger area than the object, it must contain more points than were in the original object (and so, to accomplish this, the points must expand to areas). The resolution is that there are, in fact, the same number of points in a small area as in a large one (the cardinality of these infinite sets is the same). While this lab group did not prove that point on their own (the instructor spoke with them and offered a simple proof of this), it strikes us as an important aspect of lenses and magnification that is rarely considered by the undergraduate. That students were able to link questions of magnification to questions of the cardinality of infinite sets is a profound insight.

A second question arose when students questioned how the lens accommodates to focus on closer objects. Knowing that there are muscles on the edge of the lens, and familiar with the experience of "straining" these muscles to focus on closer objects, one research group expected that this implied that a "flatter" lens, pulled taut by the muscles, must focus at shorter distances, while "rounded" lenses focus at farther distances. Figure 4 is a group's whiteboard illustrating this idea.

FIG. 4. Comparison of the divergence of rays for an object close to and farther from a lens, illustrating that the "closer" object will require a greater refraction by the lens.

Other groups (who had been working with a set of lenses donated by a theater lighting company) observed the opposite: For a fixed lens-to-screen distance, flatter lenses focused on farther objects. The instructors were stumped, too: How could the contraction of the ciliary body (the muscles responsible for accommodation) create a rounder lens? It turns out that the students had hit on an open question in physiology, with several proposed mechanisms for how the contraction of muscles changes the focal length of the lens [23].

A third question—one we were prepared to answer—arose in the context of the Advanced Inquiry course. As we modeled the potential energy of a ball attracted to a magnet, students wanted that energy to increase as the ball moved farther away, but then to "turn off" once it was too far away to feel the attractive force of the magnet. As they worked through the question of where that energy would "go" once it "turns off," a student asked, "Is it possible for the energy to always be increasing, but never reach infinity?" The instructor rephrased the question: "I think another way of asking that is, can you imagine a stairwell where you're always walking up, but the stairs don't reach infinitely high?" Students played around with sketches of possible series (1 + 1/2 + 1/3 + ... and 1 + 1/2 + 1/4 + ...) and calculators, slowly convincing themselves that for some series this would be possible. The instructor presented a sketch of nested squares (a proof of the convergence of a geometric series). Students in the course had not taken calculus or considered limits of infinite series, and yet were able to connect this question to that of potential energy as objects move infinitely far apart.

2. Content ideas

Through their inquiry, students develop ideas we would typically describe as "content knowledge." These ideas are usually qualitative conceptual ideas. This is particularly true in the Scientific Inquiry class, but is true in the Advanced Inquiry class as well. For example, students will describe a lens's focal point as related to the curvature of the lens, and generate descriptions of how rays bend (more bending the more the incident angle is away from the normal), but the same students will never have generated a mathematical relationship to predict that bend (e.g., Snell's Law).

Nonetheless, these ideas are generally normative and provide significant physical insight. As one student commented:

> Right now, our group is working on the idea of how glasses and contacts change the shape of your cornea to balance out a person's misshapen cornea. We thought we could explain it by explaining that people with near-sighted vision need glasses with thicker glass on the sides and that people with far-sighted vision need glasses with thicker glass in the center. However, we only knew what near-sighted glasses looked like. We didn't know what far-sighted glasses (e.g., reading glasses) looked like. When I was at Walgreens the other day, I saw some reading glasses and decided to investigate. And sure enough, the glasses were thicker in the center and as the intensity of the prescription increased, so did the thickness of the center. I was so proud of our group to turn out correct!

Groups have defined focus precisely, describing "focus" as occurring "when all rays that leave one point on an image are brought back to one point on the retina (or screen)." They describe the role of the lens as "organizing" the light rays so that this is possible. They find, phenomenologically, that such focus occurs for a plane. In Figure 5, a group has sketched initial ideas about why focusing at one depth will lead to blurriness at other depths.

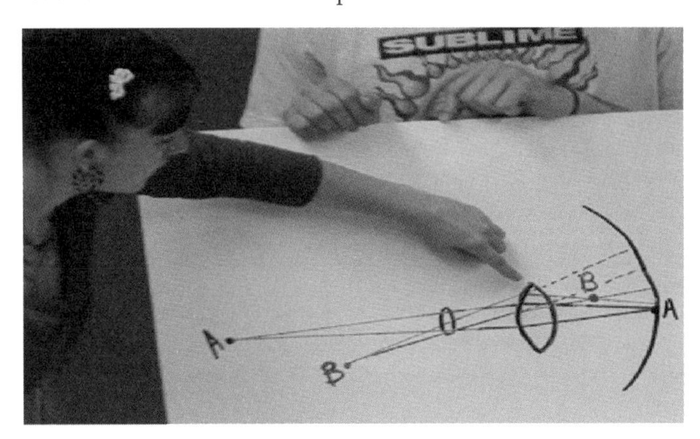

FIG. 5. A group diagrams ideas about focus. The first oval represents the pupil; the second shape is the lens; the third is the retina. Point A is in focus; rays from B are diverted from their original paths, but nonetheless are not focused.

In addition to these definitions and findings, students often explicitly describe more fundamental conceptual ideas underlying the physics. For example, one semester a student noted, "You can think of an image as a bunch of tiny dots of light, making up a picture—comparable to a Lite Brite but on a much smaller scale."

As a more quantitative example, in the Advanced Inquiry course, students found that a falling object gains 19.6 m^2/s^2 of speed squared for every meter that it falls. This description is consistent with physics, but not a common way of conceptualizing motion. Ultimately, it proved a powerful conceptualization for describing energy transfers and transformations [24].

B. Quantitative data: Survey responses

In addition to descriptions of student ideas, we have been collecting data on students' understanding of the nature of science and the degree to which the class impacts students' views about what science is and how to learn science. Most of the available surveys developed for undergraduate physics (e.g., the Epistemological Beliefs About Physics Survey) are not applicable to this course, as they ask questions about textbooks and solving book problems. We gave students the Views of Nature of Science survey to capture changes in their understanding of the nature of science. To our surprise, students' survey responses and written descriptions regarding the nature of science did not change considerably through the course. Our current research raises questions about interpretations of Nature of Science surveys [21,25]. Briefly, we find a mismatch between what students report on surveys regarding the nature of science and what they do in class as they construct scientific ideas. This raises methodological challenges for how to best assess student understanding of the nature of science: If the purpose of an inquiry class is to enable students to pursue scientific questions in productive, scientific ways, then we have evidence that this class can promote such abilities. However, if the purpose is to enable students to articulate correct ideas about how scientific ideas are constructed, this class does not address that purpose.

A second, more preliminary study investigates the degree to which this course affects students' out-of-class experiences. That is, do students actively seek out opportunities to apply ideas and questions from class, and do they value doing so? This survey is patterned on Pugh's studies of *transformative experience* (TE) [26], a measure of the degree to which students use and value ideas from class in their everyday lives. Questions on the survey are detailed in the Appendix and further details are available in [27]. Briefly, the questions ask about students' experiences using physics ideas (e.g., "I talk about light and optics just for the fun of it," and "I find it exciting to think outside of school about light and optics."). Possible responses are: strongly disagree, disagree, agree, or strongly agree. Agreeing with the statements is an indication of transformative experiences with physics ideas.

Early results suggest that students who have taken the scientific inquiry course described here have markedly higher TE scores than do students who have taken traditionally taught courses. Data from two courses—PHYS 202, a second-semester introductory physics course covering light and optics, and NSCI 321, the inquiry course described here—are shown in Figure 6. Additional data from other institutions are available in [28].

C. Student reactions to the class

Overwhelmingly, students find the class to be markedly different from other science courses they have taken, including courses that employ guided inquiry curricula. They often describe the initial weeks as frustrating, but over time, as they develop models of phenomena and find those to be productive and ultimately correct, they gain confidence in their abilities and in this approach to instruction.

Echoing many other comments, one student notes:

> This is the first science class that has ever made me feel like a scientist. Even in other classes when I had to do experiments, everything was so structured that I felt it was more of just an assignment. I have never just been given a topic, like the eye, and then let loose to try to figure out how it works. I am really glad that I took this class because it gave me a whole new outlook on science. Before, even though this is my last year, I never knew what it was like to be a scientist. At times I felt extremely frustrated and like I was going nowhere in finding out new information. But then when I finally did figure out a piece of information and had that 'ah-ha!' moment, the frustration was well worth it. I felt a sense of accomplishment and excitement every time that myself or my group discovered something new. I began to realize that the feelings I was experiencing are probably an everyday thing for actual scientists.

In addition, students often comment on the value of that frustration and struggle, noting that they have come to recognize the struggle as part of what it is to "do science" as opposed to an indication of a lack of ability in science. (References [1] and [22] help support these ideas.) As one student comments on an end-of-class survey:

> I used to believe that doing science was a fun and collaborative effort that happened naturally in a logical sequence. I now realize that this is what school has taught me. Every science class was strictly guided to give all the answers at the correct moments...I now know that is not science...This class had shown me that science can be a struggle, and should be.

This student had taken guided-inquiry-style courses in the past (e.g., [16]) that emphasized the collaborative, logical nature of scientific work but not, as she notes, the dead ends and struggles of scientific work.

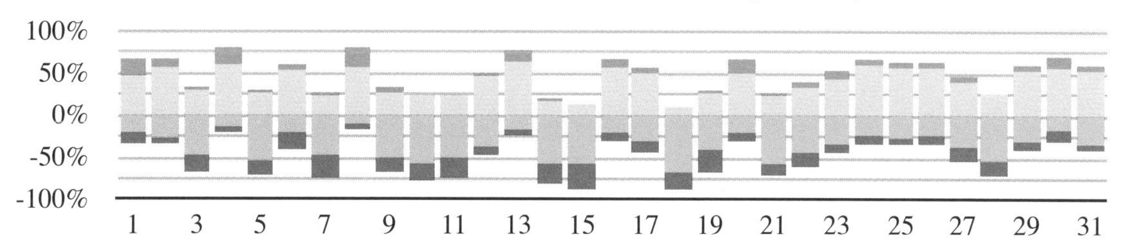

NSCI 321 (Scientific inquiry) N = 38

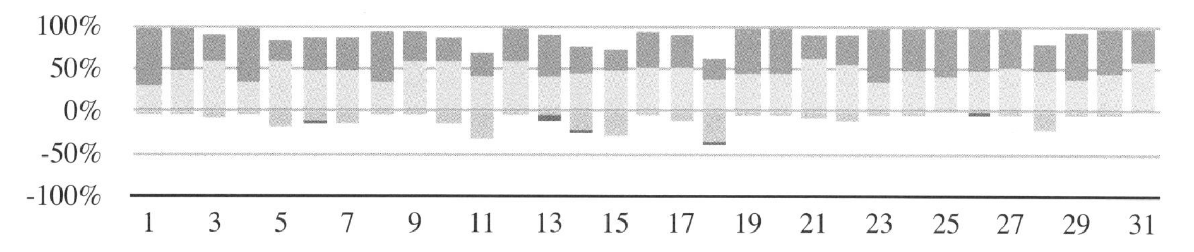

FIG. 6. The bar graphs show student responses to individual survey statements (survey questions appear in the Appendix). Each bar represents 100% of the responses to a question, and the bar's location is positioned vertically, to indicate the percentage agreement (above the horizontal axis) and the percentage disagreement (below the axis). Dark shadings are used to indicate the fraction of strong agreement or strong disagreement. For example, PHYS 202 shows a mix of bars above and below the axis, indicating a mix of agreement and disagreement, while NSCI 321 shows bars mostly above the axis, indicating a strong preference toward agreement.

V. TEACHING THE INQUIRY COURSE: PLANNING AND TIPS

A. Proposing a course

CSU, Chico has a Department of Science Education within the College of Natural Sciences. This department comprises a group of faculty who teach science content courses for undergraduate preservice teachers. The Scientific Inquiry course is housed in this department; Advanced Inquiry is housed in the physics department. Neither is part of the credentialing program at CSU, Chico. Instead, these majors address scientific content, while instruction in pedagogy is addressed in a postgraduate program. Adding a new course to either sequence can be challenging, and must be approved as a science course by the college. In proposing the course at CSU, Chico, we used the California science education standards [29], which, like many state standards from the past decade, positions "inquiry" as its own strand of content. In this way, we were able to argue that the core content of the course were these inquiry skills, that they were critical for future teachers, and that this content should be taught by science faculty rather than in the education school.

The new standards make a point of integrating inquiry more seamlessly into the traditional content goals [10]; nonetheless, one could argue for a course that foregrounds these skills, given how rarely they are core topics in traditional physics courses.

Offering to teach the course as a "writing proficiency" course (as with most universities, all majors are required to have a course that focuses on writing in the discipline) further supports the course, both in its adoption by the college and by ensuring adequate enrollment.

B. Anticipating ideas

The research literature on student ideas about topics that you might cover in such a course can be invaluable in anticipating what questions might arise, interpreting student ideas, deciding what materials to have on hand, and developing questions that might prove fruitful starting points. For the content we discuss, *Physics by Inquiry* [14] and materials developed by the Exploratorium [19] are useful starting points. We have avoided typical "Pasco"-style lab materials, opting instead for everyday objects when possible. These include Maglite flashlights, laser pointers, magnifying glasses, "bug box" lenses, electrical tape, dry erase markers, food coloring, printer inks, etc.

A caveat: We initially imagined that students' ideas might unfold in predictable ways, similar to the ways in which reform curricula proceed. This quickly proved not to be the case. One class, beginning with the *Physics by Inquiry*-inspired question regarding whether or not you can see a beam of light as a flashlight shines down a hallway, quickly became interested in how light reflects off surfaces, as opposed to our imagined discussions regarding the nature of light and seeing. The following semester, this question prompted students to characterize the parabolic reflector that sits inside the flashlight. Instructors should be prepared to follow student questions wherever they lead, which may vary significantly from anticipated directions. While we do constrain student investigations (for example, as noted above, discouraging "psychological" kinds of studies or questions like "how does a laser pointer work?"), as long as their questions and investigations in the inquiry course are related to the topics of light, how it behaves, and the implications of that, we give students leeway to study phenomena and develop models that arise in the moment.

C. Responding to ideas

Above, we note that the roles of the instructor in classroom conversations are primarily to (1) understand the ideas that students offer; (2) highlight dissenting ideas; (3) promote a conversation among students; and (4) prioritize students' everyday ideas. While the instructor may be evaluating the correctness of those ideas when considering where to focus attention, what to suggest a group do next, or how to design a class assignment, this evaluation is rarely apparent to the students. These roles are challenging, both for the instructor, who may find it uncomfortable to hear "wrong" ideas and not correct them, and for the students, who expect the instructor to tell them "right answers." We have found that, with little exception, students become more comfortable once they receive grades that reward their thinking and not the correctness of their answers. For instructors, "answer day" provides a kind of "right answer" safety net.

That is not to say that we never interject with some piece of factual information. For example, it has been surprisingly common that students use the word "refract" for diffuse reflection. (Perhaps the word is similar to "fracture" and sounds scientific.) We might say, "When you say 'refraction' I think you mean that the ray breaks up. 'Refraction' actually means that light changes direction." We also frequently have students who know that light is a "wave," and believe this means light cannot travel in straight lines, but instead travels along a wave-shaped trajectory. In some semesters this idea has become so distracting (students will not even consider that light is moving in a straight line, or that a diagram of light traveling could possibly have straight lines) that we have provided a short discussion of what scientists mean when they say something is a "wave."

We notice that one kind of discussion that we find particularly exciting and try to draw out is that of dissent. In this course, we position dissent as a discovery and a rich opportunity to "do" science. In the example of light, noted above, we often have a group of students who think of light traveling in wavelike trajectory, and another group who thinks of it as a ray. The groups don't, at first, notice the difference. "Wave" people view the "ray" diagrams as a shorthand way of drawing a wave. The "ray" people might think of the "wave" diagrams as scientists do, as conveying more than spatial information, though they have rarely explicitly considered this. Drawing out those differences presents students with two competing models of light and a chance to develop experiments that select between those models. When we find such opportunities, we work to make students' models as explicit as possible, so that their experiments are as meaningful as possible.

Finally, we have had colleagues tell us that their students have trouble coming up with "testable questions" and that many of the "experiments" students do are hardly experiments at all, but just "what happens if..." kinds of activities. Don't expect "testable questions" when students are just beginning to consider a new topic. Instead, expect them to spend some time simply playing around with the materials as they work out nascent ideas and ways of discussing and representing those ideas, and as they come to understand the differing ideas their peers have. Encourage them, particularly in the first week, to be as precise and specific as possible—but again, we often let that push for precision emerge through whole-class discussions, during which the need for precision becomes apparent. Use early observations, representations, and ideas as a way to set the stage for more rigorous, "testable" models and ideas over time.

D. Engaging with colleagues

Throughout this chapter, we have referred to "us" and "we"—the instructors in the class—as we discuss our responses and interactions with students. For the first iterations of the course, while it was grant supported, the class was co-taught. We met regularly after class to discuss and interpret student ideas and consider possible next steps. Since then we have each taught the class individually, but we have found these initial discussions invaluable in expanding our ideas and strategies for teaching this kind of course.

Another means for accomplishing this kind of collaboration with colleagues has been creating a blog and asking interested faculty to read and comment. Throughout the teaching of the course, author Atkins has maintained a password-protected blog, and received feedback on student work and suggestions for readings and experiments. Journaling student ideas and chronicling their progress has been an important reflection and a means of better understanding student progress. A list of blogs on the website PhysPort provides new bloggers with models of how to begin and a means to connect with other faculty [30].

VI. CONCLUSIONS

Speaking to the reforms that are called for in mathematics education, Lampert notes that very few Americans

> have any idea what a mathematical community is or what a conjecture is or what it would look like to do mathematical

> reasoning. Most of us have never done that...The goal that all students by some year...are going to be able to do this thing that most Americans have no sense of right now, let alone many teachers, seems like a rather ambitious goal. [31]

The same can be said for the scientific community: It seems evident that many Americans have a weak understanding of what a scientific theory is, how models are built, the messiness of real data, how the scientific community comes to consensus, and when scientist might decide to revise earlier, established claims. The Next Generation Science Standards ask teachers to engage their students in model building, theorizing, and developing and using representations in scientific ways [10]. Lecturing *about* scientific inquiry ("a theory is...") is an inadequate response, particularly for those who then need to lead students through scientific investigations. This course, which offers students the opportunity to develop their own research questions and pursue those with their peers, is a course that models (to the degree possible) the ways in which a scientific community would develop ideas. For future teachers, it provides an opportunity to experience the kinds of scientific reasoning and scientific practices that they will be asked to engage their own students in doing.

ACKNOWLEDGMENTS

The course in its current incarnation is strongly influenced by a range of colleagues and courses. Foremost among those are David Hammer [31]; Paul Hutchison [32] and the University of Maryland's Physics 115; Matty Lau and our co-taught course for high school students (Why is the Sky Blue?); Brian Frank, who teaches a version of this course at Middle Tennessee State University and has contributed to TE surveys; and Kim Jaxon from CSU, Chico's English department, who has helped us improve our writing instruction.

We also offer our gratitude to the students involved in NSCI 321 and PHYS 341, who so generously share their ideas with us and allow us to share them here.

This work was funded by National Science Foundation grants #0837058, #1140784, #1140785, and #1140860.

---

[1] E. Duckworth, *The Having of Wonderful Ideas and Other Essays on Teaching and Learning* (Teachers College Press, New York, NY, 1987).

[2] K. Catley, *How Do Teachers Work and Learn—Specifically Related to Labs*, presented to the Committee on High School Science Laboratories: Role and Vision (National Research Council, Washington, DC, June 2004).

[3] M. Windschitl, Folk theories of "inquiry:" How preservice teachers reproduce the discourse and practices of an atheoretical scientific method, J. Res. Sci. Teach. **41** (5), 481 (2004). doi:10.1002/tea.20010

[4] National Research Council, *Taking Science to School: Learning and Teaching Science in Grades K-8* (National Academies Press, Washington, DC, 2007).

[5] M. Windschitl, Inquiry projects in science teacher education: What can investigative experiences reveal about teacher thinking and eventual classroom practice?, Sci. Educ. **87** (1), 112 (2003). doi:10.1002/sce.10044

[6] A. C. Graesser and N. K. Person, Question asking during tutoring, Am. Educ. Res. J. **31** (1), 104 (1994). doi:10.3102/00028312031001104

[7] M. Windschitl, *The Reproduction of Cultural Models of "Inquiry" by Pre-Service Science Teachers: An Examination of Thought and Action*, presented at the meeting of the American Education Research Association (New Orleans, LA, April 2002).

[8] B. L. Shapiro, A case study of change in elementary student teacher thinking during an independent investigation in science: Learning about the "face of science that does not yet know," Sci. Ed. **80** (5), 535 (1996). doi:10.1002/(SICI)1098-237X(199609)80:5<535::AID-SCE3>3.0.CO;2-C

[9] National Research Council, *America's Lab Report: Investigations in High School Science* (National Academies Press, Washington, DC, 2005).

[10] NGSS Lead States, *Next Generation Science Standards: For States, By States* (National Academies Press, Washington, DC, 2013).

[11] I. Y. Salter and L. J. Atkins, Student-generated scientific inquiry for elementary education undergraduates: Course development, outcomes and implications, J. Sci. Teach. Educ. **24**, 157 (2013). doi:10.1007/s10972-011-9250-3

[12] W. M. Roth and G. M. Bowen, Knowing and interacting: A study of culture, practices, and resources in a Grade 8 open-inquiry science classroom guided by a cognitive apprenticeship metaphor, Cognition Instruct. **13**, 73 (1995). doi:10.1207/s1532690xci1301_3

[13] E. H. van Zee, D. Hammer, M. Bell, P. Roy, and J. Peter, Learning and teaching science as inquiry: A case study of elementary school teachers' investigations of light, Sci. Educ. **89** (6), 1007 (2005). doi:10.1002/sce.20084

[14] L. C. McDermott and the Physics Education Group at the University of Washington, *Physics by Inquiry* (Wiley, Hoboken, NJ, 1995).

[15] M. C. Linn, The knowledge integration perspective on learning and instruction, in *Cambridge Handbook for the Learning Sciences*, edited by R. K. Sawyer (Cambridge University Press, Cambridge, MA, 2006), pp. 243-264. doi:10.1017/CBO978 0511816833

[16] F. Goldberg, S. Robinson, and V. Otero, *Physics and Everyday Thinking* (It's About Time, Herff Jones Education Division, Armonk, NY, 2007).

[17] D. Donovan, J. Rousseau, I. Salter, L. Atkins, A. Acevedo-Gutierrez, C. Landel, V. Mullen, P. Pape-Lindstrom, and R. Kratz, *Life Science and Everyday Thinking* (It's About Time, Armonk, NY, 2014).

[18] L. J. Atkins and I. Y. Salter, Using scientists' notebooks to foster authentic scientific practices, AIP Conf. Proc. **1513**, 50 (2012). doi:10.1063/1.4789649

[19] See http://www.exploratorium.edu/education/publications, retrieved Sep. 2, 2013.

[20] L. J. Atkins, Peer assessment with online tools to improve student modeling, Phys. Teach. **50**, 489 (2012). doi:10.1119/1.4758155

[21] I. Y. Salter and L. J. Atkins, Surveys fail to measure grasp of scientific practice. AIP Conf. Proc. **1513**, 362 (2012). doi:10.1063/1.4789727

[22] M. A. Schwartz, The importance of stupidity in scientific research, J. Cell Sci. **121**, 1771 (2008). doi:10.1242/jcs.033340

[23] See http://en.wikipedia.org/wiki/Accommodation_(eye)#Theories_of_mechanism, retrieved Sep. 2, 2013.

[24] L. J. Atkins, and B. W. Frank, Examining the products of responsive inquiry, in *Responsive Teaching in Science*, edited by D. Hammer, A. D. Robertson, and R. E. Scherr (Routledge, New York, NY, to be published).

[25] I. Y. Salter and L. J. Atkins, What students say versus what they do regarding scientific inquiry, Sci. Educ. **98**, 1 (2014). doi:10.1002/sce.21084

[26] K. J. Pugh, L. Linnenbrink-Garcia, K. L. K. Koskey, V. C. Stewart, and C. Manzey, Motivation, learning, and transformative experience: A study of deep engagement in science, Sci. Educ. **94**, 1 (2010). doi:10.1002/sce.20344

[27] B. W. Frank and L. J. Atkins, Adapting transformative experience surveys to undergraduate physics, in *Proceedings of the Physics Education Research Conference* (Portland, OR, 2013), pp. 145-148. doi:10.1119/perc.2013.pr.024

[28] B. W. Frank and P. Mittura, Transformative experiences and conceptual understanding of force and motion, in *Proceedings of the Physics Education Research Conference* (Minneapolis, MN, 2014), pp. 91-94.

[29] S. Bruton, F. Ong, and G. Geeting, *Science Content Standards for California Public Schools: Kindergarten through Grade Twelve* (California Department of Education, Sacramento, CA, 2000).

[30] See https://www.physport.org/webdocs/resources.cfm, retrieved Sep. 2, 2013.

[31] D. Hammer, Discovery learning and discovery teaching. Cognition Instruct. **15** (4), 485 (1997). doi:10.2307/3233776

[32] P. S. Hutchison, *Epistemological Authenticity in Science Classrooms*, Ph.D. thesis, University of Maryland, College Park (2008). http://www.physics.umd.edu/perg/dissertations/Hutchison/Hutchison%20Dissertation.pdf

APPENDIX

Statements on the Transformative Experiences survey specify physics content and some reference specific physical contexts. Table 1 shows a list of the stems that are used. For example, in the force and motion survey, statement 2 reads, "I think about the laws of force and motion, when I see objects that are falling, colliding, or balanced." In the optics survey (taken by courses described in this chapter), statement 2 reads, "I think about the rules for optics and the nature of light when I see things like light bulbs, cameras, or glasses."

1. During class, I talk about…
2. I think about…, when I see…
3 Outside of class, I talk about…
4. During class, I think about…
5. I talk about… just for the fun of it
6. Outside of class, I think about…
7. I find myself thinking about… in everyday life.
8. During class, I use the knowledge I've learned about …
9. Outside of school, I use the knowledge I've learned about…
10. I use the stuff I've learned about…even when I don't have to.
11. I look for chances to use my knowledge of…in my everyday life
12. During class, I see things in terms of the laws I've learned about…
13. When I am working on a class assignment about… I tend to think of them in terms of…
14. If I see a really interesting situation (either in real life, in a magazine, or on TV), then I think about it in terms of…
15. I can't help but see situations in terms of …
16. During class, I notice examples of…
17. I notice examples outside of class of…
18. I look for examples outside of class of…
19. Learning about…is useful for my future studies or work.
20. Knowledge of…helps me to better understand the world around me.
21. Knowledge of… is useful in my current, everyday life.
22. I find that knowledge of…makes my current, out-of-school experience more meaningful and interesting.
23. Knowledge of…makes learning physics much more interesting.
24. In class, I find it interesting to learn about…
25. I think…is an interesting topic
26. I find it interesting in class when we talk about… in terms of…
27. I'm interested when I hear things about…outside of school
28. I find it exciting to think outside of school about…
29. The ideas we learned changed the way I see…
30. I think about…differently now that I have learned about…
31. I pay more attention to…now.

Periscope: Supporting novice university physics instructors in looking into learning in best-practices physics classrooms

Rachel E. Scherr
Department of Physics, Seattle Pacific University, Seattle, WA 98119

Renee Michelle Goertzen
American Physical Society, College Park, MD 20740

Periscope is a set of instructional materials designed to support novice university physics instructors in observing, discussing, and reflecting on best practices in university physics instruction. *Periscope* is organized into short lessons that highlight significant questions in the teaching and learning of physics, such as: "How do I ask questions that facilitate students' thinking?" and "When is it okay to leave students with the wrong answer?" Key topics in teaching and learning are introduced through captioned video episodes of introductory physics students in the classroom; episodes are chosen to prompt collaborative discussion. Video episodes from universities with exemplary physics education programs showcase a variety of research-tested instructional formats such as Modeling Instruction and *Tutorials in Introductory Physics*. Discussion questions prompt participants who view the episode to reflect on their pedagogical beliefs, on their own practices, and on the results of physics education research. *Periscope* materials are especially appropriate for undergraduate Learning Assistants and other university instructors, but may also serve preservice K-12 teachers and other populations.

I. INTRODUCTION

The experience of teaching physics using effective pedagogies can recruit undergraduates into the teaching profession. Learning Assistant (LA) programs, in which talented undergraduates help faculty transform courses to be more student centered, can be particularly valuable as preservice teacher programs. LA programs offer a coordinated experience of teaching, pedagogical instruction, and content preparation in a supported, team-based learning environment [1,2]. We describe *Periscope*, a set of materials designed to increase the effectiveness of LA and other preservice teacher programs by providing opportunities to observe, discuss, and reflect on best practices in university instruction. Watching and discussing video episodes of classroom teaching supports LAs and preservice teachers in examining authentic teaching events in the classroom and applying lessons learned about teaching to actual teaching situations.

Periscope is organized into short lessons that highlight significant questions in the teaching and learning of physics; examples of questions include "How do I ask questions that facilitate students' thinking?" and "When is it okay to leave students with the wrong answer?" The topic is introduced through a captioned video episode of introductory physics students in the classroom, chosen to prompt collaborative discussion. For example, in one video episode, frustrated tutorial students ask an LA to tell them the right answer, and the LA responds with more questions. Line-numbered transcripts and excerpts of the activity help participants engage with the specifics of the interactions, leading to discussions of questions such as: "Which student asked for the answer, and why?" and "What was the tone of the LA's response?" Subsequent discussion questions also prompt participants who view the episode to reflect on their pedagogical beliefs and on their own practice, e.g., What might the LA in the episode have been trying to accomplish? What are the potential benefits and risks of her approach? What effect did the LA's response have on the students? What else might an LA have done in that situation? Through *Periscope,* LAs and preservice teachers observe, discuss, and reflect on teaching situations similar to the ones they themselves face, developing their pedagogical content knowledge and supporting their identities as teaching professionals.

II. NEEDS OF LEARNING ASSISTANT PROGRAMS

LA programs provide potential future teachers with strongly supported and low-stress early teaching experiences that can encourage them to pursue teaching certification. The effectiveness of any LA program hinges critically on the quality of preparation that LAs receive for their university teaching assignment. LAs who are well prepared are better positioned to have the kinds of successful and rewarding teaching experiences that will recruit them into the teaching profession as well as support the course transformation in which they play a part.

Key to the success of any LA development program is the expertise of LA supervisors, the senior personnel—usually faculty—who are responsible for the preparation of LAs (and often graduate teaching assistants) at their college or university. LA supervisors need LA development materials that are immediately accessible to and of value for themselves and for the LAs they supervise. They need activities that are flexible enough to be useful in a variety of LA preparation contexts: daylong workshops, weekly meetings, or low-credit courses about teaching and learning, as appropriate to each local situation. They need materials that are relevant to the instructional formats at their institution. Finally, they need a program that invites collaboration with other faculty members, promoting faculty professional development and institutional change as well as LA development. *Periscope* supports all of these needs, enhancing and complementing any LA development program or other preservice teacher preparation program.

III. BENEFITS OF CLASSROOM VIDEO FOR PRESERVICE TEACHER DEVELOPMENT

Classrooms are complex environments. Even in small-group interactions, many different things are happening simultaneously: Students express their physics ideas, respond to questions, gesture, joke or argue with each other, look at the clock, speak animatedly or quietly, and so on. Educators cannot, and do not, pay attention to everything with equal weight. Instead, they learn to notice and interpret student behavior in a distinctive way that is answerable to the interests of their profession, much as archaeologists learn to see soil in a way that is unique to their field [3]. This development of *professional vision* is particularly critical for educators in transformed physics courses, who are expected to respond to students' ideas and interactions as they unfold moment to moment [4]. National standards emphasize formative assessment as among the most valuable tools for enriching student understanding in science [5], consistent with research demonstrating the effectiveness of formative assessment for student learning [6]. Educators who supervise highly interactive classrooms need to develop practices of moment-to-moment formative assessment: continual, responsive attention to students' processes of developing understanding as they are expressed in real time [4].

Periscope video episodes provide a rare opportunity for future teachers to stop the classroom action, share observations, and develop a repertoire of responses, thus building skills for real-time formative assessment. By watching and discussing short video episodes of classrooms similar to those in which they are presently teaching, educators enter vividly into the classroom events and explore the principles and values that inform instructor and student behavior. Video episodes provide shared, detailed experiences for collaborative discussion. They also make teaching moments available for repeated watching, allowing a type of detailed reflection not possible with individual teaching experiences shared verbally. Video-based professional development programs for teachers have been demonstrated to support the development of teachers' professional vision, specifically assisting them in attending to student ideas (instead of exclusively to teacher actions) and interpreting those student ideas in increasingly nuanced ways [7,8]. In other words, video-based professional development supports the development of pedagogical content knowledge [9]. *Periscope* adapts that research to support novice instructors in learning to notice and interpret classroom events the way an accomplished teacher does.

IV. FEATURES OF *PERISCOPE* SUPPORTING NOVICE INSTRUCTOR DEVELOPMENT

Periscope video episodes include LA-student, faculty-student, and student-student interactions. The video episodes for *Periscope* showcase a variety of exemplary instructional formats in undergraduate physics, including *Tutorials in Introductory Physics* [10], Modeling Instruction [11], and *Open Source Tutorials in Physics Sensemaking* [12,13], all of which emphasize small-group interactions. Semi-structured discussions of the video episodes illuminate the issues facing both novice and expert university educators. *Periscope* provides LAs and their supervising faculty with opportunities to observe, discuss, and reflect on teaching situations similar to their own, developing their pedagogical content knowledge and supporting their identities as teaching professionals. In addition, *Periscope* offers users a view of other institutions' transformed courses, which can support and expand the home institution's vision of its own instructional improvement and support the transfer of course developments among faculty. *Periscope* is, to our knowledge, the only resource providing LAs with video of other LAs teaching physics. *Periscope* offers value even beyond that offered by the video episodes alone, because it packages the episodes into lessons that structure activity, focus attention, and standardize implementation while allowing adaptation to local needs. *Periscope* materials are available free to educators at physport.org/periscope.

Dissemination of these materials to faculty LA supervisors has shown that the materials are also appropriate for development of other university physics educators, including four-year university faculty, two-year college faculty, and graduate teaching assistants. In dissemination workshops, faculty LA supervisors not only learn how to facilitate LA discussions of teaching and learning, but also sincerely engage in such discussions themselves. Faculty members report successfully using *Periscope* for teaching assistant (TA) and faculty development. *Periscope* meets needs specifically identified for disseminating physics education research to physics faculty [14]. First, the instructional materials are easily modifiable: The written handouts are supplied both in fixed (PDF) and editable (PowerPoint) formats, to support leaders in using the materials either with minimal preparation or modified to fit their particular contexts. Second, the instructional materials are packaged together with relevant research findings: The topics of *Periscope* lessons provide educators with access to the pedagogical ideas common to many types of transformed physics courses, such as the purpose of teacher questions and the means of facilitating cooperative learning.

Periscope is aligned with factors known to promote effective professional development for educators. A meta-analysis of research on teacher professional development has established five components essential to increasing teacher knowledge: a focus on content knowledge, opportunities to engage in active learning, the coherence of what is taught with educators' current beliefs, participation with colleagues, and a duration of more than 20 hours [15]. *Periscope* conforms to four of these five components, and is compatible with the fifth (duration). First, *Periscope* uses episodes from introductory physics classrooms, embedding the discussions of pedagogy in the context of specific physics content. *Periscope* promotes active learning through reflection on the video episode and collaborative group discussions. It is coherent with educators' teaching experiences in that the classrooms in the video episodes are similar to the ones in which the teachers themselves teach. In addition, *Periscope* increases opportunities to participate with colleagues by increasing the variety of possible colleagues: Faculty, graduate TAs, and undergraduate LAs can all participate together.

A. Sample lesson

Each lesson includes one short (one- to five-minute) captioned video episode with good-quality sound, curriculum excerpts, a transcript, and sample prompts for discussion. The presentation style is documentary. Each lesson provides 30 to 45 minutes of LA development activity.

Each lesson is associated with a one-page handout (front and back) similar to that shown in Figure 1. The front page has a photo of the student group in the video episode, the physics task the students in the video are working on, and discussion questions that target that lesson's topic. The back page has a line-numbered transcript.

The lessons that comprise *Periscope* are designed for simple implementation. This usability is particularly important since users of *Periscope* will frequently be lead educators who guide professional development for other educators (e.g., instructors of pedagogy courses for LAs). For these users, successful use of *Periscope* will mean not only engaging successfully with its content, but also seamlessly leading other educators in doing so.

B. What LAs learn about

The lessons in *Periscope* address a wide variety of questions about teaching and learning, such as:

- What ideas do students have about energy, forces, circuits, etc., and how do I address them?
- How can I facilitate students working well in groups?
- How do I bring out students' physics ideas?
- When is it okay to leave students with the wrong answer?
- Does it matter if students are unhappy in my class?
- How do I ask questions that facilitate my students' thinking?
- How can I encourage my students to engage in sense-making instead of answer-making?
- How can I support underrepresented groups in succeeding in my class?
- How can I physically arrange my classroom to facilitate student learning?
- What kinds of tasks help students work together constructively?

These topics reflect the National Research Council's Framework for K-12 Science Education and the eight scientific/engineering practices contained therein, including constructing explanations, developing and using models, engaging in argument from evidence, and obtaining, evaluating, and communicating information [16].

V. SUGGESTIONS FOR IMPLEMENTATION

Periscope is organized into sets of lessons that are intended to serve a variety of needs. For example, a facilitator leading a ten-week seminar on physics teaching and learning (such as an LA pedagogy class) might select a ten-lesson set on "Big Ideas in physics teaching and learning," or a set on "Student ideas in physics." If the goal is instead to share best practices in physics instruction with fellow faculty over a shorter timespan, a three-lesson set on "Best practices of teaching assistants" would be a better choice. The *Periscope* website (www.physport.org/periscope) offers lesson sets for a wide variety of circumstances.

Periscope lessons are designed for use in a setting that alternates whole-group discussion with smaller discussions in groups of two to four participants. The main part of a *Periscope* lesson is a cycle of watching a video episode as a whole group, discussing a question or prompt about the video in small groups, and then having groups report the results of their discussions back to the whole group. Each cycle should last 10 to 20 minutes, and there should be a minimum of two cycles. Therefore, a single lesson should be scheduled for a minimum of 30 to 45 minutes.

A. Establishing a respectful atmosphere for discussions

Discussions about teaching often involve values that run deep for the participants. Maintaining a respectful and safe atmosphere is crucial not only for developing a learning community among participants, but also for enabling participants to identify and share their values, examine them

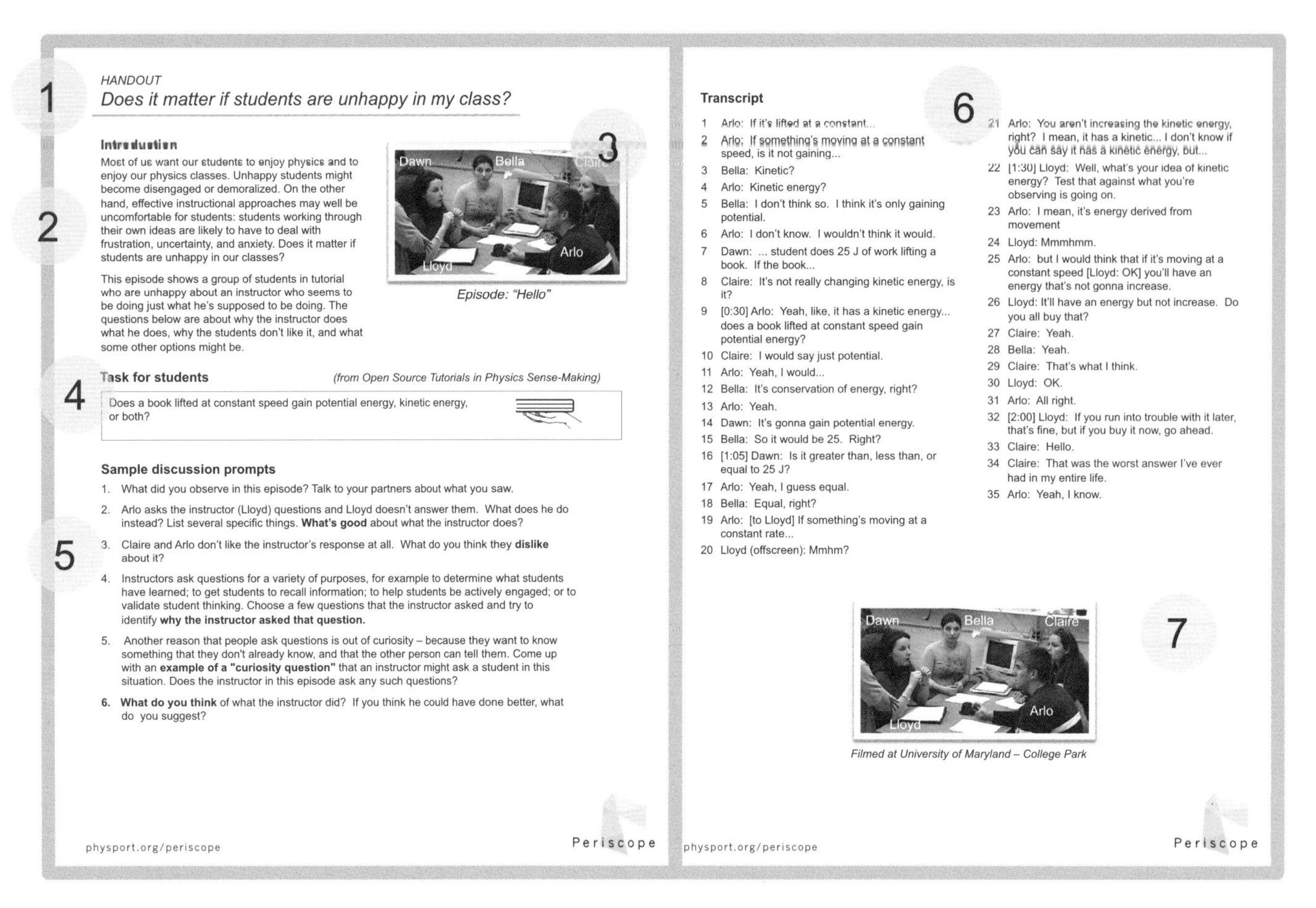

HANDOUT

Does it matter if students are unhappy in my class?

Introduction

Most of us want our students to enjoy physics and to enjoy our physics classes. Unhappy students might become disengaged or demoralized. On the other hand, effective instructional approaches may well be uncomfortable for students: students working through their own ideas are likely to have to deal with frustration, uncertainty, and anxiety. Does it matter if students are unhappy in our classes?

This episode shows a group of students in tutorial who are unhappy about an instructor who seems to be doing just what he's supposed to be doing. The questions below are about why the instructor does what he does, why the students don't like it, and what some other options might be.

Episode: "Hello"

Task for students *(from Open Source Tutorials in Physics Sense-Making)*

Does a book lifted at constant speed gain potential energy, kinetic energy, or both?

Sample discussion prompts

1. What did you observe in this episode? Talk to your partners about what you saw.
2. Arlo asks the instructor (Lloyd) questions and Lloyd doesn't answer them. What does he do instead? List several specific things. **What's good** about what the instructor does?
3. Claire and Arlo don't like the instructor's response at all. What do you think they **dislike** about it?
4. Instructors ask questions for a variety of purposes, for example to determine what students have learned; to get students to recall information; to help students be actively engaged; or to validate student thinking. Choose a few questions that the instructor asked and try to identify **why the instructor asked that question.**
5. Another reason that people ask questions is out of curiosity – because they want to know something that they don't already know, and that the other person can tell them. Come up with an **example of a "curiosity question"** that an instructor might ask a student in this situation. Does the instructor in this episode ask any such questions?
6. **What do you think** of what the instructor did? If you think he could have done better, what do you suggest?

physport.org/periscope Periscope

Transcript

1 Arlo: If it's lifted at a constant...
2 Arlo: If something's moving at a constant speed, is it not gaining...
3 Bella: Kinetic?
4 Arlo: Kinetic energy?
5 Bella: I don't think so. I think it's only gaining potential.
6 Arlo: I don't know. I wouldn't think it would.
7 Dawn: ... student does 25 J of work lifting a book. If the book...
8 Claire: It's not really changing kinetic energy, is it?
9 [0:30] Arlo: Yeah, like, it has a kinetic energy... does a book lifted at constant speed gain potential energy?
10 Claire: I would say just potential.
11 Arlo: Yeah, I would...
12 Bella: It's conservation of energy, right?
13 Arlo: Yeah.
14 Dawn: It's gonna gain potential energy.
15 Bella: So it would be 25. Right?
16 [1:05] Dawn: Is it greater than, less than, or equal to 25 J?
17 Arlo: Yeah, I guess equal.
18 Bella: Equal, right?
19 Arlo: [to Lloyd] If something's moving at a constant rate...
20 Lloyd (offscreen): Mmhm?
21 Arlo: You aren't increasing the kinetic energy, right? I mean, it has a kinetic... I don't know if you can say it has a kinetic energy, but...
22 [1:30] Lloyd: Well, what's your idea of kinetic energy? Test that against what you're observing is going on.
23 Arlo: I mean, it's energy derived from movement
24 Lloyd: Mmmhmm.
25 Arlo: but I would think that if it's moving at a constant speed [Lloyd: OK] you'll have an energy that's not gonna increase.
26 Lloyd: It'll have an energy but not increase. Do you all buy that?
27 Claire: Yeah.
28 Bella: Yeah.
29 Claire: That's what I think.
30 Lloyd: OK.
31 Arlo: All right.
32 [2:00] Lloyd: If you run into trouble with it later, that's fine, but if you buy it now, go ahead.
33 Claire: Hello.
34 Claire: That was the worst answer I've ever had in my entire life.
35 Arlo: Yeah, I know.

Filmed at University of Maryland – College Park

physport.org/periscope Periscope

FIG. 1. Sample handout for a *Periscope* lesson. Features include: (1) Questions about teaching and learning that are accessible and compelling for both novice and expert physics educators. (2) Introduction to pedagogical issue. (3) Photo associating lesson with video. (4) Task reproduced from research-based and research-tested curricula. (5) Discussion questions highlighting lesson topic. (6) Line-numbered transcript supporting detailed discussion. (7) Editability: Users can change anything they want, or use a fixed format for convenience.

thoughtfully, and consider other possible perspectives. The first time using a *Periscope* lesson with a particular group, an agreement such as one of the following is useful:

- Strive to characterize what's going on in the episode according to the people in it. Describe events in a way that the participants themselves would likely agree with if they were present.
- Limit discussion to what we see happening in the episode (observable evidence) and what we think it means (evidence-based interpretation). Set aside opinion, judgment, and critique.
- Recognize that while we will likely all agree on observations (e.g., "The LA never spoke"), and we may persuade each other of interpretations ("Those two students have the same idea"), value statements (such as "The LA should not have done that") are personal: they provide an opportunity to learn about the person speaking, and may reveal commitments and priorities that are not universally shared.

B. Using the three-stage cycle of communal viewing, small-group discussion, and whole-group discussion

The main part of a *Periscope* lesson is a 10 to 20 minute cycle of communal viewing, small-group discussion, and whole-group discussion that repeats at least twice. Each cycle of viewing and discussion should be focused on a particular question or prompt.

The first stage of each cycle is for the large group to watch the episode together. Communal viewing gets the whole group thinking about the same teaching and learning event at the same time and is technically simpler than having individuals watch the episode on separate screens. When starting the episode, say something simple such as "Okay, let's watch," without any special instructions. Have participants watch the captioned episode rather than following along with the transcript, so that they can see the action as well as hear what is said.

The second stage of each cycle is for small groups of two to four participants to discuss what they saw in the episode. The small-group discussions give all individuals a chance to process their immediate reactions to the episode. It can also help them to focus their observations on the specific prompt, if there was one. While participants talk to each other, facilitators may participate in a small-group discussion, float to different groups and listen in, or just wait. After one to five minutes, depending on the schedule and the richness of participants' discussions, transition from small-group to whole-group discussion with a sentence like, "Okay, I'm interested to hear what you observed."

An alternative is to have participants write individually in response to the prompt, before or instead of discussing the prompt in a small group. We suggest trying both of these techniques in order to give people with different interactional styles opportunities to participate in different ways.

The third stage of each cycle is a whole-group discussion. The purpose of the whole-group discussion is to expose participants to observations and interpretations that had not arisen in their small group and (in some cases) to build consensus about a question or prompt. The following guidelines may be useful for facilitating whole-group discussions. In general, a facilitator should:

- Encourage participants to ground their statements in evidence from the episode. For example, when a participant says, "Great group dynamic," a facilitator might say, "What do you see that makes you say that?"
- Encourage participants to respond to each other and let the discussion develop. E.g., "You see Deb as doing a thought experiment. Is that how other people interpreted line 15?"
- Revoice participant contributions by rephrasing what participants say. For example, if a participant says, "They have the idea that electrons jump from one tape to the other," a facilitator might say, "You see them talking in terms of a transfer of electrons."
- Stay aware of when participants are making claims and inferences, in order to help them stay connected to the evidence. Claims and inferences can be welcomed, but identified—as in, "You're thinking that Caleb is the only one to use scientific vocabulary. That's a claim. Did anyone else make any observations about that?" or "You see Deb as being the leader. What observation led to your making that inference?"
- While listening to and revoicing participants' contributions, attempt to recognize issues relevant to the lesson question, or recognize if the participant is raising a new issue. For example, "You saw Deb disagreeing with Bridget. What ideas does each of them have about electrostatic charge?"
- When there's a lull in the talk, draw out additional comments by saying, "What more did you see?"

The complete cycle (communal viewing, small-group discussion or writing, large-group discussion) should repeat at least once, usually with a different prompt each time. With more than one cycle of viewing, participants experience seeing different things in an episode than they saw the first time or reconsider inferences that they had made. Both of these experiences are important for the development of professional vision.

C. Useful discussion prompts

Each cycle of viewing and discussion addresses a particular question or prompt. There are many possible sources of questions and prompts to stimulate discussion of *Periscope* episodes. The first prompt, however, should always be completely open ended.

1. First prompt (Open-ended prompt)

We have found repeatedly that participants cannot focus on a specific question about a video until they have had a chance to process what they have seen in their own way. Thus, after the first time watching the episode with participants, use an open-ended prompt such as one of the following:

- "What did you notice? Talk to your neighbor about what you noticed."
- "What did you observe? Tell your group what you saw."
- "What struck you? Talk to the person next to you about what stuck out."

An open-ended prompt such as this has the special benefit of helping facilitators learn about the participants, and helping participants learn about one another. Different audiences will tend to focus on different aspects of the episode, e.g., the physics ideas, the instructional format, the group interactions, or issues of equity and inclusion. Facilitators can also get a sense of participants' expertise with best-practices instruction, with video analysis, and with physics content from their responses to this prompt. These natural interests and areas of expertise or development can shape the rest of the discussion.

2. Lesson question prompt

A prompt for the second cycle of viewing, and subsequent cycles, should focus participants' attention on a particular issue or question. The lesson question (which is the title of each lesson handout) makes a good second prompt, along the lines of, "Let's watch this again, and this time I want you to think about the students' electrostatics ideas."

3. Handout prompts

Another source of prompts is the "sample discussion prompts" printed on the lesson handout. Usually there are more sample prompts than should be used in a single session. A facilitator using a handout prompt would transition from the whole-group discussion of the previous prompt by saying something along the lines of, "Take a look at question three on the handout: 'Caleb proposes a mechanism for how charge gets from one object to another. What is the mechanism that he proposes?' Let's watch the video again, and this time see what you observe that addresses that question."

There is great value in getting different groups to share their responses to the same prompt. We designed these prompts to have more than one reasonable answer—sometimes even opposite reasonable answers. We prefer these kinds of prompts not only because they seem to be more inviting to participants, but also because we think they reflect the real complexity of teaching and learning events. We hope that different participants will see the events in the video episode differently, perhaps even taking different sides on a question. Subsequent viewings may either develop consensus or affirm distinctive viewpoints.

4. Prompts generated and refined in class

Other sources of discussion prompts are the questions and issues raised by the participants themselves. To organize these sources, facilitators should write participants' contributions on the board as they make them. While writing, facilitators should refrain from response, judgment, or follow-up questioning. Once the contributions are listed on the board, the facilitators choose (or co-choose with the participants) which contribution(s) to discuss in greater depth. This will be the prompt for the next round of viewing.

Facilitators may find it valuable to classify (and lead the group in classifying) the contributions that are written on the board into categories. Some frequently useful categories are questions, observations, claims or inferences, and value statements.

5. Tried-and-true advanced prompts

Another class of prompts is especially worthwhile for relatively advanced inquiry—when the lesson objectives have already been achieved, or with a particularly experienced group. The following general prompts can lead to increased insight about any episode:

- Prompt participants to attend to only one "channel" of communication—only gestures, only prosody (the music of the voice), only facial expressions, only body movements, etc.—and see what they learn from narrowing their attention in that way. A facilitator might ask each participant or each small group to select a different channel.
- Prompt participants to take the perspective of a single student in the episode—to try to live through the episode as that person, foregrounding what they say, see, and do in their attention. A facilitator might ask each participant or each team to select a different focal person.

VI. GOALS SERVED BY *PERISCOPE*

A. Support and enhance LA programs

Periscope promotes individual and group reflection on high-quality teaching and learning practices, provides learning opportunities about key pedagogical concepts in physics education, and promotes effective implementation of a variety of research-based and research-validated instructional materials.

B. Promote quality instructional practices

Periscope showcases research-based course transformations from around the country, offering vivid, realistic, and intellectually compelling examples of proven curricular transformations in action, and helping instructors value such approaches. The lessons provide a forum for instructors at all levels to talk substantively about teaching, providing a means for transferring course developments among faculty as well as for shaping educators' values.

C. Provide unobtrusive faculty development

Faculty development can be a sensitive matter, especially since some faculty do not see themselves as needing to improve their teaching practices. When faculty initiate LA programs as part of a course transformation effort at their home institution, however, there is a valuable opportunity for faculty development. Most faculty members agree that LAs need support in order to function as effective instructional assistants in undergraduate courses. Faculty members who prepare to train LAs can engage with important issues in teaching and learning while minimizing risk to their professional identities. *Periscope* helps faculty members encounter transformed teaching in a way they may not have previously experienced. Faculty members who are new to course transformation may be struck by the relatively long teacher-student interactions they observe in the video episodes, the dilemma of whether and when students get to the right answer, and the challenge of involving every student in the discussion. Discussion questions and scholarly readings support faculty in appreciating effective teaching practices. Faculty development thus happens authentically in a context that faculty members accept as legitimate and appropriate. As a result of this development, faculty may become more engaged with preservice teacher preparation in physics departments, a high priority in the national landscape of physics teaching and learning [17]. *Periscope* also provides a means for faculty to explore potential course transformations privately—for instance, to get a sense of what *Tutorials* or Modeling Instruction look and feel like—without having to visit another institution or engage the department in supporting such exploration.

VII. EVIDENCE OF EFFECTIVENESS

Periscope is a resource to support and enhance LA programs based on compelling classroom video of best-practice university physics instruction. Preliminary evidence indicates that *Periscope* is effective for this purpose. The nationally acclaimed LA program at the University of Colorado Boulder uses *Periscope* for development of LAs and faculty. Faculty at the University of Maine in Orono have used *Periscope* with LAs, TAs, and teachers for at least five years. Faculty at California State Polytechnic University, Pomona have used *Periscope* not only in the U.S. but also to engage faculty in France about physics education reform [18]. Workshops disseminating *Periscope* have been organized at conferences of the Physics Teacher Education Coalition (PhysTEC), Learning Assistant workshops, and meetings of the American Association of Physics Teachers. Over 150 users have acquired *Periscope* materials so far. Feedback from facilitators indicates high engagement with a variety of issues in teaching and learning. One new user at the University of Colorado said,

> I was impressed with the variety of things that [participants] noticed—a great mix of things ranging from body language, LA and student behaviors, power dynamics, physics content, gender stuff, the structure of the rooms, student attitudes... Maybe the coolest moment for me was when someone observed that [a particular LA] didn't seem to be following our Colorado suggestion of asking Socratic questions about the physics; instead, they thought that LA seemed to be focusing on drawing out student positions and getting them to debate. I also got a comment that they had never realized there was all this other stuff to "attend to" besides just student ideas. That alone seemed worth the price of admission.

The online delivery of *Periscope* allows the developers to continually improve and update the resource in response to user feedback and requests.

ACKNOWLEDGMENTS

This material is based upon work supported by the Physics Teacher Education Coalition and by the National Science Foundation under Grant No. 1323699. We are particularly grateful to Monica Plisch for her advocacy and to early users, including Mackenzie Stetzer, Laurie Langdon, Steve Iona, Cyrill Slezak, Bud Talbot, and Alexander Rudolph, for their intellectual contributions. *Periscope* was inspired by insightful suggestions from Stamatis Vokos.

[1] V. Otero, N. Finkelstein, R. McCray, and S. Pollock, Who is responsible for preparing science teachers?, Science **313** (5786), 445 (2006). doi:10.1126/science.1129648

[2] V. Otero, S. Pollock, and N. Finkelstein, A physics department's role in preparing physics teachers: The Colorado learning assistant model, Am. J. Phys. **78** (11), 1218 (2010). doi:10.1119/1.3471291

[3] C. Goodwin, Professional vision, Am. Anthropol. **96** (3), 606 (1994). http://www.cogsci.ucsd.edu/~johnson/COGS102B/Goodwin94.pdf

[4] F. Erickson, Some thoughts on "proximal" formative assessment of student learning, Yearb. Natl. Soc. Stud. Educ. **106**, 186 (2007). doi:10.1111/j.1744-7984.2007.00102.x

[5] National Research Council, *National Science Education Standards* (National Academy Press, Washington, DC, 1996). http://www.nap.edu/openbook.php?record_id=4962

[6] J. M. Atkin, P. Black, and J. Coffey, eds., *Classroom Assessment and the National Science Education Standards* (National Academy Press, Washington, DC, 2001). http://www.nap.edu/openbook.php?record_id=9847

[7] M. G. Sherin, Effects of video club participation on teachers' professional vision, J. Teach. Educ. **60**, 20 (2009). doi:10.1177/0022487108328155

[8] M. G. Sherin, The development of teachers' professional vision in video clubs, in *Video Research in the Learning Sciences*, edited by R. Goldman, R. D. Pea, B. Barron, and S. J. Derry (Earlbaum, Hillsdale, NJ, 2007), pp. 383–396.

[9] L. S. Shulman, Knowledge and teaching: Foundations of the new reform, Harvard Educ. Rev. **57**, 1 (1987). http://her.hepg.org/content/j463w79r56455411/

[10] L. C. McDermott, P. S. Shaffer, and the Physics Education Group at the University of Washington, *Tutorials in Introductory Physics* (Prentice Hall, Upper Saddle River, NJ, 2002).

[11] E. Brewe, Modeling theory applied: Modeling Instruction in introductory physics, Am. J. Phys. **76** (12), 1155 (2008). doi:10.1119/1.2983148

[12] A. Elby, R. E. Scherr, T. McCaskey, R. Hodges, E. F. Redish, D. Hammer, and T. Bing, *Open Source Tutorials in Physics Sensemaking: Suite I* (2007). http://umdperg.pbworks.com/w/page/10511218/Open%20Source%20Tutorials

[13] R. E. Scherr, A. Elby, R. M. Goertzen, and L. D. Conlin, *Open Source Tutorials in Physics Sensemaking: Suite II* (2010). http://umdperg.pbworks.com/w/page/10511218/Open%20Source%20Tutorials

[14] M. H. Dancy and C. Henderson, Barriers and promises in STEM reform, Commissioned Paper for National Academies of Science Workshop on Linking Evidence and Promising Practices in STEM Undergraduate Education (2008). http://homepages.wmich.edu/~chenders/Publications/Dancy_Henderson_CommissionedPaper2008.pdf

[15] L. M. Desimone, Improving impact studies of teachers' professional development: Toward better conceptualizations and measures, Educ. Res. **38** (3), 181 (2009) doi:10.3102/0013189X08331140

[16] H. Quinn, H. Schweingruber, T. Keller, and National Research Council, eds., *A Framework for K-12 Science Education: Practices, Crosscutting Concepts, and Core Ideas* (National Academies Press, Washington, DC, 2012). http://www.nap.edu/catalog/13165/a-framework-for-k-12-science-education-practices-crosscutting-concepts

[17] D. E. Meltzer, M. Plisch, and S. Vokos, eds., *Transforming the Preparation of Physics Teachers: A Call to Action. A Report by the Task Force on Teacher Education in Physics* (T-TEP) (American Physical Society, College Park, MD, 2012). http://www.ptec.org/webdocs/2013TTEP.pdf

[18] A. L. Rudolph, B. Lamine, M. Joyce, H. Vignolles, and D. Consiglio, Introduction of interactive learning into French university physics classrooms, Phys. Rev. ST Phys. Educ. Res. **10**, (2014). http://arxiv.org/abs/1311.3622

Mentoring, Collaboration, and Community Building

The PhysTEC Teacher-in-Residence: What an expert high school teacher can bring to a physics education program

Monica Plisch
American Physical Society, One Physics Ellipse, College Park, MD 20740

Jacob Clark Blickenstaff
Pacific Science Center, 200 Second Ave. N., Seattle, WA 98109

Jon Anderson
Centennial High School, 4757 North Road, Circle Pines, MN 55014

Bringing an experienced high school teacher into a university physics department for a one-year Teacher-in-Residence (TIR) position can support the transformation of that department into a site for high-quality physics teacher preparation. The Physics Teacher Education Coalition (PhysTEC) project has supported the placement of more than 50 TIRs at universities across the nation. These expert teachers have been involved in nearly all aspects of physics teacher preparation programs, including recruiting, advising, and mentoring future teachers; teaching courses; organizing professional communities; and building connections with local schools. This article outlines design principles for the TIR position, documents TIR activities, describes full-time and part-time positions, provides evidence of the impact of TIRs, and offers practical guidelines for departments planning to hire a TIR.

I. INTRODUCTION AND CONTEXT

The best preparation for teachers requires close connection between theory taught in pedagogy courses and the practice of teaching in real classrooms. Many programs link theory and practice through practicum and student teaching experiences, but typically these experiences come at the end of the licensure program. A variety of strategies have been developed to achieve better integration between theory and practice, including the creation of laboratory schools on university campuses [1], and residency programs in teacher preparation [2,3]. Another strategy is to bring an expert classroom teacher with recent teaching experience to the university campus, so students can benefit from the teacher's practical knowledge and connections with local schools throughout the program. This is the vision behind the Physics Teacher Education Coalition (PhysTEC) Teacher-in-Residence (TIR).

The idea to bring an experienced K-12 teacher to a university campus as part of a teacher preparation program was not invented by PhysTEC. Other projects have supported reassigning Teachers-in-Residence, or Teachers on Special Assignment (TOSA), to a university campus for one or two academic years. For example, the Carnegie Corporation of New York funded a program called Teachers for a New Era [4], which supported Teachers-in-Residence at a number of sites. TIR or TOSA programs attempt to bring the schoolteacher's practitioner knowledge of teaching and learning to the university teacher preparation program. TIRs sometimes participate in recruiting, methods courses, and pedagogy courses; they can also serve in the teaching practicum experiences as coaches [5]. Some teacher preparation programs hire what are essentially permanent full-time TIRs. For example, the UTeach program at the University of Texas at Austin and about 40 replication sites nationwide employ master teachers who teach certification courses, arrange for early teaching experiences, and mentor future science, technology, engineering, and mathematics (STEM) teachers [6,7].

PhysTEC TIRs are unique in that their assignments are not to colleges of education, but to physics departments. This integrates the TIR into a departmental culture that usually has very little contact with K-12 education. Most university physics faculty have little or no direct experience of high school physics classrooms beyond that acquired when they were students; nor do physics faculty typically have much contact with high school physics teachers. To help change departmental culture in a way that PhysTEC values—i.e., to help physics departments see their key roles in teacher preparation—interaction among faculty, students, and high school teachers is essential. Potential future teachers (and faculty) need to interact with people who share their passion for physics and find the teaching profession rewarding and challenging. In addition, physics faculty need to hear first-hand from a teacher what knowledge and skills are needed to be successful in a high school classroom.

The following sections of this article will detail the current state of the Teacher-in-Residence at PhysTEC Supported Sites. We will detail the differences between full- and part-time TIR positions and the Visiting Master Teacher (VMT) position, and describe the range of activities in which TIRs participate. Brief case studies of particular sites will serve to illustrate successful models for the TIR. We will also make suggestions for the recruiting, hiring, and long-term support of TIRs, and provide evidence of the impact TIRs have on departments and future teachers. This article is intended both for PhysTEC Supported Sites as well as for institutions seeking to improve their physics teacher education programs through the addition of a staff member with expertise in high school physics teaching.

II. DESIGN PRINCIPLES

The TIR position was originally envisioned as a one-year full-time appointment in a university physics department, after which the TIR returns to the classroom. The TIR works closely with the faculty lead on nearly all aspects of the physics teacher preparation program. The TIR position has evolved considerably over the lifetime of the PhysTEC project, in response to accumulated project experience with TIRs as well as the redefinition of the project itself over time. As described in more detail below, PhysTEC has experimented with part-time TIRs as well as multi-year appointments for TIRs.

An experienced high school physics teacher brings valuable expertise to the preparation program that is typically not found on a university campus. The TIR can also provide a reality check on program design, as well as influence faculty and student attitudes about the teaching profession. With his or her recent classroom experience, the TIR is uniquely qualified to lead the recruitment and mentoring of future teachers and build bridges with local schools. The TIR also possesses considerable knowledge and skill as a teacher, and sites are able put this expertise to good use in the college classroom.

A. Defining the TIR position

Early PhysTEC Supported Sites were given considerable latitude in defining the TIR position within their own programs. Some emphasized professional development for the TIR, offering training in pedagogical methods or conducting classroom observations of teachers using a research-based protocol like the Reformed Teaching Observation Protocol (RTOP)[8]. Other sites assigned the TIR to work on introductory course reform and develop instructional materials. These early trials and implementations at later supported sites led to significant evolution of the TIR position as project leaders gained experience and learned what was effective.

The TIR position is now formalized, with specific TIR duties articulated in a memorandum of understanding (MOU) between funded sites and the PhysTEC project (see Appendix: PhysTEC MOU Template). Each supported site completes an annual MOU detailing the activities planned for the upcoming year. The MOU template is prefilled with suggested activities based on what was effective at other sites. Institutions add and remove items to fit their specific contexts and goals, but all sites start with the MOU template as a guide. The portion of the MOU template listing specific activities is provided in the appendix. In 45 of the 96 (i.e., nearly half) of the activities listed, the TIR has full or partial responsibility for executing the task, underscoring his or her central role in the program. TIR activities are grouped in the following sections: Recruitment Plans, TIR Activities, Learning Assistants, Mentoring, Teacher Advisory Groups, Early Field Experiences, Institutional Change, and Assessment.

B. Variations

PhysTEC teacher education programs have included the TIR as a core element since the beginning of the project in 2001. In the earliest years, some sites decided to focus on the preparation of future elementary and middle school teachers, and a number of these sites hired elementary and middle school teachers to serve as TIRs. Since 2005, the focus of PhysTEC has shifted primarily to high school physics teacher preparation, and all TIRs hired since then have been high school teachers. This article focuses on high school TIRs, though some cumulative statistics are included.

As mentioned above, the TIR role was originally envisioned as a one-year position, after which the TIR would return to the classroom. In practice, several PhysTEC sites kept their TIR for more than one year, and there are a number of advantages and disadvantages to doing so. A new TIR every year allowed programs to build strong relationships with a greater number of teachers; such relationships are critical to a successful program. Changing the TIR every year also brought fresh ideas and connections to the program as well as current knowledge of the classroom. In addition, it allowed programs to gain experience with a number of different individuals, some of whom were a better fit than others for the TIR role.

However, changing the TIR every year added more work for the lead faculty member responsible for conducting the search and completing the hiring process, which sometimes involved extensive negotiations with the TIR's school district. Individuals who remained in the TIR role for more than one year had the opportunity to build stronger relationships with students and faculty, and developed greater institutional knowledge, both of which tended to increase their effectiveness over time. Another consideration that was especially relevant to rural institutions was a limited pool of strong candidates for the TIR position, which in some instances prevented hiring a new TIR every year. The decision to retain or change the TIR was made on a case-by-case basis, after weighing the factors listed above and any other relevant considerations. Most PhysTEC Supported Sites that sustained a full-time TIR position retained the same individual after project funding ended, pointing to the considerable advantages of multiple-year positions.

More recently, a number of PhysTEC Supported Sites have experimented with making the TIR position part time rather than full time. This was typically driven by limitations in funding and/or limited needs with respect to the teacher preparation program. Smaller programs did not necessarily need a full-time TIR, in contrast to larger programs where a full-time TIR was typically needed to meet program goals. In addition, research-intensive physics departments frequently employed a full-time TIR due to very limited time on the part of faculty to engage in teacher preparation activities. While part-time TIRs necessarily had more limited duties than full-time TIRs, they still had a significant impact on teacher preparation programs. Further information on part-time and full-time TIR positions, including outcomes and case studies, is provided below.

C. TIR coordinator

The PhysTEC project engages a TIR coordinator to facilitate the exchange of ideas among TIRs at different sites, provide orientation to new TIRs, and serve as a resource to site leaders. TIR coordinators are typically experienced leaders in education as well as former TIRs. Duties of the TIR coordinator include organizing an annual one-day workshop for TIRs and VMTs. This event provides an opportunity for new TIRs to interact with veteran TIRs and PhysTEC project management, and to get oriented to their new roles. In addition, the TIR coordinator organizes several videoconferences throughout the academic year for TIRs to discuss successes and challenges at their respective sites and exchange ideas. The TIR coordinator also administers project surveys and collects data on TIR activities.

Jon Anderson (a coauthor of this chapter) has served as TIR coordinator since 2009. Anderson has more than 20 years of experience in the classroom and has served as a Modeling Instruction leader and QuarkNet Lead Teacher; he also was a TIR for two years at the University of Minnesota. Previous TIR coordinators Paul Hickman and Drew Isola brought similar classroom and teacher leadership experience to the position.

III. TIR ACTIVITIES

Physics faculty members not familiar with the TIR concept are sometimes unsure what such a position can do for a teacher education program. This section details roles that TIRs typically fill in a teacher preparation program and how TIRs spend their time. TIRs often prove to be invaluable to physics teacher preparation programs, because of the wide variety of relevant activities they can lead and support.

A. TIR roles

The PhysTEC project management, TIR coordinator, and assessment consultant developed a list of specific roles that TIRs fill, based on collective experience working with TIRs. This list of roles formed the basis of a survey that was sent to all TIRs and site leaders since the beginning of the project. As discussed in more detail below, the survey results validated the list as an accurate description of what TIRs do. Below is the list of roles and their descriptions.

- **Recruiter:** Discuss with students and faculty the nature of the teaching profession, including its rewards and challenges, and promote teaching as a viable career option.
- **Advisor:** Engage future teachers in one-on-one interactions and track the progress of students over time, providing individualized attention and support.
- **Instructor:** Teach or co-teach methods or content courses, bringing practical insights from the classroom and providing a stronger physics emphasis to general science methods classes.
- **Course and curriculum developer:** Improve laboratory, content, and methods courses with insights from the classroom.
- **Learning Assistant/Teaching Assistant leader:** Develop the teaching skills of Learning Assistants (LAs) and TAs, through individual interactions or by teaching a pedagogy course; take on administrative tasks in organizing LA or TA training programs.
- **Mentor:** Provide critical mentoring support for future and new teachers; serve as a university supervisor for student teachers.
- **Professional community leader:** Bring together prospective and practicing teachers to meet and communicate on a regular basis.
- **Program coordinator:** Take on responsibilities necessary for a teacher education program that faculty cannot take on, due to lack of time or experience; improve a program by identifying elements that are missing and then developing solutions.
- **Professional development provider:** Design and deliver workshops to support local teachers, and develop demonstration shows to excite and teach school children about physics.
- **Ambassador to school of education:** Build stronger ties between physics departments and schools of education through, for example, discussions or joint meetings.
- **Ambassador to school districts:** Build relationships with local teachers and districts through, for example, the securing of student teaching placements and exchanges of teaching ideas.

Figure 1 shows the percentages of TIRs and site leaders reporting that the TIR performed each of the roles listed

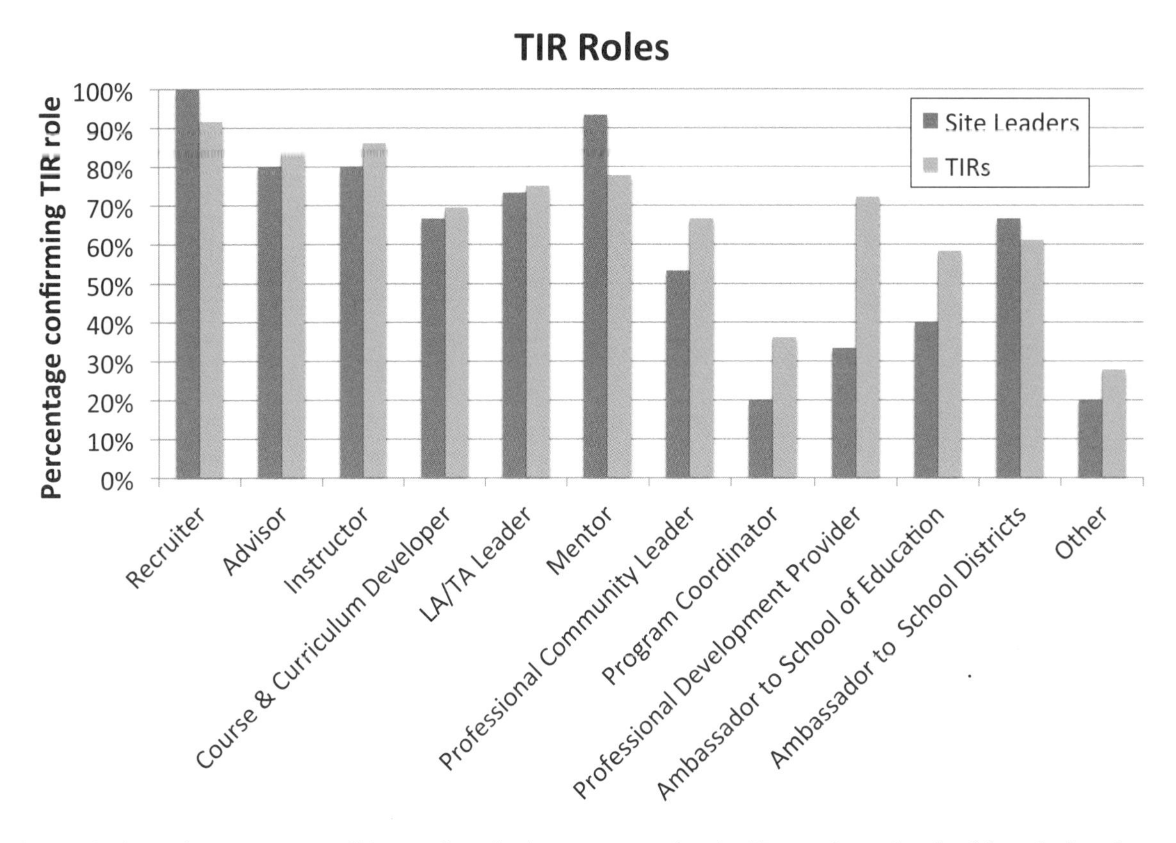

FIG. 1. The graph shows the percentages of TIRs and site leaders reporting that the TIR performed each of the roles listed.

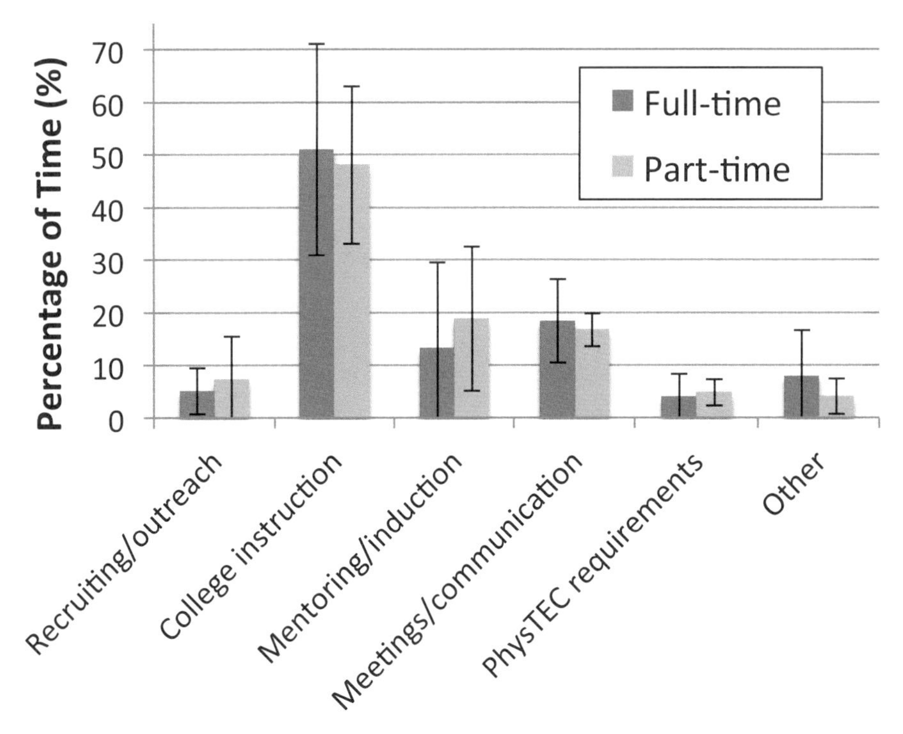

FIG. 2. The graph shows time allocations for full-time and part-time TIRs. Error bars indicate the standard deviations of individual responses.

above. It is based on responses from 15 of 20 site leaders (75% response rate) and 36 of 53 former TIRs (68% response rate) at the time that the survey was administered (academic years 2010–2011 and 2011–2012).

The percentages in Figure 1 show that the categories on the survey are a good match for what the TIRs are actually doing, since every item except Program Coordinator was selected by more than 50% of the TIRs. Fewer than 30% of the TIRs and only 20% of the site leaders selected Other, indicating the list was reasonably complete. The five roles TIRs were most likely to perform are recruiter, instructor, advisor, mentor, and LA/TA leader; these are also the roles that involve the strongest interactions with students.

The similarity between the site leader and TIR responses (most within 10 percentage points) shows that there is generally good agreement on which roles the TIRs held. Some of the difference in responses is likely due to TIRs engaging in a different mix of roles at different institutions, and not having a perfect institutional match between the sample of responding site leaders and TIRs. One area that stands out with a substantial difference is professional development provider, with the TIRs reporting engaging in this role much more frequently than the site leaders. We speculate this might be due to a difference in how the two groups interpret the term "professional development."

B. Time allocation

Figure 2 shows how TIRs allocate their time, based on the monthly time logs submitted by TIRs employed during the 2011–2012 and 2012–2013 academic years. The data are averaged from time logs of 13 full-time TIRs at nine different institutions and five part-time TIRs at three different institutions.[1]

Categories on the time log were grouped to make a few broad categories shown on the graph.[2] Error bars indicate the standard deviations of the individual responses.

TIRs spend about half of their time in instruction and instruction-related activities, including tutoring, working with LAs, and course revision. This is not entirely surprising, given that universities are structured around classes, and TIRs are expert teachers. The first three categories account for about 70 to 75% of TIRs' time, indicating that TIRs spend most of their time working directly with students. This is consistent with the discussion above on TIR roles. They spend less time on interaction with other high school teachers, university faculty, or school administrators. There is some variation among TIRs, as seen by the error bars.

On average, TIRs reported spending little time on direct recruiting of future teachers. This may seem surprising, given that nearly all TIRs engage in recruiting. Moreover, on the roles survey, a substantial proportion of site leaders (60%) and TIRs (36%) indicated it was one of the three categories in which TIRs spent the most time. A likely explanation for this apparent conflict is that TIRs spend a large fraction of their time working with students, for example as instructors, where informal recruiting can occur. With overlapping roles in many activities, and the perceived importance of recruiting, such results are not necessarily inconsistent.

Full-time and part-time TIRs allocate their time similarly, even though part-time TIRs work about one-fifth as many hours. (Full-time TIRs reported an average of 146 +/- 19 hours per month, whereas the part-time TIRs reported an average of 30 +/- 8 hours per month.[3]) A possible explanation is that factors that determine the nature of the TIR position, including program priorities as well as skills and knowledge of a typical physics teacher, do not change depending on whether the position is part-time or full-time.

IV. FULL-TIME AND PART-TIME TIR POSITIONS

Since the start of the PhysTEC project there have been 58 TIRs. Most (53) served as full-time TIRs, and a few (14) served as part-time TIRs; this number includes nine TIRs who were full-time one year and part-time another year. Below is a description of both types of positions, with specific cases to illustrate how these positions fit within the overall context of the various teacher preparation programs.

A. Full-time positions

Full-time TIR positions have been the standard for PhysTEC Supported Sites since the beginning of the project, and continue to be the most common type of TIR position. Full-time TIRs typically engage in the full spectrum of activities listed above in the discussion of TIR roles. Given the substantial work of running a physics teacher preparation program, a full-time position is critical for achieving

---

[1]Full-time and part-time TIRs were distinguished based on full-time versus part-time employment at the university, regardless of the fraction of time spent working directly with future teachers.

[2]"Recruiting/outreach" includes the time log categories "Organize/conduct future teacher recruiting" and "PhysTEC outreach/PR/Ambassador." "College instruction" includes "Course revision (physics or education)," "Tutor/office hours/advising," "Teach/co-teach a course lab or seminar," and "Interview/hire/train Learning Assistants." "Mentoring/induction" includes "Mentor pre/inservice teachers in high schools," "Organize HS observe/volunteer times," and "Organize/conduct TAG meetings." "Meetings/communications" includes "Meet with physics and/or education faculty" and "Email/communication." "PhysTEC requirements" includes "PhysTEC administrative requirements" and "National PhysTEC meetings/videoconferences."

[3]The term after the "+/-" sign is the standard deviation of the average hours worked by individual TIRs.

program goals in many cases, especially at larger institutions. The following example illustrates the variety of activities that can engage a full-time TIR.

Virginia Polytechnic Institute and State University, or Virginia Tech (VT), became a PhysTEC Supported Site in 2011. Virginia Tech's first TIR was also a graduate of the VT physics teacher education program. She spent eight years teaching all levels of physics at Wakefield High School in Arlington, Virginia, before returning to VT as TIR. As a result of her familiarity with the physics department and the teacher licensure program, she was in a unique position to serve as TIR. The TIR taught two courses (Physics Outreach and Physics Teaching and Learning), led the physics department outreach program, served as the Society of Physics Students (SPS) advisor, and organized professional development opportunities for high school teachers and for physics faculty. She also attended the Learning Assistant workshop at the University of Colorado Boulder and worked with faculty to launch an LA program. As a result of her efforts and her ability to build connections with students, the TIR successfully recruited several future physics teachers. During the first three years as a PhysTEC Supported Site (2011-2014), Virginia Tech graduated an average of three physics teachers per year, substantially due to the TIR's efforts.

B. Part-time positions

Part-time TIR positions are a relatively new type of configuration in the PhysTEC project. This type of position can be appropriate for smaller teacher preparation programs, or when the needs of the program are more limited in scope. In some cases a part-time TIR position was driven by limited funding available in the initial award. In other cases, PhysTEC Supported Sites chose to create a part-time position since this is what the institution could sustain after PhysTEC funding ended. Part-time TIRs necessarily have more limited duties than full-time TIRs, but can still have a significant impact on teacher preparation programs. Below are descriptions of three part-time TIR positions at different sites.

California State University, Long Beach (CSULB) became a PhysTEC Supported Site in 2010. In three years the site had three different TIRs, and all three continued teaching full-time in their respective high schools while devoting about 10 hours per week to the TIR role. The CSULB TIR's primary responsibility was co-teaching a physics pedagogy course for preservice and inservice physics teachers. TIRs also spent time organizing a monthly demo sharing event for teachers, assisting with the physics department open house, mentoring teachers in local schools, and publishing a monthly newsletter for local physics teachers.

Since the State University of New York at Geneseo (SUNY Geneseo) became a PhysTEC Supported Site in 2010, a local physics teacher has served as a part-time TIR each year of the project. The teacher had a special connection with the SUNY Geneseo physics department: His father was a founding department faculty member and former chair. The TIR continued to teach full-time at his high school, and was on campus once or twice each week after the school day ended. During his five- to 10-hour weekly commitment, he maintained an office hour, worked in the student physics assistance center, and gave promotional talks on physics teaching in introductory physics course sections. In addition, he organized the "Build it, Teach it, Leave it" program, an opportunity for undergraduates to work with high schools students. Through the program, undergraduates build physics demonstration equipment, take it to a high school classroom and teach a lesson, and then leave it for the classroom teacher.

Seattle Pacific University (SPU) piloted the concept of a part-time TIR by creating a position called the Visiting Master Teacher (VMT).[4] SPU already had a full-time TIR who worked primarily with the elementary teacher preparation program, and the university needed an expert high school physics teacher on a very part-time basis to work with the relatively small number of future physics teachers. The current VMT at SPU is a retired high school physics teacher who has one other part-time job. His decades-long participation in the Seattle high school physics teaching community means that he is familiar with the teachers and schools in the region and can provide substantial mentoring support to future physics teachers at SPU. The VMT co-teaches the physics pedagogy course at SPU, and assists during student teaching by conducting classroom observations of student teachers and new teachers. He meets regularly with future teachers, both individually and in small groups, often at a local coffee shop. He also works with SPU graduates to help them find their initial job placements in the classroom, and has even made visits to first-year teachers to help them make sense of the items in their science stock rooms.

V. IMPACT OF TIRs

Teachers-in-Residence bring an understanding of the high school classroom to physics departments that likely have had limited contact with high school teachers. TIRs are also in an excellent position to mentor physics students considering teaching as a career. In these central roles, TIRs have a direct impact on students, faculty, and physics teachers in the local community. Even the TIRs themselves are changed by the experience.

[4]The Visiting Master Teacher position should not be confused with a Visiting Professor position; the VMT is very part-time (only a few hours per week) whereas a Visiting Professor is typically full-time.

A. Shifting attitudes about teaching

On site visits to PhysTEC Supported Sites, it has often been our experience that one does not have to ask too many questions to uncover some negative attitudes about teaching as a potential career. These attitudes include, for example, that high school teaching is appropriate only for students who are not talented enough for graduate school, that working conditions are so bad that teaching is not a suitable career for anyone, and that teaching is what you do after having a real career first. Future teachers sometimes report hearing these types of comments from faculty, their peers, their parents, and other influential sources. One future teacher at a top-tier institution shared that his peers asked, "Why are you going here to become just a teacher?"

The TIR serves as a positive role model for future physics teachers, and can also influence faculty attitudes about teaching as a possible career. TIRs have recent and extensive classroom experience, and can bring some reality to discussions about the profession of teaching physics. They can share their love of teaching, communicate both the rewarding and challenging aspects of their positions as high school teachers, and answer questions students and faculty may have about day-to-day realities of the profession. TIRs can also arrange for undergraduate students contemplating teaching to go into schools and observe high school physics classrooms and experience for themselves what the environment is like. Sometimes such conversations between TIRs and students take place over time. For example, the TIR at Florida International University manages the Learning Assistant program and works closely with students serving as LAs. During a site visit to the campus, several students noted: "[Being a LA] opened up teaching to me as a real worthwhile pursuit." It is likely that the early teaching experience, guided by the TIR, led to the shift in attitude.

B. Mentoring

PhysTEC TIRs have mentored hundreds of preservice and inservice physics teachers. Research shows that providing high levels of support for beginning teachers through mentoring or teacher-induction programs can lead to higher rates of teacher retention [9,10]. To better understand the mentoring relationship, PhysTEC implemented an annual survey of VMTs and PhysTEC Noyce Scholars, with both Likert-style and open-ended questions. While the detailed mentoring survey was restricted to institutions in the PhysTEC Noyce program, it is likely representative of all TIR-teacher mentoring relationships within the project.

VMTs and Noyce Scholars alike felt that mentoring by the VMTs had a very positive impact on the scholars and provided a professional relationship that would continue to be useful to the scholars after they entered the teaching profession. One scholar noted, "The experience of the mentor has been invaluable to me." Several Noyce Scholars commented on the value of getting feedback from the VMT on their teaching practices and the value of being able to observe the VMT's class as a component of their education classes. Several VMTs expressed the importance of physics teachers networking together as a community of practice, since most schools only have one or two physics teachers. VMTs reported that their mentoring interactions occurred nearly weekly and were initiated by both the VMT and the mentee. Email was the primary means of communication, along with less frequent phone conversations or face-to-face meetings, which took place in the mentee's classroom, a local coffee shop, or at the university.

C. Building community

At most PhysTEC Supported Sites, TIRs have played a central role in building a community of teachers, both preservice and inservice. For example, at SUNY Geneseo the TIR meets regularly with students to lead them in structured outreach activities to local schools. This has created a community of future teachers and students interested in teaching; the community has been growing as students recruit each other to join the group. Prior to the presence of the TIR, students interested in teaching had no place to go to get together and explore their common interests. The number of students enrolled in the physics teacher education program at SUNY Geneseo has been growing rapidly, and faculty and students attribute this in large part to the TIR's activities, which center on the outreach program.

The TIR often takes on the role of organizing a Teacher Advisory Group (TAG) at PhysTEC Supported Sites. The TAG is a group of area high school physics teachers who meet regularly to advise and engage with the project. Some TAGs organize professional development activities; for example, the CSULB TAG holds regular demonstration-sharing events as a way for preservice physics teachers at the institution to meet inservice teachers in the area. The TAG begins to build a local community of physics teachers around the PhysTEC site, which also benefits inservice teachers, many of whom are the only physics teachers in their buildings.

D. Transforming teaching practices

Faculty often recognize the expertise of the TIR in the domain of the high school classroom. It can take a little longer for faculty to appreciate that the teaching skills of the TIR also apply to the university classroom. When this happens, the TIR is in a good position to help transform teaching practices in the physics department. For example, the TIR at Towson University gained the respect of several faculty members regarding teaching practices. Faculty came to him for assistance on developing and implementing lecture demonstrations as well as creating outreach activities. As part of her duties, the TIR at Virginia Tech taught a section of introductory physics. When her students outperformed all the other sections, including those led by faculty, the faculty took notice and started coming to her for advice about effective pedagogy.

E. Impact on TIRs

TIRs and VMTs reported being enriched by the experience of working on a university campus, and they appreciated feeling that they were giving something back to the teaching profession. The following quotations are typical of comments made by TIRs and VMTs on end-of-year surveys:

> [Being a TIR] has helped me broaden my contacts at the university and with other high school teachers. Additionally I found that it has helped me evaluate my present teaching methods.

> This association with the mentees has rewarded me by continuing and extending my connection with a variety of school districts. Also the work has been valuable for my personal growth (as a veteran teacher with time to give back) and professionally as a link to the School of Education.

A somewhat unexpected impact on TIRs was that some were hired away from the high school classroom at the end of their service. Approximately 10 percent of TIRs remained employed at the university and did not return to the classroom. Though this was likely a net benefit to the physics teacher education enterprise, it could be seen as a problem by school districts facing a shortage of qualified physics teachers.

VI. IMPLEMENTATION ISSUES

The TIR position was new to almost all physics departments starting a PhysTEC program. Moreover, most school districts had little to no experience with sending a teacher to work at a university for a year, and administrative barriers were not uncommon. We describe successful strategies used by PhysTEC Supported Sites to address common challenges in recruiting, hiring, managing, and sustaining the TIR.

A. Recruiting TIRs

Universities found excellent TIRs in many different places. It was critical to start the search for a TIR as early as possible, as the best teachers tended to line up commitments far in advance, and negotiations with school district administrators unfamiliar with the TIR concept often took more time than expected. Below is a list of strategies that have worked for one or more PhysTEC Supported Sites.

- **Look for a TIR among people already involved with your institution.** Look for outstanding participants in teacher workshops or summer institutes. Other candidates may include those who work as adjuncts or in outreach programs at your university. Also consider past graduates of your teacher education program.
- **Organize a professional learning community or teacher advisory board.** Developing a community of teachers around your institution will give you the connections you need to find good candidates.
- **Network through national organizations to identify local leaders.** Contact the American Association of Physics Teachers (AAPT) to learn who the Physics Teaching Resource Agents (PTRAs) are in your area [11]. Look on AAPT and National Science Teacher Association (NSTA) websites for lists of local section leaders. Also, if there is a local QuarkNet program [12], ask about the physics teachers who are involved.
- **Ask your students if they had a good high school physics teacher.** If your institution draws a significant number of students that are local or from surrounding counties, your students will be able to help identify good TIR candidates.
- **Use your current TIR to recruit the next one.** Experienced teachers usually have lots of contacts in the local physics teacher community, and may be able to suggest good candidates.
- **Supplement with conventional recruiting strategies.** Website job postings, mailings, and word of mouth can also be effective TIR recruiting strategies.

B. Hiring TIRs

When PhysTEC Supported Sites attempted to hire full-time TIRs who were currently teaching, it was critical to allow plenty of time for negotiations. Understandably, many school districts were reluctant to release an expert teacher from the classroom, and a convincing case needed to be made that the arrangement would bring benefits back to the district. Also, it was useful to first ask teachers about their school districts, to learn which administrators to approach and in what order. The decision-making process required to release a teacher was highly dependent on the particular district.

Below is a list of strategies that one or more PhysTEC Supported Sites used in successful negotiations with school districts.

- **Help the district save money on salaries.** Frequently, the school district pays the salary of the replacement teacher while the university pays the salary of the TIR, which typically results in a cost savings for the district. (Some universities negotiate splitting the difference in salaries, which still results in savings for the district. A few universities negotiate paying for the replacement teacher while the TIR remains on the district payroll, in which case the district breaks even.)

- **Offer a recent graduate as a replacement for the TIR.** This can help greatly with building relationships with local school districts.
- **Have the TIR mentor his or her replacement.** The TIR can work with his or her replacement to make that teacher as effective as possible.
- **Grant the TIR access to resources.** Allow TIRs (current and former) to borrow lab and demonstration materials and use department facilities.
- **Point out that the TIR will build relationships that can benefit high school students.** The TIR will return with good connections to the university, especially with the physics department and education school. This can help students, for example, get into university labs and gain research experience while in high school. Also, letters of recommendation written by the TIR may carry more weight as a result of stronger relationships.
- **Support the TIR in leading professional development for teachers in his or her district.** Curriculum changes including the Next Generation Science Standards [13] and the new Advanced Placement physics curriculum [14] will require significant professional development. The TIR is a natural leader who can help his or her colleagues implement such changes as well as improve existing practices.

It was also important to ensure that the TIR did not lose status, pay, or even his or her job after serving as TIR and returning to the school system. In many cases the TIR remained an employee of the school district (on special assignment), which avoided negative impacts on the teacher's years of service, retirement and other benefits, and placement on the salary schedule. It worked best to get a commitment in writing that the TIR would return with the same status (department head, for example), same teaching options, same room, and other important conditions of employment.

Some school districts were easier to work with than others; in a few cases a district was all but impossible to work with. In such cases, sites hired a retired teacher or a teacher willing to leave his or her position to solve the problem. Retired teachers[5] have benefited PhysTEC programs by bringing a wealth of cumulative experience and relationships. In a few instances, retired teachers did not work as well, for reasons that included decreasing relevance of the teacher's experience the longer the teacher was out of the classroom, diminished overall energy, and barriers in connecting with students across age differences. Keeping these kinds of possible concerns in mind can be useful during the interview process.

For part-time TIRs who remained in the classroom, negotiations with the school district were simpler; in some cases they bypassed the district altogether. TIRs who remained teaching a full load in their schools typically just informed their principal and/or superintendent about the additional commitment. In a few cases, the university paid the school district a partial salary for the TIR, who was granted a partial release from the standard teaching load.

C. Managing the job description

The biggest challenge with managing the TIR position at many PhysTEC Supported Sites was often time management. Typically it was easy to generate a long list of items to do, and the key was to keep the list prioritized. Regular, frequent interaction between the TIR and the site leader was critical to establishing and adjusting priorities as needed. It was helpful for the TIR to discuss the job description with the site leader before starting, in order to get the big picture. Some degree of flexibility was also useful, as a number of TIR positions evolved considerably. In addition, recognizing the unique skills of the TIR was critical to keeping activities focused on the TIR's strengths, including expertise in the precollege classroom, connections with local schools, and teaching skills that frequently transferred to college classrooms.

D. Sustaining the TIR

Most PhysTEC Supported Sites that completed PhysTEC funding (i.e., legacy sites) found a way to sustain (or partly sustain) a TIR. In many cases, PhysTEC legacy sites secured internal funding for the TIR through teaching courses, and in a few cases internal funding was provided for the TIR to do administrative work related to teacher education. Some PhysTEC legacy sites found external funding to sustain a TIR; sources included UTeach replication grants, Noyce grants, and private donations.

The most readily available source of internal funding for sustaining TIRs was through having the TIR teach courses. TIRs have taught or co-taught a variety of courses, mostly in physics departments, including lab and recitation sections of introductory physics, physics courses for elementary teachers, physics pedagogy courses, and other science methods courses. Virginia Tech sustained a full-time TIR position by having the TIR teach sections of introductory physics, a course for students doing physics outreach projects, and a pedagogy course for Learning Assistants. CSULB sustained its part-time TIR by having the teacher teach two courses: a pedagogy course for Learning Assistants and a special topics course on physics pedagogy for preservice and inservice teachers. At Florida International University, TIRs were sustained internally through a mix of administrative

[5]To date, 11 of the 58 TIRs (19%) employed by the project were retired teachers.

and teaching assignments. One full-time TIR was hired as the director of the secondary science education program; the other full-time TIR became the supervisor for the Learning Assistant program in physics and taught sections of introductory physics.

A number of PhysTEC Supported Sites received external funding to become UTeach replication sites. The UTeach master teacher positions provided an opportunity for hiring and sustaining an expert physics teacher to instruct and mentor future teachers. At Middle Tennessee State University, the PhysTEC site leader was part of the UTeach leadership team and worked with colleagues to hire a physics master teacher. In addition to teaching UTeach certification courses, this master teacher also coordinated with the PhysTEC program and taught the Modeling curriculum [15] in a nearby high school, so that future physics teachers could experience an excellent classroom. Some PhysTEC Supported Sites received external funding from other sources, such as the six sites that participated in a National Science Foundation (NSF) Noyce Scholarship grant which funded Visiting Master Teachers, described above.

The most recent PhysTEC solicitations for new sites encouraged a robust sustainability plan as part of the initial proposal. Proposal writers used the promise of external funds to leverage internal funding to continue the project after the three years of PhysTEC support. Most successful proposals had commitments to sustain the TIR for an additional three years, for a total of six years of funding, after which one could reasonably hope that the position would become institutionalized. In most cases the commitment for internal funds to sustain the TIR came from the dean level, and in a few cases from the provost level or department level. Some institutions had two or more units contribute a partial salary to reach the total desired amount. At the University of Missouri, the College of Arts and Science and the College of Education each contributed 0.25 FTE to create a 0.5-FTE teaching professor position for the TIR to continue after PhysTEC funding ends. At Georgia State University, the College of Education promised funds for a 0.5-FTE position to be filled by a science education Ph.D. student (typically an expert science teacher).

VII. SUMMARY AND CONCLUSIONS

The presence of a high school physics teacher in a university physics department can bring about a paradigm shift for physics faculty and students. Faculty start to see the role that physics departments play in the preparation of future teachers, and have the opportunity to observe first-hand the professionalism of high school teachers. Students see that being a high school teacher is an intellectually stimulating and rewarding career path, not just a backup plan for physics majors who are not interested in or prepared for graduate school.

Institutions that hope to become leaders in physics teacher education should consider hiring a Teacher-in-Residence. Experienced physics teachers have the flexibility to fill many key roles in a teacher preparation program, and bring a unique skill set that is typically not found elsewhere on campus. Both full-time and part-time positions can be effective, depending on the needs and resources of the program.

ACKNOWLEDGMENTS

We gratefully acknowledge contributions from Gay Stewart on strategies for hiring TIRs and from David Meltzer and Paul Hickman on developing the survey on TIR roles. This material is based upon work supported by the National Science Foundation under Grant Nos. 0108787, 0808790, and 0833210. Any opinions, findings, and conclusions or recommendations expressed in this material are those of the authors and do not necessarily reflect the views of the National Science Foundation.

[1] C. Hinton and K. W. Fischer, Research schools: Grounding research in educational practice, Mind Brain Educ. **2** (4), 157 (2008). doi:10.1111/j.1751-228X.2008.00048.x

[2] Organisation for Economic Development and Cooperation, *Building a High-Quality Teaching Profession: Lessons from Around the World* (2011). http://www2.ed.gov/about/inits/ed/internationaled/background.pdf

[3] See http://www.bostonteacherresidency.org

[4] Carnegie Corporation of New York, *Teachers for a New Era: Transforming Teacher Education* (New York, NY, 2006). http://carnegie.org/fileadmin/Media/Publications/PDF/Carnegie.pdf

[5] K. Zeichner, Rethinking the connections between campus courses and field experiences in college- and university-based teacher education, J. Teach. Educ. **61**, 89 (2010). doi:10.1177/0022487109347671

[6] See https://uteach.utexas.edu

[7] See http://www.uteach-institute.org

[8] M. Piburn, D. Sawada, K. Falconer, J. Turley, R. Benford, and I. Bloom, Reformed Teaching Observation Protocol (RTOP): Reference Manual, in *ACEPT Technical Report IN003* (2000). https://mathed.asu.edu/instruments/rtop/RTOP_Reference_Manual.pdf

[9] C. Brewster and J. Railsback, *Supporting Beginning Teachers: How Administrators, Teachers, and Policymakers Can Help New Teachers Succeed* (Northwest Regional Education Laboratory, Portland, OR, 2001). http://files.eric.ed.gov/fulltext/ED455619.pdf

[10] L. Olsen, Finding and keeping competent teachers, Educ. Week **19** (18), 12 (2000).

[11] See http://www.aapt.org/ptra

[12] See https://quarknet.i2u2.org

[13] NGSS Lead States, *Next Generation Science Standards: For States, By States* (The National Academies Press, Washington, DC, 2013).

[14] See http://apcentral.collegeboard.com/apc/public/courses/teachers_corner/2262.html

[15] See http://modelinginstruction.org/about/modeling-method-synopsis

APPENDIX: PHYSTEC MOU TEMPLATE

RECRUITMENT PLANS			
What Efforts?	**Who is Responsible?**	**Timeline**	**What Outcomes?**
Announce in each intro algebra- and calc-based physics class the option of teaching and provide a contact person. TIR is present during announcement to provide a "face" to the contact name. Make available sheet with pertinent information for students to take after class.	Physics faculty	15 Sep 15 Feb	Signup sheets collected
Bring in a HS physics teacher to give a student seminar on HS Physics teaching	TIR	Fall	Signup sheet from students who want more information
Have majors involved in judging at science fairs and helping in other school events	PI (xxxx) **NOTE: In all parentheses in this document, please provide name of person or persons responsible.**	Spring	Contact information from students met, photographs
Create new or adapt existing posters to recruit people into teaching. Place posters throughout science building.	TIR	15 Oct	Posters up in halls, copy of original to PC
Contact admissions office to see if "interest in teaching" can be placed on postcards sent to admitted students	PI ()	Fall	New postcard
Meet with local SPS or NSTA student chapters; show video and discuss recruiting strategies	TIR	Fall	List of actions
Meet with physics teachers at regional meetings (e.g. AAPT or NSTA section meetings), raise awareness about PhysTEC, and discuss recruiting strategies	TIR	Fall	List of actions
Hire external PR person to develop campaign to recruit more teacher candidates and majors	PI (), Chair	Fall	Develop contract with PhysTEC leadership

Meet with Engineering chair to discuss recruiting in engineering classes (this could involve many of the same actions taken within the physics department)	PI ()	15 Sep	Action item list
Help with physics open house to promote secondary teacher education, building in a recruitment pitch into schedule	TIR	10 Oct	Signup list from interested students
Ask physics advisors to explicitly mention teaching as a career option	PI ()		List of students interested in physics teaching option
Contact student newspaper to describe effort and get story on project in paper	PI ()	1 Oct	Story appears. Send copy to PC
Give talk to SPS chapter on teaching careers	TIR	1 Nov	Signup list for more information
Form a "doubling" committee to work toward increasing the number of majors	Physics faculty	All year	List of actions
Write a Noyce teacher scholarship proposal	PI ()		Submitted proposal
Have TIR meet with graduate students	TIR		

• TIR ACTIVITIES			
What Efforts?	**Who is Responsible?**	**Timeline**	**What Outcomes?**
Develop a TIR plan, refine plan at summer AAPT meeting	PI ()	15 Sep	Plan approved by PhysTEC TIR coordinator
Entrance and exit surveys given to TIR	PI ()	15 Aug 15 Jun	Surveys sent to PC
TIR contacts previous TIRs (from this or other sites) to learn about issues	TIR	15 Sep	Prepare a list of activities and review with PI. Send copy to TIR coordinator
Present at regional AAPT meeting on impact of TIRs and PhysTEC	TIR, PI ()	Nov	Send copy of presentation to PC

Weekly meetings with PI to discuss effective ways to further PhysTEC goals with the TIR	TIR, PI ()	Entire year	TIR will keep notes for annual report
TIR orients new TIR by reviewing ongoing activities, departmental challenges, support networks and resources	TIR, TIR (next year)	Summer	New TIR writes and keeps notes on issues
TIR attends weekly LA and TA meetings and offers suggestions to aid LA/TA thinking about pedagogical issues	TIR	Fall, Spring	
Co-teach science methods course and mentor 5th year preservice teachers	TIR	Fall, Spring	Course evaluations available to project
Attend TIR professional development workshop prior to AAPT meeting	TIR	July before and after year	Evaluation form from workshop
Recruit next year's TIR	PI (), TIR	Fall	Short list of possible candidates
Hire next year's TIR	PI ()	February	Agreement in place
Draft plans to engage new TIR during current year	TIR	January	Draft plan
Meet with Noyce Fellows to understand their needs	TIR	1 Nov 1 Mar	List of actions
Classroom observations of recent graduates	TIR	Ongoing	Discussion with teachers on issues seen
Serve as liaison between physics faculty, education faculty, prospective teachers, and local HS teachers	TIR	Ongoing	

• LEARNING ASSISTANTS			
What Efforts?	**Who is Responsible?**	**Timeline**	**What Outcomes?**
Recruit LAs for coming semester (announce in classes, emails to top students, pizza information party)	PI (), TIR	1 Nov	List of possible LA candidates
Conduct LA interviews, make hiring decisions	Faculty (), TIR	1 Dec 1 Apr	Final list of prospective LAs

Make LA offers, handle paperwork		7 Dec 7 Apr	Final list of LAs
Teach pedagogy course for LAs (mandatory) and TAs (optional)	Faculty member (), TIR	Spring	Course enrollment and evaluations provided to project
Weekly LA meetings to review content for coming week, discuss issues of past week	TIR	Weekly	TIR's notes on issues to bring back to PI or others
Semester-end evaluation of LAs	TIR, PI ()	15 Apr	List of issues to address in next year's LA cohort

• MENTORING (BOTH IN THE UNIVERSITY AND IN THE SCHOOL DISTRICTS)			
What Efforts?	**Who is Responsible?**	**Timeline**	**What Outcomes?**
Create a 1-2 page document that outlines each of the pathways for student to become a teacher	PI ()	Oct	Document sent to PC, up on site website, given to prospective teacher candidates
Meeting between PhysTEC physics and education faculty to review classroom observations of new PhysTEC teachers	PI (), Education faculty ()	Spring	Brief paragraph for student's portfolio
Take mentees to regional AAPT meeting	TIR	Nov	Collect evaluation of experience from each
TIR takes state mentoring workshop to achieve state certification in mentoring	TIR	Feb	Reports experience to PI
Meet with identified PhysTEC future teachers once/semester to inform them of internships, research experiences and other opportunities	TIR	1 Nov 1 Mar	Signup list handed out to track interest
Develop a mentoring plan for PhysTEC graduates and other interested new local teachers	TIR	15 Sep	Review plan with PhysTEC TIR coordinator
Mentees complete year-end exit survey	TIR	15 May	Send survey to TIR coordinator
TIR identifies future mentees and arranges contact with them	TIR	15 May	Sends list of future mentees contact info to PI

Contact new local HS physics teachers to open discussion on how site might provide support for them	TIR	15 Oct	List of potential support efforts

• TEACHER ADVISORY GROUPS			
What Efforts?	**Who is Responsible?**	**Timeline**	**What Outcomes?**
Send word out to local teachers about a TAG meeting	TIR	1 Oct 1 Feb	
Hold TAG meeting	TIR, Faculty ()	1 Nov 1 Mar	Signup list of local teachers, evaluation forms from participants
Meet to discuss improvements for future TAG meetings (improved content, attracting more teachers, better involvement from teachers and university faculty)	TIR, Faculty ()	15 Nov 15 Mar	List of improvements for next meeting
Meet with TAG to discuss needs of local teachers	PI (), TIR	1 Feb	List of actions to address teacher's concerns
Find best local teachers by asking current students if they had a "great" physics teacher. Get to know this person, and include them in TAG and other activities.	PI	1 Nov	Assemble names of these teachers for TAG, future placements, etc.

• EARLY FIELD EXPERIENCES			
What Efforts?	**Who is Responsible?**	**Timeline**	**What Outcomes?**
Have prospective teachers visit high school physics classes, meet with TIR subsequently to discuss experience	TIR	Spring	Note for student's portfolio if needed
Visit classrooms of PhysTEC future teachers observe early field experience	PI ()	Fall Spring	Note for student's portfolio if needed
Group meeting with LAs on teaching experience, address concerns about teaching and teachers	PI (), TIR	1 Apr	Gather names of those interested in teaching

Prospective teachers / LAs help with physics demonstration show	TIR	Spring	

• INSTITUTIONAL CHANGE (INCLUDING PLANS FOR BOTH CURRICULAR AND ORGANIZATIONAL CHANGE)			
What Efforts?	**Who is Responsible?**	**Timeline**	**What Outcomes?**
Hold regular meetings with Education faculty to discuss interaction between secondary methods and content courses	PI (), Education faculty ()	Spring	Task list of items to address to improve interaction
Ask Education dean/chair to distribute flyer/email to all Education faculty explaining PhysTEC's goals and how TIRs can present in courses in order to promote better prepared science teachers	TIR	15 Aug	Flyer distributed
Present at College of Education Luncheon regarding elementary issues: PET – Physics and Everyday Thinking; role of physics department	TIR, PI ()	Spring	List of questions to address from education faculty
Report on PhysTEC progress to physics faculty at departmental meetings	PI (), TIR	Fall, Spring	List of questions to address
Send new physics department hire to AAPT / APS / AAS New Faculty Workshop	Chair	Fall	
Invite HS teacher to give departmental colloquium	TIR, PI ()	November	
Invite PER faculty to give departmental colloquium	TIR, PI ()	February	
Send one faculty member to national workshop on LAs at UC Boulder	PI ()	Fall	Faculty writes 1-page synopsis of experience and possible changes to local program, and sends along with travel reimbursement to PC
Introduce new HS Teaching track into physics major	PI ()	Fall	File paperwork with university curriculum committee
Introduce new PER track into physics major	PI ()	Fall	File paperwork with university curriculum committee

Extend implementation of PET to accommodate all elementary education majors	PI ()	15 Mar	Change in requirements for ElEd degree

• **ASSESSMENT (INCLUDING PROVIDING HISTORICAL DATA)**			
What Efforts?	**Who is Responsible?**	**Timeline**	**What Outcomes?**
Develop a tracking plan to follow all HS teachers	PI ()	1 Oct	Plan sent to PC for approval
Track all PhysTEC high-school teachers	TIR	1 Jun	Results included in annual report
Give FCI exam as pre and post in all sections of calc-based and algebra-based intro physics	Physics faculty ()	30 Dec	Results reported to PI, sent to PC
Give CSEM exam as pre and post in 2nd semester of intro physics	Physics faculty ()	15 Jun	Results reported to PI, sent to PC
Analysis of all pre/post data from final year of project	PI ()	30 Jun	Results included in annual report
Provide PC baseline data as requested in PhysTEC template	PI ()	1 Sep	Data sent to project staff for review
Review a set of lesson plans from three PhysTEC future teachers	TIR	15 May	Write note for student's portfolio
Document, for eventual publication, specific actions or stories that describe unique nature of PhysTEC project at your institution.	Faculty ()	Ongoing	Provide outline to PIs for review along with list of data needed to collect for story.

• **SUSTAINABILITY**			
What Efforts?	**Who is Responsible?**	**Timeline**	**What Outcomes?**
Write NSF grant to study XXX in LAs/teachers/curriculum	PI ()		Submitted grant
Generate talking points for dean/provost on University's efforts to improve teacher education for the community in STEM education fields	PI ()		Talking points given to Dean/Provost

Invite dean/provost to event for prospective teachers	PI ()		
Send brief (1-page) report to dean, provost on advances made during year on teacher education in physics. Review with chair before sending.	PI ()	May	One page report
Draft letter to university President from project	PI ()	15 May	Letter sent by project to inform university of progress
Meet with college development officer about fundraising for STEM teacher education	PI ()		
Get list of foundations to consider approaching from APS	PI ()	1 Apr	Use list to open discussion with university relations
Discussion with chair about continued funding for TIR past grant period	PI ()	1 Apr	Notes from discussion, actions to inform next steps
Meet with "strategic initiatives" university committee to discuss role of teacher education	PI ()	Spring	Notes/actions from meeting

• **PROJECT ACTIVITIES**			
What Efforts?	**Who is Responsible?**	**Timeline**	**What Outcomes?**
Attend PhysTEC annual meeting and half-day PhysTEC project meeting. Team will include PI, TIR, additional interested faculty	PI ()	3 Feb	Notes on actions to take in coming year
Attend full-day PhysTEC project meeting before start of project	PI ()	July	Notes on actions to take in coming year
IRB paperwork in place	PI ()	1 Oct	Send message to PC indicating this and any limitations project must exercise over data it receives
Write cumulative annual report according to PhysTEC template	PI ()	1 Jun	First draft in by 1 June, final by 15 June
Submit data with annual report according to PhysTEC template	PI ()	1 Jun	First draft in by 1 June, final by 15 June

Contact PhysTEC graduates and ask them to submit information requested by the project	PI ()	Annually	PhysTEC graduates contacted
Attend monthly project videoconferences	PI ()	Monthly	Occasional reporting out
Build local website to include PhysTEC and teacher education materials	PI ()	1 Nov	New content up on web
Hire photographer to take publicity pictures for website and articles. Get release forms from all photographed	PI ()	15 Nov	Send picture CD, release forms to PC
Hold teleconference with Project Director on MOU progress	PI ()	1 Dec 1 May	Revisit MOU. Prepare list of actions to take
Monthly meeting of project team to discuss progress and set priorities	Team	Monthly	Action items from meeting
Host PhysTEC site visit	PI ()		Action list from visiting team
Participate in a PhysTEC site visit at another university	PI () or other faculty member ()		Half-page summary of visit

Cultivating outstanding physics teacher mentorship

Jeffrey Nordine
Leibniz-Institute for Science and Mathematics Education (IPN), Kiel, 24118, Germany

Angela Breidenstein
Trinity University, San Antonio, Texas, 78212, USA

Amanda Chapman and Penny McCool
Robert E. Lee High School, San Antonio, Texas, 78213, USA

Physics mentor teachers wield tremendous influence over the learning trajectories of preservice physics teachers. As the primary example of what effective high school physics teaching looks like in practice, mentors model how to design and implement physics instruction that can inspire an increasingly diverse group of learners to understand and appreciate physics. Although university-based teacher education relies heavily on mentor teachers, physics mentors teachers are often selected solely based on their willingness, location, or availability, and are often given little support in performing their vital roles. In our work at Trinity University, we have learned that outstanding physics teacher mentors are not simply discovered; they are cultivated through sustained school-university partnerships and intentional support from university faculty. This chapter describes critical practices in physics teacher mentoring and how these practices are systematically supported in the Trinity teacher education program. While not all teacher education programs should look like Trinity's, the mentorship practices and support systems we describe hold promise for fostering outstanding physics teacher mentorship in a wide range of teacher education contexts.

I. INTRODUCTION

Physics mentor teachers play a critical role in developing the beliefs and practices of preservice physics teachers. Mentor teachers in the schools, also known as cooperating teachers, are current classroom teachers who work with university students for a significant segment of their education studies. Although this partnership varies in different education programs, the mentors usually share their classrooms and students with the preservice teachers. As preservice teachers spend a large portion of their time working with mentors and observing them teach, mentors wield enormous influence over the learning trajectories of beginning physics teachers. Mentors help preservice teachers transition from learning physics in predominantly lecture-based university classes, usually with academic requirements for student entry and chosen student interest in the subject, to teaching in a high school physics classroom with a diverse student body and an emphasis on student engagement. The mentor can be critical in determining how the preservice teacher develops the knowledge, skills, and dispositions not only to teach the physics curriculum but also to inspire students to appreciate and continue their study of physics.

We write this chapter as two university-based teacher education faculty (Jeff, a science educator formerly in the Trinity University Department of Education, and Angela, a high school teacher educator currently in the Trinity Department of Education) and two experienced physics mentor teachers (Amanda and Penny). In the chapter, we describe the Trinity University teacher preparation program and discuss how physics teacher mentors are supported through a long-term school-university partnership and through specific support systems offered by university-based faculty. Throughout our work together, we have learned that outstanding physics teacher mentorship doesn't just happen on its own; rather, it is fostered through a series of systemic collaborations between university faculty and teachers in the schools. Our aim in this chapter is to provide a picture of how physics teacher mentors are identified and supported at Trinity, so that colleagues in other university and school settings may benefit from our experiences as they work to support physics teacher mentors.

II. THE TEACHER EDUCATION PROGRAM AT TRINITY UNIVERSITY

A. General overview

Trinity University's teacher preparation program is designed as a five-year sequence in which teacher candidates earn a B.S. or B.A. degree after four years and a Master of Arts in Teaching (MAT) in year five. It is possible for students to enter the Trinity MAT program without attending Trinity as an undergraduate. In these cases, Trinity faculty work with incoming MAT students to design an individualized plan for them to explore the teaching profession and gain field experiences with K-12 students in advance of beginning the MAT program.

The MAT program, which has been in place since 1991, is designed according to principles of best practice in teacher education: strong content knowledge for teaching, extended and intensive internships, integration of theory and practice, and carefully designed contexts for authentic professional learning experiences. For Trinity these contexts are school-university partnerships, known as professional development schools (PDS), intended to be mutually beneficial to both partners [1].

The MAT program is small by design, graduating 12 to 20 teachers each year who are certified to teach secondary school. Prospective secondary teachers of all content areas study together in the same cohort. Since 2006, the Trinity MAT program has graduated seven physics teachers (roughly one physics teacher per year). Of these seven physics teachers, four completed their B.S. or B.A. in the Trinity physics department (which graduates roughly six physics majors per year) and one completed a B.S. in Trinity's engineering science department with a minor in physics. The other two physics MAT graduates earned a B.S. in physics and a B.S. and MS in chemical engineering from other institutions. With roughly one graduate per year, Trinity's production of physics teachers is relatively strong compared to many universities in the United States [2].

Prospective physics teachers at Trinity benefit from the attention of faculty in both the physics and education departments. (Trinity employs a departmental structure.) Physics professors serve as advisors in prospective teachers' majors and as instructors in their content courses. Education faculty teach methods courses and other education-related courses, and provide education-focused advising support. A physics faculty member holds a place on the university's Council for Teacher Education, which is responsible for oversight of the MAT program; this faculty member participates, along with education faculty, in all other key certification areas and in admissions decisions for the MAT program. This council also provides programmatic feedback and consultation, including developing a set of recommended courses that address the state certification exam and teaching curriculum.

This physics-education partnership extends as students continue into the MAT year and into their professional practice. Physics faculty offer both teaching candidates and mentor teachers content-related ideas, resources such as laboratory equipment, and sometimes opportunities to bring high school students to the university for laboratory or other experiences (e.g., astronomy nights). Education faculty collaborate with physics faculty to hold undergraduate seminars in physics education, recruit prospective teachers to the MAT program, and facilitate collaborative projects.

B. Learning sequence in the Trinity University teacher education program

The Trinity teacher education program is designed as a five-year sequence in which teacher candidates first earn bachelor's degrees in their teaching fields and then complete a one-year, 36-credit-hour master's degree program that culminates in the MAT degree. During their undergraduate years, physics candidates will complete a B.S. or B.A. in physics while taking between eight and 18 hours of core courses in education to prepare for entry into the MAT year. Included in these cores courses are site-based practicum courses in which undergraduates gain experience working with students and teachers in K-12 schools, as well as courses in human growth and development, special education, science and mathematics teaching methods, and urban education. At this early stage, it is critical to expose prospective teachers to teachers who are engaging, innovative, and positive, as these teachers can be the best advertisements for entering the teaching field [3]. This is particularly true in science and mathematics, because prospective teachers maybe feel pressure to pursue higher-paying or more prestigious careers.

The Trinity MAT program takes place within one intensive year, during which prospective teachers complete graduate work and complete an extended teaching placement. The intensity and extended nature of the teacher education program is reflected by the use of the term "interns," rather than "student teachers," to describe MAT students. Trinity interns begin their work in mid-July by taking courses focusing on curriculum development and pedagogy. Also in the summer, interns meet their formally assigned mentor teachers for an intern-mentor orientation. Often, this mentor is someone with whom the intern has worked during a practicum course. Interns then join their mentors on the first day that teachers return from summer break for professional development prior to the beginning of the academic year. The internship lasts until interns phase out of the classroom in April, eight months later. In the fall semester, interns are with their mentor teachers in the school four days a week; on the fifth day the interns are at the university to participate in two graduate courses, in which they continue to learn about curriculum, pedagogy, and approaches for teaching the diverse groups of learners in their classrooms. Interns work closely with faculty in the MAT program and their mentor teachers to complete three key projects as part of their fall semester work. Prior to and during all of these projects, MAT program faculty frequently meet with mentors to introduce the projects and explain the mentor teachers' role in them.

Interns take over all lead teaching responsibilities from their mentors in the spring semester. In addition, interns continue their graduate coursework by participating in a graduate seminar in which they discuss their ongoing classroom work with peers, and by taking a course focusing on teacher leadership. After a lead teaching experience of at least nine weeks, interns begin to phase out of the classroom in April and turn their focus to writing their internship portfolio, in which they document and reflect upon evidence for their progress relative to the professional teaching standards that guide the Trinity MAT program.

The MAT year is both an intensive and an extensive experience for interns, and requires substantial growth in a number of intellectual and dispositional dimensions. Physics interns must enhance their existing knowledge of physics principles to develop pedagogical content knowledge (PCK), which includes knowledge of how to represent and connect physics principles for learners, a deep understanding of common student misconceptions, and content-specific strategies for helping students develop their naïve ideas into more accurate and sophisticated conceptions [4].

We believe that the function of a teacher preparation program is to set teacher candidates on a trajectory along which they begin to learn to teach, and that teaching expertise is developed over years rather than months. Further, we acknowledge that without careful attention to the day-to-day realities of schools and the integration of theory and practice during their teacher education program, beginning teachers often struggle to transfer what they have learned in their undergraduate physics coursework and in their education coursework to their practice as teachers. Thus they tend to forget, or rather not rely upon, what they learned during their preservice teacher education [5,6]. Since pedagogical content knowledge and professional teaching dispositions are practiced in context (i.e., not solely in the university classroom), these core competencies are truly developed through interns' experiences in the high school classroom with their mentor teachers, professional development school colleagues, students, and university faculty. Thus, high-quality mentor teachers and their support for interns' induction into the profession are among the important components of a successful teacher preparation program. Trinity's program has been recognized in the teacher education literature as an exemplary program for its programmatic design, use of professional development schools as contexts for clinical practice, and commitment to mentor development and support [7,8]. Research on the Trinity program has revealed high levels of retention, positive self-efficacy assessments, and strong reviews from principals in whose schools interns teach [9,10]. In the next sections, we identify critical practices of outstanding physics teacher mentors and describe how we systematically support the development of these practices.

III. CRITICAL PRACTICES IN PHYSICS TEACHER MENTORING

Our mentoring program was designed with consideration for how mentor teachers are capable of offering unique support and guidance to their interns as well as with an understanding that learning to teach is a long-term developmental process that takes place over years [11]. In this section, we identify critical areas of physics teacher mentorship and describe how mentors are uniquely positioned to guide preservice physics teachers as they begin the process of developing expertise in physics teaching. These areas are modeling physics pedagogy for all students, modeling reflective practice, and supporting pedagogical content knowledge.

A. Modeling physics pedagogy for all students

One of the challenges facing many high school and undergraduate physics classrooms today is that traditional lecture-based instruction dominates, and often fails to effectively reach all students [12,13]. Approaches that encourage greater student engagement and active participation show greater promise for broadening student success in physics [12,14,15].

Mentor teachers play a critical role in modeling instruction that expands access to physics for all students. In particular, they assist in moving interns beyond the lecture-based pedagogy that is effective for only a narrow band of learners. This modeling is especially critical for interns who were successful in traditional lecture-based undergraduate physics classrooms, as these students may have come to identify lecture-based pedagogy as the fundamental means of learning physics, and therefore struggle to see the need for rethinking this pedagogy. For example, one of Penny's interns, Thomas (all intern and student names are pseudonyms), clung to the belief that most of his students had learned a concept after he verbally explained it in class. In conferences and course meetings, he repeatedly expressed a belief that, despite what he was learning in his education courses, inquiry-oriented physics teaching was unnecessary. After many weeks of observing Penny implement inquiry-oriented lessons and collaborating with Penny and university-based faculty, Thomas began to express a new belief: that Penny's inquiry-oriented instruction involved students in thinking more deeply about physics than his didactic explanations had. He then began to create and teach his own inquiry-oriented lessons. We have seen that mentors are critically important for validating the legitimacy of educational methods, especially inquiry-oriented physics instruction, for interns.

Challenges inevitably arise for interns as they make their initial attempts at inquiry-oriented physics pedagogy. By "inquiry-oriented pedagogy," we mean teaching that engages learners in questioning, gives priority to evidence, emphasizes learners' construction of explanations, connects explanations to scientific knowledge, and engages learners in communicating their explanations [16]. After observing their mentors and other teachers conduct inquiry lessons, interns can be inclined to think that creating such lessons occurs fairly easily, as a smoothly run inquiry lesson may obscure the careful planning and experience that went into it. Interns' own first attempts at inquiry-oriented pedagogy often are not as seamless as their mentor's teaching, and reveal gaps that hinder students' learning. These challenges can arouse frustration, anxiety, and a tendency to blame students, which may lead to efforts to suppress future use of such pedagogy [17]. Interns look to their mentors to model positive dispositions toward inquiry (e.g., encouraging divergent questions from students) and for strategies to adapt lessons and classroom management practices to overcome these challenges. For example, when Sarah (a physics intern) used a sound editing program to teach waves, students initially spent more lesson time using the technology than learning about waves. Sarah also spent the majority of instructional time solving students' computer problems instead of helping direct students' inquiry about waves. At first, Sarah expressed frustration. During a lesson debrief, she explained to her mentor that she wanted to abandon the lesson for something less student directed. However, after working with her mentor and MAT faculty, she developed a strategy to change the lesson for the next day. Sarah identified the common computer problems students were encountering, and created various handouts with clear illustrations and directions explaining how to solve them. She placed these handouts at each of the lab tables for students to use when needed. Students were able to quickly fix their technology problems and spend the majority of their time investigating the behavior of waves.

Other common challenges include problems with pacing lessons, balancing guidance and freedom during independent work, and ensuring that all students have opportunities to substantively engage during collaborative work time. Mentor teachers model specific solutions to these challenges because they know the students individually, know the class dynamic, and have an understanding of how curricular activities must fit within their school day, school year, scope, and sequence. They can also refer to experience from past years of teaching similar concepts and skills.

When Matt (a physics intern) taught an inquiry unit on sound, he received curriculum-specific feedback via observation notes when his mentor noted, "Your essential questions are 'How do musical instruments work?' and 'Why does the same note played on different instruments sound differently?' How do you plan to assess if they have answered these questions?" Matt's mentor was able to provide expertise in teaching and physics instruction: "Continue to work on and think about your wait time. Maybe 'take a break' and have the students talk to their neighbor for one minute about why different instruments sound differently." Matt also received student-specific feedback when he was asked, "Isaac is still having a very hard time understanding the relationships among velocity, frequency, and wavelength. How can we help him?"

By being able to offer feedback that addressed several different levels of analysis, Matt's mentor teacher was able to help him to understand how students, classrooms, and curriculum must all be considered when designing effective inquiry-oriented instruction. Mentor teachers can improve interns' ability to give descriptive, specific, and substantive feedback, both in writing and in conversation.

B. Modeling reflective practice

During the MAT year, interns learn to continually reflect upon and revise their practice. Interns reflect formally by completing journal entries, participating in classroom discussions, and debriefing orally after a lesson observation with education faculty and their mentor. During these reflection opportunities, interns consider whether and how their practice (e.g., lesson planning and classroom management) aligns with the principles of curriculum development and pedagogy that they learn about in their MAT coursework. Mentor teachers, who are with the intern full-time, have a unique opportunity to prompt interns to reflect and revise before, during, and after teaching and other professional experiences. It is very helpful when mentors can model the type of content-specific and job-embedded reflection that they want interns to engage in. This demonstrates to interns the importance of constant refinement in physics teaching practices.

As opposed to simply asking interns to reflect on what they see or do in a classroom, mentors can involve interns in the professional reflection that they already do. For example, after a mentor teaches a lesson in which students struggled with completing a lab, he or she might include the intern in discussing what went wrong and how to revise it before the next lesson. Successful mentors are intentional about including interns in the normal process of post-lesson reflection and revision. Penny and Amanda have found that something as simple as eating lunch with interns after a laboratory activity ensures an opportunity for productive informal reflection. Interns learn a great deal when mentors include them in their own professional reflection as they plan new instructional units, revise existing units, examine student assessment data, and consider how student learning may be improved.

C. Supporting pedagogical content knowledge

Strong pedagogical content knowledge is essential to being an outstanding physics teacher mentor. This kind of knowledge extends beyond knowing physics content and generalized pedagogical strategies to include an understanding of curriculum, student ideas, effective instructional strategies, and assessments that are unique to physics teaching [4,18]. Mentor teachers not only demonstrate this knowledge to interns but also demonstrate how it grows through reflective practice and experimentation.

For example, Penny decided to reorder the traditional sequence of physics instruction at her school and included her intern, Sarah, in that process. Based on Penny's prior experience with students in the science, technology, engineering, and mathematics academy where she teaches, she chose to begin her school year with a study of waves instead of kinematics. The kinematics unit she teaches requires stronger mathematical skills than the unit she teaches about waves, and she noticed that many of her students had weaker mathematical skills at the beginning of the school year. Additionally, beginning the year with kinematics often resulted in students gaining the impression that learning physics was mostly about selecting the right equation. As the wave unit was traditionally one of her students' favorites, Penny hypothesized that it would allow students to begin the course with a more engaging and more accurate impression of physics.

In making this change, Penny needed to reconsider how to account for students' initial ideas, how she was building an instructional sequence throughout the year, and how she would assess students' understanding. During this process, she engaged in continual reflection with her intern to evaluate the benefits and limitations of this approach and to make instructional adjustments along the way. Such curricular experimentation reaffirms for interns that teaching expertise develops over the course of years, not months, and that their main task during their MAT experience is to become reflective practitioners with a capacity for continued growth. While some mentor teachers are more disposed toward this kind of inquiry and experimentation than others, we have found that a structured program of support for mentor teachers helps all mentors to be more comfortable with uncertainty and curricular experimentation while mentoring.

IV. SUPPORTING PHYSICS MENTOR TEACHERS

As two university-based teacher educators and two experienced physics mentor teachers, we have learned that outstanding physics teacher mentors are rarely discovered *in situ*; rather, they are developed through continual support and collaboration between school-based and university-based faculty. Not all outstanding teachers make outstanding mentor teachers. Opening one's classroom to a novice teacher and to university faculty can be seen as a risky proposition; many teachers hesitate to open their practice to what they might perceive as potential scrutiny from others. Also, mentorship requires allowing a novice teacher to instruct the students for whom mentors are still ultimately responsible. In particular, many mentors feel under pressure to produce high Advanced Placement test scores and statewide standardized test scores. Rather than viewing an intern as an asset to the classroom, some teachers, administrators, and parents might see an intern as a competing demand on the teacher's attention.

Thus, accepting the role of mentor, with its potential risk of vulnerability and the demands of a new role, requires support from university collaborators, teacher colleagues, and school administrators. Despite the risks of additional time commitments and opening one's practice to review from other educators, outstanding teacher mentorship holds tremendous rewards for all stakeholders, including exchanges of new ideas and development of a pipeline of physics teacher educators on a PDS campus. Given the critical importance of quality mentoring and the potential benefits for all involved—prospective teacher, mentor teacher, students, and school colleagues—we invest a great deal of time in identifying and supporting mentor teachers as they work with interns in the Trinity MAT program. In this section, we discuss how potential physics teacher mentors are identified as well as the programmatic components that are implemented to ensure that mentors are better prepared to engage in the critical practices in mentoring.

A. Identifying potential mentors

At the high school level, we work extensively with two schools on the same campus. (One is a magnet school and the other is a comprehensive high school.) Trinity's long-term partnership with these professional development schools has proven to be fertile ground for building relationships with potential mentor teachers. By placing undergraduates who are not yet in their intern years in teachers' classrooms and by working with teachers in school-related activities, Trinity faculty can begin to identify outstanding teachers who may have the disposition for mentoring. This disposition includes openness to professional conversation and shared inquiry, embrace of ambiguity and uncertainty, willingness to give and receive constructive criticism, and professionalism both inside and outside the classroom. Over time, this consistent partnership has resulted in the schools hiring many MAT graduates. These alumni then become strong candidates to be mentor teachers. (Both teacher authors of this paper are Trinity alumni and physics mentors.)

Once a potential mentor is identified, he or she typically begins in smaller mentorship roles, such as working with an undergraduate who is in a practicum course or completing a service-learning project. Such collaboration gives the teacher a picture of what future mentorship in the MAT program may look like. As an example of this type of collaboration, during the 2012–2013 school year an MAT graduate who was teaching physics was matched with two undergraduate physics majors enrolled in a science teaching methods course. The physics majors collaborated with the teacher to identify which standards his students commonly struggled with, learned about the teacher's existing curricular activities, conducted classroom observations, and held content-based assessment conversations with his students. The physics majors then developed a set of curricular activities that the teacher could incorporate into his classroom. During this experience, the faculty member teaching the course (Jeff) was able to observe the interpersonal relationships between the potential mentor and students and to assess not only whether the students were on track for the MAT program but also whether the teacher might be a potential mentor for the program. This type of collaboration among a physics teacher, university student, and a university faculty member could happen in many different settings, with some attention to building a relationship with the teacher and making sure the teacher did not feel that the collaboration or intervention was designed because he or she was a weak or ineffective teacher.

Mentor support activities are helpful to facilitate the move from an initial and smaller experience with Trinity undergraduates to a more extensive working relationship between an intern and mentor. We also know that mentoring practice evolves, and we therefore anticipate developmental support for mentors, particularly as mentors move from being first-time to more experienced mentors. To systematically support mentors and provide consistent collective and individualized support, we rely on the following programmatic components:

- providing individual and collective support (informational, emotional, and professional) for mentors from university faculty;
- coaching mentors in how to form and enact relationships with interns; and
- fostering professional learning communities among mentors.

In the following sections, we discuss each of these components and how they contribute to a cohesive experience for mentors that ultimately supports interns.

B. Providing support for mentors from dedicated MAT program faculty

Perhaps the most important resource for ensuring quality mentoring for a program such as the MAT program is a dedicated leader who coordinates between and among the university faculty and the mentors. At Trinity, this liaison position is a tenured faculty line (Angela, an author, is the high school liaison), and is regarded as a core—not a peripheral—position within Trinity's education department. Because the work of the liaison is fundamentally grounded in the practice of secondary school teaching, this faculty member spends roughly equal time each week at the PDS campus and at the university. When interns complete their lead teaching during the third nine weeks of the school year, the liaison spends the majority of time in the schools. The faculty liaison teaches education courses for interns, meets regularly with mentors (both formally and informally), and communicates with other stakeholders to ensure continuity of experience for interns and a coherent PDS relationship. The dedicated faculty liaison leads a team of MAT program faculty that includes a university-based science educator (formerly Jeff), a university-based special educator, and a school-based professional development school coordinator.

C. Coaching mentors in how to form and enact relationships with interns

A critical role of the faculty liaison is coaching mentors to coach interns. This includes enhancing mentor teachers' knowledge and skills for teaching adult learners as opposed to high school students. (Interns are typically 22 to 30 years old.) Throughout the year, carefully designed intern assignments that include the mentor's involvement in specific ways are leveraged to help scaffold the mentor-intern coaching relationship. Mentors are also given direct guidance through regular mentor meetings that are run by the faculty liaison and held at the PDS campus, as well as through informal feedback provided via email, telephone, or conversations at the PDS campus. Figure 1 illustrates the major components of the mentoring program and how mentors and MAT program faculty work to support interns in their development throughout the course of their MAT years.

1. Mentor orientation meeting

The first mentor meeting occurs in the summer at Trinity, several weeks prior to the beginning of the new school year. As this is outside the school year for teachers, they receive professional development credit (continuing professional education hours required by the state) for this time. At this meeting, all mentors, whether veterans or novices, meet each other and the MAT faculty with whom they will be working. Mentors are given a handbook of resources for mentoring. This handbook includes a description of the Trinity MAT program, descriptions of key assignments and checkpoints, coaching guidelines, and resources for collaborative lesson planning. This same resource is given to interns and helps ensure transparency and consistency across stakeholders in the program.

In the orientation meeting, mentors read a book chapter [19] to prompt a structured conversation about developmental intention in adult learning as well as informational versus transformational learning. They are guided through an exploration of the cognitive apprenticeship model for teaching and mentoring [20]. The apprenticeship continuum that describes interns' learning trajectories includes observing, emerging, practicing, and refining, whereas the mentorship continuum that describes the roles and responsibilities of the mentor includes modeling, coaching, and fading. After learning about the cognitive apprenticeship model, mentor teachers are explicitly introduced to how the MAT year is structured with the constructive-developmental and cognitive apprenticeship models in mind.

After mapping the year, mentors are introduced to strategies for having their first conversations with their interns through an intern-mentor interview that provides prompts targeted at sharing educational biographies, exploring their own learning profiles, discussing dispositions for collaboration and coaching, and uncovering assumptions and questions. Mentors then begin to explore the First Days of School assignment and discuss their roles in supporting interns in this project.

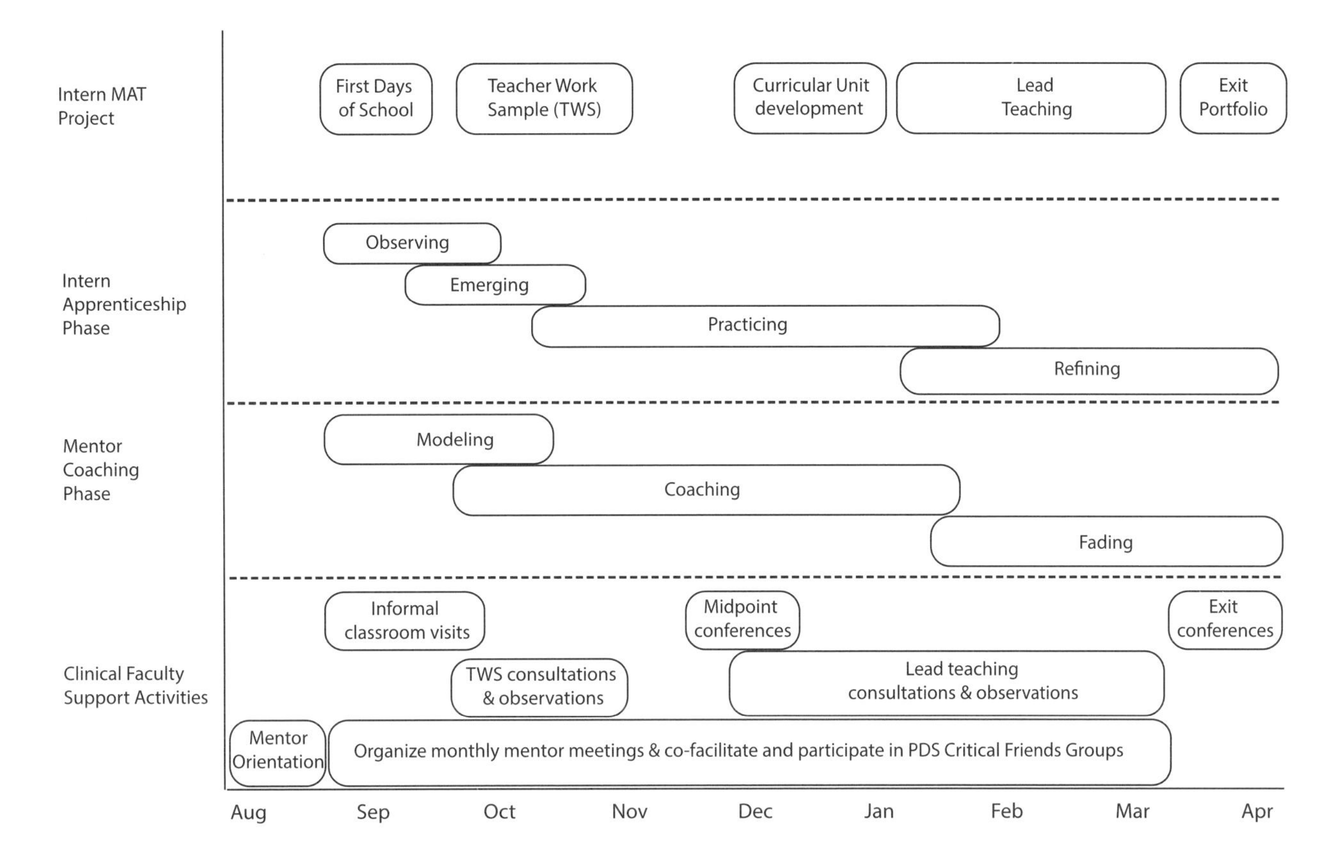

FIG. 1. This diagram illustrates how MAT projects, mentor supports, and MAT faculty work evolve over the course of the academic year. The components and assignments identified in the diagram are discussed in the sections below.

2. *First Days of School study*

The First Days of School study is the interns' first substantial assignment in the internship; it involves a systematic and substantive examination of how mentor teachers set up their classrooms, build classroom culture, begin the process of learning about their students, and introduce students to physics (both as a high school course and as a scientific discipline). This assignment also requires interns to interview other teachers (physics and non-physics) and gather artifacts beyond their primary placements. This reinforces the significance of the broader professional development school setting, in that the learning happens across the campus and is shared among the whole community of teachers.

Unlike in many "sink-or-swim" student teaching experiences, the mentor maintains primary teaching responsibility in this time period and is guided by MAT faculty as the mentors engage in the "modeling" phase of cognitive apprenticeship. During this time, the intern's primary responsibilities are observing and reflecting on what he or she sees. Interns participate in lessons by, for example, helping students conduct investigations in groups, leading a short warm-up at the beginning of the day, or guiding students to solve problems. This time period allows interns to participate with their mentors in all beginning-of-school activities, which include the first meeting with teacher colleagues during the professional development days that take place before the start of school, the first meeting with students, the first department meeting, and the first meetings with students' families. For many beginning teachers, the first days of school can easily be lost trying to accomplish all of the tasks required of them (e.g., distributing school forms, distributing textbooks, etc.). The First Days of School study gives interns the opportunity to consider how these days can also be used intentionally to build a successful physics learning community in their classroom and to specifically identify the mentor as a model.

3. *Teacher Work Sample*

Much attention and thought is given to interns' first experiences developing and teaching their own physics lessons in the classroom. Mentors play a central role in supporting the development of high-quality physics curricula. About four weeks after the beginning of the school year, mentors are introduced to the next major intern assignment, the Teacher Work Sample (TWS). In this project, interns plan a week-long sequence of lessons, design a pre- and post-assessment, teach the lessons they planned, and reflect on the teaching experience and assessment artifacts. This assignment gives interns their first taste of lead teaching responsibility and represents the "emerging" phase of cognitive apprenticeship. Simultaneously, mentors are beginning their "coaching" phase. Interns and their mentors collaborate to select a set of standards that the interns will address in their TWSs, and during these interactions, MAT faculty are intentional about serving as model coaches. Critical in this modeling is demonstrating how mentors can direct interns' thinking through intentional questioning rather than overt redirection. As mentors have taught the physics standards successfully in the past, MAT faculty guide them to avoid simply giving their past lesson plans to interns. The learning potential for interns is much greater when they develop their own plans or at least recreate plans, as opposed to when they simply enact what their mentors have always done.

The TWS represents a level of risk for the mentor teacher, since it is one of the first times the mentor hands over the class to the intern for an extended period of time. This risk is mitigated by the shared development of the assessment plan and lesson plans. MAT faculty work with the mentors to recognize and allow minor mistakes interns might make during their planning and to address possible major mistakes. For example, Toby (a physics intern) prepared his TWS on Newton's laws of motion. In a one-on-one conversation, his mentor expressed concern to the university liaison (Angela) about two problems in Toby's lessons. The first concern was that he created lesson plans with slightly more material than could be covered in one class. Second, he had not created lessons that helped students to develop an understanding of Newton's second law to the degree they would need for later physics units. Angela worked with Toby's mentor, and they decided the first mistake was not major and should remain in the lesson plans so that Toby could develop a personal sense of pacing for a class period. However, the second mistake had to be addressed before Toby could teach the unit. The mentor teacher then guided Toby to make that revision.

When interns teach the TWS, mentors are asked to be in the room the entire time. Additionally, MAT faculty members observe at least one lesson taught by each intern during this time period. Both MAT faculty and the mentor record notes during the lesson; the notes are later shared during the debriefing. At mentor meetings (described in more depth below), mentors are given different strategies for observing and recording feedback for their interns. Each observation of the intern consists of a shared observation, with the mentor and MAT faculty member co-observing, and a post-observation debrief with the intern, mentor, and MAT faculty member. Prior to the post-observation debrief, the mentor and MAT faculty plan the key points of the debrief conversation. This consultation step allows the mentor and MAT faculty member to exchange impressions of strengths and challenges to "get on the same page" and also prioritize the key feedback, as too much feedback often can be overwhelming for the intern. The observation process is another opportunity for MAT faculty to model feedback and coaching for mentor teachers.

The post-observation debriefing begins by asking the intern to reflect on the experience and identify strengths as well as areas for improvement. Thereafter, the mentor teacher shares feedback based on his or her observations and the points developed with the faculty. Lastly, the MAT faculty adds anything not already identified by the intern and mentor. This format puts the intern in a place to speak first and develop his or her own insights prior to receiving external feedback, and it allows the mentor and faculty member to hear where the intern is in terms of the content of and ability for self-assessment. It positions the mentor teacher in a central role as a curricular and pedagogical coach, and makes the MAT faculty part of the conversation as opposed to the immediate "critic." Additionally, it allows the MAT faculty to take a facilitative role, so that the intern and mentor can participate fully in the debriefing conversation without worrying about the facilitation.

Instructional missteps will happen in the TWS. This project represents a key experience during which mentors and interns experience sustained support from MAT faculty to deal with these missteps, learn from them, and recognize that the risk of opening one's classroom can be successfully managed. The outcome is intended to lead to (1) intern learning in terms of curriculum development, pedagogy, and classroom facilitation; (2) mentor learning in terms of coaching and experiencing the shift of teaching to the intern; and (3) a stronger relationship among the intern, mentor, and MAT faculty. It might even lead to instructional and curricular innovations, as interns work with mentors to build on mentors' existing curriculum and practices.

4. Lead Teaching Curricular Unit and lead teaching

The most intensive project of the first semester in the MAT program is the Lead Teaching Curricular Unit, during which interns work with university faculty (including MAT faculty and physics faculty), mentor teachers, and other colleagues to design a unit that sustains several weeks of instruction. The intensity of this project provides fertile ground for collaboration between interns, mentors, and university faculty, and it brings to the surface myriad natural opportunities for university faculty to embed intentional support for mentors as they work with their interns.

Before interns begin writing their units, MAT faculty work with mentors to select appropriate topics for the units. Mentors must consider what concepts or topics are needed for their school-based curricular requirements and what topics easily allow for the development of units. For example, Amanda collaborated with university faculty in choosing energy as the topic for her intern's unit. The topic was chosen for the following reasons: The unit's concepts were needed before the state standardized testing; it is a broad topic allowing for a variety of lesson models and strategies; and the school had a variety of lab equipment the intern could use to develop inquiry-based lessons.

As interns begin to plan their units, they take their first independent foray into translating the physics knowledge that they learned in the collegiate setting and first semester of the internship into a sustained, coherent, inquiry-oriented unit targeted at high school learners. Interns begin with a set of state-required content standards and unpack them to discern critical physics knowledge and skills for their students. The mentor teacher plays an important role in ensuring that the intern focuses on the physics skills and knowledge his or her students must have by the end of the unit, and university faculty support the mentor in this process by collaborating to identify key ideas in the content standards and providing a curriculum design framework that helps interns and mentors focus on key ideas.

Interns use the *Understanding by Design* approach to curriculum design [21]. Every step of the curriculum design process includes ongoing collaboration among interns, mentors, and MAT faculty. At this stage in the MAT program, interns are expected to be able to lead the design of their unit while collaborating with others. This replicates curriculum planning as it happens in many schools and districts, where teachers are called upon to be both independent and collaborative. To support interns and mentors, MAT faculty host cohort-wide work sessions and facilitate consultations that include interns, mentors, and university faculty. Mentors offer interns lab equipment and other curricular resources, and provide their knowledge of students and the curriculum. Each intern has multiple opportunities to participate in consultation and feedback meetings with both university faculty and mentor teachers, and MAT faculty guide and coordinate this process for each intern. Personalization is very important at this point, since each intern is likely at a different point in his or her readiness to develop inquiry-oriented physics curriculum. MAT faculty work with mentor teachers to personalize their support for interns in light of their development.

The Lead Teaching Curricular Unit project represents interns' initial movement from the "emerging" into the "practicing" phase of cognitive apprenticeship, and places the mentors solidly in the "coaching" phase. When interns take over lead teaching responsibility after the winter break, with the enactment of the unit they have developed, they will have moved solidly into the "practicing" phase, while mentors continue in their coaching role. During lead teaching, interns are formally observed at least one or two times per week by their mentor teachers and MAT faculty. Each observation by an MAT faculty member includes the observation and post-observation conference in which mentors, interns, and MAT faculty meet together. These post-observation conferences support mentor teachers by providing another lens on interns' teaching and reinforcing the feedback that mentors give interns.

Throughout the interns' experiences on the PDS campus, the Trinity faculty liaison (Angela) coordinates interactions among stakeholders and holds regular meetings with mentors to foster a professional learning community organized around supporting interns, mentors, and high-quality instruction throughout the school.

5. Mentor meetings

All mentor teachers from all content areas participate in mentor meetings approximately every three weeks throughout the school year. Meetings take place before or after school or during lunch when mentors are the lead teachers; when interns are the lead teachers, the meeting may occur during the school day. These meetings are not structured as information delivery sessions; rather, mentors are asked to be thought partners with each other. This means bearing shared responsibility for raising issues, discussing interns' progress, and brainstorming ways forward. To facilitate this kind of interaction, MAT faculty often use protocol-driven discussions that are designed to help participants function as full, equal partners in conversations around their shared practice and to make sure all mentor voices are heard [22]. These meetings are guided by the faculty liaison and framed by the cognitive apprenticeship model. At each meeting, mentors are given explicit guidance about the current assignment interns are completing, a timeline for its completion, and expectations for the mentor's role. This guidance is overlaid with explicit discussion of how the assignment is intended to help interns progress through the cognitive apprenticeship phases. Typical discussions are guided by prompts such as the following, taken from mentor meeting agendas:

- Which questions from your intern might you answer, and which might you push the intern to take a stab at before you share your thoughts?

- Where do you think your intern might be developmentally about now? Why?

- How is your intern developing resilience—physical, emotional, and intellectual—and where do you see continuing development needed? How can we support that development as mentors?

- Where are *you* as a mentor? How are you doing and how can we help you so that you can then in turn help your intern? (This provides a focus on the mentor, not the intern, as mentors often want to talk about the intern as opposed to about their coaching of the intern.)

The constant shared communication builds support for and among mentors, as well as among mentors, faculty, and interns. Mentor meetings provide space and time for this communication to happen.

V. PROFESSIONAL LEARNING COMMUNITIES AND RECIPROCAL BENEFITS

The university and the MAT program benefit tremendously from the work and thoughtfulness of participating mentor teachers. This happens at three levels: in contributions to interns' growth and development, within the mentor professional learning community, and in ongoing feedback and guidance given to MAT program faculty. Benefits for mentors include collaboration with an intern, a second adult in the classroom, opportunities for sustained learning with other mentor teachers and MAT faculty, an experience within which to focus on issues of learning and teaching, and opportunities to shape the future teaching force and help to induct new colleagues into the profession.

These opportunities help counter the unfortunate isolation that many teachers experience, with few opportunities for professional collaboration and even fewer that are school- or even classroom-based [23,24]. By serving as a mentor, teachers have the opportunity and structure (in time and space) for collaborations with other mentor teachers, university faculty (including MAT and physics faculty), and interns, as the teachers engage in meaningful discussions around teaching, learning, and professional practice. In end-of-year focus group interviews, mentors frequently express appreciation for the opportunity to work with other teachers throughout the year and for the new ideas that interns brought to their classrooms.

In addition, mentors see value in being engaged not only in collaboration around planning and teaching but also in the additional responsibility of transparently reflecting upon the strategies they use in their classrooms. In an MAT program feedback form, Amanda relates, "Mentoring a physics intern pushes me in my own practice. It forces me to justify what I do in the classroom and be ready to explain my thinking." Amanda believes that seeing her practice mirrored back to her by an intern, and the challenge of talking about what is and is not working, push her to be a better teacher.

An often unanticipated benefit of mentorship is the ability for mentors to watch their own students engaged in the learning process in a way that is not possible when mentors are the lead teachers. As interns teach, mentors get to watch their students "at work" in the learning, and they can also spend individual time with students and focus on supporting struggling learners in a fundamentally different way than they can when they are teaching the lesson. This opportunity helps mentors continue to develop their own practice and really understand their own students as individuals.

VI. CONCLUSIONS AND TRANSFERABLE LESSONS LEARNED

Physics teacher mentors are perhaps the most essential components of successful physics teacher preparation programs, since they provide a practice-focused bridge between the university and the high school classroom. They are essential to modeling high-quality curricular development, pedagogical innovation, professional collaboration, and reflective practice, which are all critical for effective physics teaching. We have learned, through our personal experiences as university-based teacher educators and as physics mentor teachers, that good physics mentors are developed through ongoing support and effective school-university partnerships. It is important to note that Trinity's school-university partnerships were initiated over 20 years ago and were much more modest at the start. The descriptions provided in this chapter are meant to give a picture of what can be built through long-term relationships, and to give ideas for getting started in early partnerships or deepening already existing ones.

What if another context does not feature elements found at Trinity, such as an extended-year teacher education program, a collaborative undergraduate physics department, mutually beneficially professional development school partnerships with committed mentor teachers, or the resources to support all these elements? What can a well-intended physics professor do to advocate for strong physics teacher mentorship? Our years of trying new ideas, self-reflecting, and revising have reinforced our commitment to continued experimentation within the Trinity program, and we have found that supporting and improving the mentoring aspect of our program usually begins with one-on-one conversations and collaboration.

Physics faculty can begin the conversation by reaching out to the education faculty on their campus to learn about the nature of the teacher preparation program. Asking specifically about how mentor teachers are supported in their work with preservice physics teachers may lead to powerful collaborations. Offering to support mentor teachers by providing content input or collaboration for curriculum design, lending (often-needed) equipment to support inquiry-oriented instruction at the high school, or even visiting high school classrooms as a guest speaker would likely be deeply appreciated by mentor teachers and education faculty, especially since physics is an area in which teacher education faculty often feel less confident in their own content expertise. Further, offering direct support for mentor teachers can enrich ongoing collaboration between the schools in which preservice teachers are placed and the university that is sending these teachers. As we have described above, we have found that the richness of Trinity's long-term partnerships with schools is a critical factor for providing meaningful support for both mentors and interns. By offering support for physics mentor teachers, physics faculty can contribute to lasting relationships that enhance this critical feature of outstanding preservice physics teacher education.

A strong mentorship program is built upon the relationships and trust that come from demonstrating to school partners that the university is willing to commit time and resources in service of supporting teachers and students at the professional campus. By reaching out to teacher education faculty at their university and to high schools where teacher candidates are placed with mentors, physics faculty demonstrate a commitment to physics teacher education and support two of the most important dimensions in physics teacher preparation. Though the Trinity program has evolved over the course of two decades, we have found that it remains fundamentally reliant on the willingness of school and university faculty to initiate and continue partnerships formed around supporting high school students and teacher candidates. With intentionality, these initially small partnerships grow over time and provide substantive support for mentor teachers in their critical work.

---

[1] A. Breidenstein, Re-forming teacher education: The promise of extended programs, Kappa Delta Pi Rec. **36** (3), 111 (2000). doi:10.1080/00228958.2000.10532033

[2] D. E. Meltzer, M. Plisch, and S. Vokos, eds., *Transforming the Preparation of Physics Teachers: A Call to Action. A Report by the Task Force on Teacher Education in Physics* (T-TEP) (American Physical Society, College Park, MD, 2012). http://www.ptec.org/webdocs/2013TTEP.pdf

[3] P. Norman and A. Breidenstein, Developing learning teachers: A curriculum for teacher education, Teach. Educ. Pract. **16**, 85 (2003).

[4] L. S. Shulman, Those who understand: Knowledge growth in teaching, Educ. Res. **15** (2), 4 (1986). doi:10.3102/0013189X015002004

[5] S. Feiman-Nemser, From preparation to practice: Designing a continuum to strengthen and sustain teaching, Teach. Coll. Rec. **103** (6), 1013 (2001). http://www.tcrecord.org/content.asp?contentid=10824

[6] K. M. Zeichner and B. R. Tabachnick, Are the effects of university teacher education 'washed out' by school experience?, J. Teach. Educ. **32** (3), 7 (1981). doi:10.1177/002248718103200302

[7] L. Darling-Hammond, *Powerful Teacher Education: Lessons from Exemplary Programs*, 1st ed. (Jossey Bass, San Francisco, CA, 2006).

[8] J. Koppich, K. K. Merseth, American Association of Colleges for Teacher Education, and National Commission on Teaching & America's Future, *Studies of Excellence in Teacher Education: Preparation in a Five-Year Program* (AACTE Publications, Washington, DC, 2000). http://eric.ed.gov/?id=ED468995

[9] L. Van Zandt, Assessing the effects of reform in teacher education: An evaluation of the 5-year MAT program at Trinity University, J. Teach. Educ. **49** (2), 120 (1998). doi:10.1177/0022487198049002005

[10] A. Breidenstein, Researching teaching, researching self: Qualitative research and beginning teacher development, The Clearing House **75** (6), 314 (2002). http://eric.ed.gov/?id=EJ655292

[11] L. Darling-Hammond and J. Bransford, *Preparing Teachers for a Changing World: What Teachers Should Learn and Be Able to Do*, 1st ed. (Jossey-Bass, San Francisco, CA, 2005).

[12] L. Deslauriers, E. Schelew, and C. Wieman, Improved learning in a large-enrollment physics class, Science **332** (6031), 862 (2011). doi:10.1126/science.1201783

[13] S. L. Hanson, *Swimming Against the Tide: African American Girls and Science Education* (Temple University Press, Philadelphia, PA, 2009).

[14] C. H. Crouch and E. Mazur, Peer Instruction: Ten years of experience and results, Am. J. Phys. **69** (9), 970 (2001). doi:10.1119/1.1374249

[15] B. A. Adegoke, Impact of interactive engagement on reducing the gender gap in quantum physics learning outcomes among senior secondary school students, Phys. Educ. **47** (4), 462 (2012). doi:10.1088/0031-9120/47/4/462

[16] S. Olson and S. Loucks-Horsley, eds., *Inquiry and the National Science Education Standards: A Guide for Teaching and Learning* (National Academy Press, Washington, DC, 2000).

[17] S. M. Ritchie, K. Tobin, M. Sandhu, S. Sandhu, S. Henderson, and W. M. Roth, Emotional arousal of beginning physics teachers during extended experimental investigations, J. Res. Sci. Teach. **50** (2), 137 (2013). doi:10.1002/tea.21060

[18] E. Etkina, Pedagogical content knowledge and preparation of high school physics teachers, Phys. Rev. ST Phys. Educ. Res. **6** (2), 020110 (2010). doi:10.1103/PhysRevSTPER.6.020110

[19] K. Taylor, Teaching with developmental intention, in *Learning as Transformation*, edited by J. Mezirow and Associates (Jossey-Bass Inc., San Francisco, CA, 2000), pp. 151-180.

[20] J. S. Brown, A. Collins, and P. Duguid, Situated cognition and the culture of learning, Educ. Res. **18**, 32 (1989). doi:10.3102/0013189X018001032

[21] G. P. Wiggins and J. McTighe, *The Understanding by Design Guide to Creating High-Quality Units* (Association for Supervision and Curriculum Development, Alexandria, VA, 2011).

[22] School Reform Initiative, Inc. See http://www.schoolreforminitiative.org/, retrieved Sep. 8, 2013.

[23] A. Hargreaves and D. Shirley, The persistence of presentism, Teach. Coll. Rec. **111** (11), 2505 (2009). http://www.tcrecord.org/content.asp?contentid=15438

[24] A. Lieberman and L. Miller, eds., *Teachers in Professional Communities: Improving Teaching and Learning* (Teachers College Press, New York, NY, 2008).

Using early teaching experiences and a professional community to prepare preservice teachers for everyday classroom challenges, to create habits of student-centered instruction, and to prevent attrition

Eugenia Etkina
Graduate School of Education, Rutgers University, New Brunswick, NJ 08901

This chapter addresses two themes of this book: (1) structuring effective preservice early teaching experiences and (2) mentoring and community building. It shows how these themes contribute to solving three major issues in teacher education: preparing preservice teachers for the challenges of everyday instruction, creating habits of student-centered learning, and preventing attrition from the profession. The chapter describes the Rutgers physics teacher preparation program, which uses the learning community approach to connect early teaching experiences in a physics teacher preparation program with the professional development and retention of program graduates. Finally, it provides evidence that the Rutgers approach to early teaching experiences and community building leads to the production of physics teachers prepared for classroom instruction and remaining on the job.

I. INTRODUCTION

Teacher preparation is a complex and multifaceted field. It includes teacher recruitment, education, early teaching experiences, and retention. It is possible that physics teacher retention depends not only on teachers' knowledge of physics content and ways to engage their students in learning, but also on teachers' ability to apply this knowledge to design effective instruction and on productive interactions with other teachers. This argument stems from the three issues that have been found to be crucial in teacher preparation:

(1) According to D. Boyd and colleagues, "The only element of teacher preparation programs that can predict how new teachers will actually teach is the amount of experience with everyday instruction relevant to the first year of their teaching received in the program" [1].

(2) According to P. E. Simmons and colleagues, "Teachers experience difficulty implementing student-centered teaching, despite being taught to do so during preservice preparation" [2].

(3) According to L. Darling-Hammond, the feeling of isolation is one of the primary reasons that teachers leave the profession [3].

These issues are especially relevant for teachers of physics. Due to the small number of preservice physics teachers in any single university, very few institutions have a sequence of courses or even one course that prepares physics teacher candidates for the challenges of everyday instruction. The Task Force on Teacher Education in Physics (T-TEP) found that most universities have one or zero courses dedicated to this material [4].

In addition, physics education research shows that many university physics courses are taught using traditional methods, whereby students listen passively to lectures, watch instructors solve problems on the board, and perform "cookbook" lab experiments. Undergraduate physics courses rarely provide teacher candidates with the experience of student-centered instruction, whereby the student is actively engaged in the construction of knowledge through interactions with peers, explorations, discussions, etc. As teachers tend to teach the way they have been taught, they have difficulties implementing such instruction even if they learn about its benefits in their education courses.

In addition, very few physics teacher preparation programs keep contact with graduates or have well-organized professional development programs for beginning teachers after they leave the university [4]. This lack of continuity contributes to the isolation of physics teachers.

The purpose of this chapter is to describe a physics teacher preparation program that connects early teaching experiences during preparation to professional development, resulting in the retention of program graduates as teachers. It also provides an analysis of learning communities and their role in teacher preparation and retention. Finally, it provides evidence of the program's results.

II. BACKGROUND

In the same way a person cannot become a physicist just by taking physics courses, a person cannot become a teacher just by learning the subject matter and taking courses describing how to teach it. People in both professional positions need to practice their crafts: A physicist needs to "do" physics and a teacher needs to "do" teaching. The classroom experience wherein a future teacher (known as a preservice teacher or a PST in this chapter) engages in any kind of instruction is called an early teaching experience. Such experience is crucial in helping PSTs learn how to teach in ways that differ from how they have been taught, and to acquire the "habits" of good teaching. Research shows that PSTs enter teacher preparation programs with a very clear vision of what they think of as "good teaching" and what it means to be a good teacher. This vision is usually based on their years of being a student and observing teacher behaviors [5]. However, it is often not supported by our knowledge of how people learn.

When prospective students come to my office to discuss their possible enrollment in the program, I always ask them, "Why do you want to be a physics teacher?" Their answers have two consistent themes: "I love physics," and "I love explaining things to people." A common image of teachers is of one explaining difficult material to students and of experiencing triumphant satisfaction when the eyes of the listeners light up, indicating that they finally understood something. Although this image of teaching is very attractive, we know that students do not learn from listening and that good explanations often do not translate to understanding. The "light bulb" goes on for an instant, and 10 minutes later the difficulty that the student had before remains unresolved [6]. Therefore teacher preparation involves changing this image of good teaching.

In addition, research shows that the intensity of instruction is such that many decisions teachers make in the classroom are based on automatic responses [7]. This means that a good teacher must develop specific habits of response that encourage learning. How do teachers form those habits? This is the first critical question. Even if PSTs form "good" habits during the training process, there is still a danger that, when PSTs become teachers, they will not be able to consistently implement them in practice.

Finally, the feeling of isolation in the school, the need to convince parents of a new approach to learning, and resistance of colleagues and administrators to nontraditional instruction often push the best teachers to quit. Therefore finding ways for good teachers to stay on the job and continue to grow professionally is very important.

A. Early teaching experiences: Goals and opportunities

If a student cannot learn physics just by listening and reading, but in fact needs to engage in the active process of knowledge construction, the same should apply to learning to teach physics. One can only become a teacher by actively constructing knowledge of teaching physics in the process of teaching. If our goal is for PSTs to be skilled in student-centered instruction, then learning to be a teacher starts with experiencing high-quality student-centered instruction with subsequent reflection on these teaching experiences [8,9]. These experiences can happen in courses in which PSTs learn physics or in courses in which they learn how to teach physics. The next step is for PSTs to practice the skill of student-centered instruction. One way to do this is in the form of *microteaching*. Microteaching is a technique used in teacher preparation, whereby preservice teachers teach a K-12 lesson to their peers, who act as students.

Such practice can be extremely useful. PSTs learn to plan a lesson, to interact with students, and to reflect on the lesson after its implementation. This structure also allows multiple opportunities for the mentor (i.e., the course instructor) to provide formative feedback to the PST, and to give the PST an opportunity to revise a particular moment of the lesson and "replay" it again (explained in Section IV.A.2).

Another way to engage future teachers in student-centered teaching is to give them opportunities to either (1) serve as lab or recitation instructors in college physics courses that follow reformed curricula, or (2) become Learning Assistants. Learning Assistants are talented undergraduate science majors with a demonstrated interest in teaching who are hired to facilitate interactive, student-centered instructional approaches in large-scale introductory courses [10].

B. Student teaching

Student teaching is an early teaching experience that lasts for one or two semesters. PSTs are placed in schools under the supervision of teachers specializing in the same content area (e.g., physics). Each PST spends a week or two observing classroom instruction before he or she slowly assumes the responsibilities of the teacher. These responsibilities include lesson planning and teaching, assigning homework, grading, and meeting with parents. In different preparation programs PSTs have different teaching loads during student teaching: Sometimes they teach the full load of the cooperating teacher and sometimes they teach only a few lessons per day. While teacher educators agree that student teaching is the most crucial aspect of teacher preparation [11–13], student teaching has inherent positive aspects [14] as well as deficiencies [15]. One of the most important deficiencies is the frequent disconnect between what PSTs see happening in the classroom and what they learned in their teacher preparation programs [16].

For both microteaching and student teaching to contribute to the growth of PSTs, physics teacher educators need to provide future teachers with help and feedback. This feedback needs to be intensive at the beginning and should then slowly decrease as the PSTs become more skilled [17,18].

C. Learning communities

Linda Darling-Hammond wrote in 2003 that the cause of teacher shortage in the U.S. is not a shortage of teacher preparation but the attrition of teachers from the profession [19]. Some researchers find the main contributors to teacher attrition to be salaries, working conditions, preparation, and mentor support [20]. Others think it is the feeing of isolation that contributes most to the attrition of young teachers and prevents the professional growth of those who stay

[19]. While one can argue that salaries are not in the hands of teacher educators, the working conditions, quality of teacher preparation, and mentor support are. A good teacher preparation program can address these issues by focusing on solving the problem of teacher isolation, specifically by creating teacher learning communities. Imagine a strong, supportive community of graduates, united by the same goals and similar philosophies of teaching. This community might then affect the participants' working conditions. This is particularly true if more than one program graduate works in the same school, or if the community provides materials, equipment, and teaching ideas. The community continuously provides professional development as more seasoned teachers use their experience to help novice teachers, and novice teachers inspire seasoned teachers by sharing new ideas and approaches. Such a community needs to be built while the PSTs are still in training, which is one of the reasons many teacher preparation programs create cohort experiences for their students. This community needs to extend after graduation, allowing novice teachers to slowly transition into the roles of experienced and seasoned teachers.

III. PHYSICS TEACHER PREPARATION AT RUTGERS: PROGRAM OVERVIEW

This section briefly summarizes the program structure and explains how Rutgers integrates coursework, early teaching experiences, and a learning community into its teacher preparation program.

The Rutgers physics teacher preparation program is housed in the Graduate School of Education. It is a 45-credit graduate program that can be completed either as a five-year program in combination with an undergraduate degree in physics or as a separate two-year post-baccalaureate program [21]. Students progress through the program in cohorts, spending two years plus one summer in the program. For students in the five-year program, the first year in the GSE is the last year of their undergraduate studies. They finish their undergraduate degrees and start taking courses in the teacher preparation program simultaneously. Student teaching occurs in the fall semester of a student's second year in the program. During the last 10 years, the Rutgers physics program has graduated an average of six PSTs per year.

It is important to mention that the program functions effectively due to the collaboration of the Graduate School of Education and the Department of Physics and Astronomy at Rutgers. The program coordinator (the author of this chapter) is a faculty member in the Graduate School of Education who teaches most of the physics methods courses and conducts the meetings of the learning community (see Section III.C). The course coordinator for one of the key physics courses for early teaching experiences, Physics for the Sciences, is a staff member in the Department of Physics and Astronomy who is knowledgeable about physics education research and is supportive of the goals of the program. These two people collaborate extensively to make sure that early teaching experiences reinforce the knowledge that PSTs gain through coursework.

A. Coursework

A large portion of PSTs' work in the program is dedicated to physics methods courses. These are the courses in which PSTs develop their physics content knowledge for teaching (which is a combination of content knowledge and knowledge of how to help students learn physics) and participate in early teaching experiences [22]. Only a small portion of the coursework is dedicated to general education courses. Table I provides an overview of the coursework with the details of physics-specific teaching methods courses. Unless specified, the courses are each three credits. As mentioned above, the program graduates about six physics teachers every year, but the enrollment in these courses is typically between nine and 16 students. Other enrolled students are already-certified physics and physical science teachers who are working on their master's degrees, graduate students from the physics department who wish to learn more about physics teaching methods, and science education doctoral students from the Graduate School of Education.

B. Early teaching experiences

Early teaching experiences consist of three elements: (a) microteaching in physics teaching methods courses throughout the program, (b) teaching in a reformed introductory algebra-based physics course (Physics for the Sciences) during the first year of the program and spring of the second year, and (c) student teaching in the fall of the second year. Each of these elements is described in detail in the following sections.

C. Community development

The PSTs develop the cohort-based learning community and integrate themselves into a broader community of program participants and graduates as they progress through the two years. The community framework starts to take shape with group work during class meetings in physics methods courses, group projects, group-based microteaching, and group preparation for oral exams. It grows with teaching as a cohort in the introductory physics course. The PSTs begin integrating into the community of program graduates during student teaching, when they are paired with cooperating teachers who went through the program four to nine years ago. Finally they join this broader community after graduation by participating in online discussions and face-to-face meetings.

TABLE I. Summary of coursework in the Rutgers physics teacher preparation program.

Year/semester	General education courses	Physics and physics methods courses	Course goals
1/Fall (combined with the undergraduate major courses)	Educational Psychology Individual and Cultural Diversity	Development of Ideas in Physical Science	To learn how physicists developed the ideas and laws that are a part of the high school physics curriculum.
1/Spring (combined with the undergrad major courses)		Teaching Physical Science	To learn how to build student understanding of crucial concepts, and how to develop and implement curriculum units and lesson plans.
		Demonstrations and Technology in Physics Education	To learn how to use technology to develop and apply physics concepts and assist in course management.
		Upper-level physics elective	For example, Physics of Modern Devices helps students learn about applications such as solar cells, cell phones, MRI, etc.
1/Summer	Assessment for Teachers (2 credits)	Engineering Education	To introduce preservice teachers to ways of including engineering projects in a physics course.
2/Fall	Classroom Management (1 credit)	Teaching Internship Seminar for physics students	To simultaneously support preservice teachers who are doing student teaching and to explore teaching approaches to the new concepts in the curriculum.
		Teaching Internship (9 credits)	These are the credits that preservice teachers sign up for while doing student teaching.
2/Spring	Inclusive Education	Multiple Representations in Physical Science	To integrate different representations of physics knowledge into problem solving.
	Teacher as a Professional		

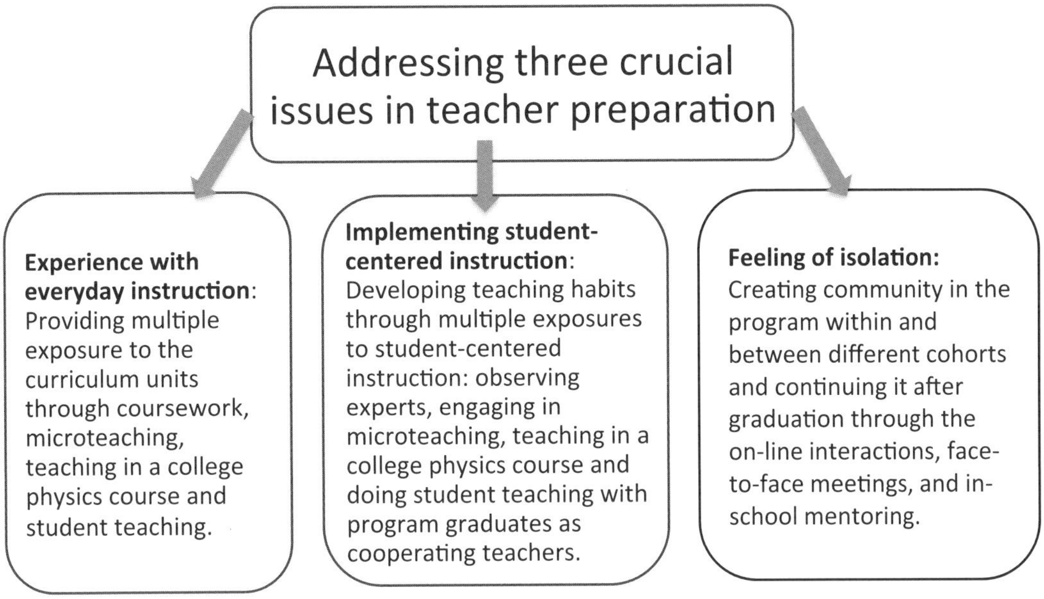

Addressing three crucial issues in teacher preparation

Experience with everyday instruction: Providing multiple exposure to the curriculum units through coursework, microteaching, teaching in a college physics course and student teaching.

Implementing student-centered instruction: Developing teaching habits through multiple exposures to student-centered instruction: observing experts, engaging in microteaching, teaching in a college physics course and doing student teaching with program graduates as cooperating teachers.

Feeling of isolation: Creating community in the program within and between different cohorts and continuing it after graduation through the on-line interactions, face-to-face meetings, and in-school mentoring.

FIG. 1. Rutgers program and three issues in teacher preparation.

IV. HOW THE RUTGERS PROGRAM ADDRESSES THREE CRUCIAL ISSUES IN TEACHER PREPARATION

Figure 1 summarizes Rutgers' approach to three critical issues in teacher preparation. The details of addressing each issue are provided in the subsequent subsections.

A. Preparing future teachers for everyday instruction

1. Coursework

The first step in preparing PSTs for everyday instruction is to help them develop three types of knowledge: knowledge of the progression of curriculum topics, knowledge of student ideas, and knowledge of ways to assess student learning. At Rutgers, PSTs build this knowledge slowly over two years through coursework in physics-specific methods courses. The physics topics addressed in each course are shown in Table II. The topics that are in every high school curriculum (motion, forces, energy, momentum, fields) are addressed twice, and those that might not appear (optics, waves) are addressed once.

2. Early teaching experiences in physics methods courses

The structure of each physics methods course is important. During the first five or six weeks in every physics methods course, 130 to 140 minutes of each class meeting mimics a student-centered high school lesson. During this time the PSTs play the roles of students and the course instructor plays the role of a high school teacher. The PSTs work in groups exploring physical phenomena (e.g., collecting and analyzing data, devising and testing explanations), presenting their findings on whiteboards, and participating in discussions, in the same way that high school students would. In the last 40 minutes of each three-hour class meeting, the PSTs reflect through whole-class discussions on the teaching methods used during the lesson, and how the same lesson can be implemented in high school. From week to week these lessons follow the progression of a standard physics curriculum, so the PSTs can see how ideas develop and build on each other over time. In the second part of the course (weeks seven through 14), students microteach. In groups of two they prepare and teach 140-minute lessons to their peers, each of which is roughly equivalent to one week of high school instruction (i.e., high school students typically have 180 to 225 minutes of physics per week, and the PSTs can teach the same material to their peers in 140 minutes).

Each microteaching experience can be broken into three stages:

(a) Preparation. During the preparation stage the PSTs work together to draft a lesson plan according to the requirements articulated in the syllabus, and post it as a document on Google Docs, giving access to the course instructor. The course instructor can provide immediate feedback and see the students' corrections. Working on the document takes one to two weeks, after which the instructor and students meet for a discussion of the lesson, equipment needed, and other logistical considerations. After this meeting the PSTs usually take one more week to polish the lesson before they teach it in class.

(b) Implementation. While the PSTs are teaching the lesson, the instructor uses a technique called "stop and rewind." When the "teachers" miss a question, misunderstand a

TABLE II. Relationship of course content to everyday physics instruction in high school

Physics-specific methods course	Physics content and ways of teaching that are explored
Development of Ideas in Physical Science	Motion, forces, momentum, energy, specific heat, kinetic-molecular theory, electric charge (including its quantum nature), electric current, magnetism, photon, atom, and nucleus
Teaching Physical Science	Motion, forces (including buoyant force), vibrations, energy, electric field, complex circuits, power, and magnetic field
Demonstrations and Technology in Physics	Topics depend on student choice but these are usually connected with available computer sensors (e.g., motion detector, force probe, microphone)
Upper-level physics elective	Modern physics including semiconductor physics
Engineering Education	Most projects involve mechanics and DC circuits
Teaching Internship Seminar (physics)	Motion, forces, momentum, statics, mechanical waves, and wave optics
Multiple Representations in Physical Science	Electric and magnetic fields, complex DC circuits, electromagnetic induction, geometrical optics, and photoelectric effect

student comment, miss a teaching opportunity presented by a student comment, or continue the lesson according to plan while students are either "ahead of the game" or having trouble with the material, the instructor says, "Stop and rewind." This means that the lesson stops and the "teachers" and "students" have an opportunity to reflect on why the lesson was stopped and how the lesson might be revised. The PST "teachers" have the first opportunity to offer amendments. If they cannot, then the PST "students" provide suggestions. After this rewind and adjustment, the lesson continues in the new direction. Overall, during a 120-minute lesson the instructor uses "stop and rewind" an average of two to four times. This number decreases gradually as the PSTs progress through the program, the first lesson requiring four or five and the last lesson zero or one.

(c) Reflection. After the lesson ends, a whole-class reflective discussion begins. The reflection prompts are different for the "teachers" and the "students." The teachers answer questions such as: What were the goals of the lesson? How do you know if the goals were achieved? What did you learn during the preparation for the lesson? What did you learn while teaching? The "students" have the following prompts: What physics concepts did you learn in this lesson and how did you learn them? What teaching methods did you notice? How were they used? What was successful in the lesson? What needs improvement? The "teachers" are the first speakers, followed by the "students." The course instructor summarizes the reflections, specifically discussing the role of "stop and rewind." These reflections help the PSTs remember what happened in the lesson and why the "stop and rewind" cases occurred, and it teaches them to listen to students during the lesson and be flexible in changing the direction if needed. The positive effects of this approach can be seen as each group's lesson is better than the previous group's lesson, due to the fact that the PSTs reflect on previous instructor feedback and incorporate this feedback into their lesson design and implementation. This same growth in pedagogical skill is also observed over the length of the program.

At the end of the semester, the materials that the PSTs produced for their microteaching as well as lesson and unit plans are shared with the rest of the class. This leaves everyone with a rich set of unit and lesson plans that can be used in the high school classroom. The details of work on lesson and unit plans are described elsewhere [21].

3. *Student teaching*

Student teaching is an integral part of preparing PSTs for the challenges of everyday instruction. During student teaching the PSTs are placed with graduates of the program who practice student-centered teaching, the same methodology PSTs have been learning for a year and implementing during microteaching. It is important for student teaching to be done in the fall (as opposed to the spring semester) so that the PSTs see how the experienced teachers start the school year and build student learning from the beginning.

B. Helping future teachers implement student-centered teaching

1. Pre-program experience

Before entering the program, PSTs have an option to serve as Learning Assistants in a reformed introductory physics course. (About 50% elect to do this.) The course, Physics for the Sciences, is an introductory algebra-based large-enrollment (200 students) physics course that employs methods of student-centered instruction and follows the Investigative Science Learning Environment (ISLE) approach [23]. Each week, students attend two large-room meetings (during which students work in pairs on ISLE activities constructing and testing new ideas [24]), one 80-minute problem-solving session, and one three-hour lab. In the problem-solving sessions and labs there are about 25 students. In these smaller sessions, students work in groups on activities based on the findings of physics education research [23]. In the labs, students design their own experiments [25]. Overall, the course structure, teaching methods, and teacher-student interactions resemble the environment that we would like the physics teachers to create in their classrooms. LA responsibilities include facilitating group activities during problem-solving sessions and leading after-hours study groups.

Those PSTs who indicate a desire to become physics teachers early in their undergraduate physics major enroll in the Physics for the Sciences course instead of the general physics course for majors. Although it is an algebra-based course, we believe that experiencing student-centered learning is beneficial for PSTs while they are undergraduates. After they have passed the course, in the following year the PSTs serve as Learning Assistants in problem-solving recitations or lead study-group sessions.

2. First-year experience

In addition to microteaching, PSTs engage in two other kinds of early teaching experiences in their first year of the program: observations in schools and teaching in college physics courses.

(a) Observations in schools. Simultaneously with microteaching, in spring of the first year, PSTs spend 10 days (one day per week) in high schools observing teachers who graduated from the program three to 10 years before. These teaching observations are a required component of the Teaching Physical Science course. Eventually, the PSTs plan and implement one "real" lesson. The lesson preparation and after-lesson reflection proceed the same as during microteaching, but without "stop and rewind" episodes, giving PSTs an opportunity to see how these methods work in the classroom.

(b) Teaching experiences in a college physics course. During the two semesters prior to student teaching and the one semester after student teaching, the PSTs work as lab or problem-solving instructors in the Physics for the Sciences course, as described above. Their teaching load is usually limited to one lab and/or one problem-solving session per week, which includes two to three contact hours, office hours, grading of homework or exams, and attendance at training meetings. The PSTs individually are fully responsible for their students' learning of introductory physics in their sections. However, the PSTs do not choose the activities for their recitations, nor do they design lab materials or write course exams. These materials are provided by the course coordinator. The PSTs need to plan what will happen during class sessions: what difficulties the students might have, how to structure students' reflections on what they learned at the end of each session, what additional examples to provide, and how to grade student classwork and homework. Thus their college physics teaching is a simplified and sheltered version of high school teaching wherein a teacher creates lesson plans, assembles equipment, writes tests, and assigns course grades. The PSTs' major responsibility is to implement student-centered instruction and later reflect on what happened in class.

The influence of this teaching experience within the graduate program can be seen from a few quotations provided by former PSTs who are currently New Jersey high school physics teachers:

> I feel that teaching in Physics for the Sciences (PFTS) was incredibly useful in preparing me for both student teaching and my first teaching position. I definitely felt considerably more at ease in front of the class by the end of the PFTS year than I did the first time. Also, the format of [PFTS], where I was forced to work with students in small groups, who were working at different individual paces, prepared me for being a better teacher. Otherwise I might have resorted to significantly more whole-class, teacher-led instruction because it seems easier, but the experience in PFTS showed me that managing small group work isn't as hard as it seems and has incredibly strong benefits for the students.
>
> I felt that being in charge of a classroom so early on was most helpful for me in two ways: (1) allowing me to experience the kinds of common ideas and questions students had and allowing me the opportunity to deal with those questions, and (2) giving me experience in dealing with the paperwork and non-teaching aspects of being in charge of a class. Those are two things that you can't fully understand unless you experience them and collaborate with others to deal with them.

3. Second-year experience

In the fall of their second year in the program, PSTs do a semester-long student teaching internship. In the spring they return to teaching in the college physics course and become mentors to the first-year PSTs teaching in the same course.

For the teaching internship, the PSTs are placed with cooperating teachers who are graduates of the program (usually the same teachers who were observed by the PSTs in the spring of the previous year). These placements are possible only because of the continuous interaction of the program coordinator with graduates (see the section on the learning community). Placing the interns with program graduates allows the interns to practice what they learned and avoid conflict between how they are "supposed to teach" and "how real teachers teach." The following comment from a graduate illustrates this point:

> One of the most important factors for me was having an incredibly skilled and supportive cooperating teacher, who encouraged me to take risks and try things in the classroom that I would probably have been afraid to try on my own during my first year in my independent teaching. It was also crucial to me that my cooperating teacher was such a strong proponent of the ISLE method. So I could really observe it in action during the first few weeks and then have support in implementing it throughout the rest of the semester.

During the student teaching internship, the PSTs plan and execute lessons under the supervision of the cooperating teacher and university supervisor. The university supervisor is usually a retired physics or chemistry teacher who is trained by the program coordinator on the philosophy of the program and the expectations of the PSTs. The training helps ensure that university supervisor provides the PSTs with feedback that is consistent with the goals and teaching approaches advocated by the program. The supervisor visits the PSTs about once every two weeks (usually eight observations during the semester). In addition, the instructor of the Student Teaching Internship Seminar course (usually the program coordinator) observes each PST once during student teaching. These observations strongly influence the discussions in the Student Teaching Internship Seminar, which PSTs attend once per week. In this class, in addition to discussing classroom observations, PSTs reflect on what happened during the week, learn to interpret and assess student work, and plan new lessons. To guide the PSTs, the program developed a self-assessment rubric to support lesson planning and post-teaching reflections. Typically, at the start of each week the PSTs choose four to six items from the rubric (out of a total of 17) to work on. At week's end, they summarize their progress in writing and choose a new set of rubric items to guide their self-assessment in subsequent weeks. As the course progresses, after the PSTs go through the entire set of rubric items, they continue working on those items that are the most difficult. An excerpt of the syllabus for the seminar, including the rubric, is provided in Appendix A.

The following quotation illustrates how beginning teachers think of their student teaching experience in retrospect:

> In retrospect I was very prepared for student teaching. Having a well-prepared unit plan for dynamics[1] was crucial. Also, I felt I had a very good grasp of not just the content material but also the specific points that students might struggle with. However, I must admit that at the start of student teaching I was terrified! (Although every student teacher I've talked to has said the same thing.) I quickly realized that the Rutgers program had prepared me so well, it wasn't nearly as difficult as I had anticipated, and actually was really fun.

[1]Each PST creates a dynamics unit plan during the Teaching Physical Science class in the spring before student teaching.

C. Preventing the feeling of isolation

This section describes the process through which the learning community of PSTs and program graduates is built: specific aspects of the coursework and clinical practice that help develop the camaraderie of the PSTs within a cohort, across the cohorts, and with the graduates of the program who are already teaching.

1. First-year experience

The building of the learning community is a purposeful activity that begins on the very first day of PSTs being admitted to the program. In September of their first year, PSTs start teaching in the Physics for the Sciences course. They come to training meetings and share reflections on teaching on the course Google discussion group [5]. This process helps PSTs get to know each other. At the same time, in the Development of Ideas in Physical Science course, the PSTs work in groups, changing group partners every week. After three to four weeks they start working on a group project that lasts for two months, which is the first real bonding that occurs during the program. Similar group interactions occur in the spring courses, and so by the end of the first year each PST has worked with every other PST in the cohort several times. In addition to the formal coursework, each PST has worked with the rest of the cohort on lab and problem-solving activities related to teaching in the introductory physics course (detailed in Section IV.B.2).

Helping PSTs establish a learning community prior to student teaching is important, since the community will help the PSTs get through difficult times. To strengthen the community connections that form the first year, PSTs are required to participate in an oral exam at the end of the Teaching Physical Science Course (see exam questions in Appendix B). Five weeks before the spring semester ends, the course professor posts 30 to 35 questions that the PSTs should be able to answer during the oral exam. In addition to answering these questions, during the exam the PSTs also need to solve problems from an algebra-based physics textbook, solve problems involving real-time data from videotaped experiments, and demonstrate an interesting experiment for the class. The number and depth of the exam questions and problems are designed in such a way that it is impossible for the PSTs to answer all the questions and solve all the problems on their own; they have to collaborate with others and share the work. PSTs realize this quickly and start working on the questions and problems together. This preparation for the oral exam bonds them together even more because they know that nobody can pass the exam by working alone.

2. *Second-year experience*

The first semester of the second year is dedicated to the student teaching internship (described in Section IV.B.3). Since PSTs are placed with graduates of the program as cooperating teachers, the PSTs and graduates start building connections. The PSTs also meet once per week for the Student Teaching Internship Seminar. During these meetings they reflect on the past week, share their achievements, failures, and concerns, and support each other. They also join the community of program graduates on the Internet (described in Section IV.C.3) and pose questions about their teaching practices. In the past three years the interns created their own Facebook group.

In the second semester of the second year, the PSTs return as instructors to the Physics for the Sciences course (second semester of the course), where they meet the first-year cohort. As both cohorts attend the same training meetings and reflect on their teaching using the same Google group page, the first- and second-year cohorts bond and learn from each other. This experience is invaluable for the PSTs from the first-year cohort because they can ask questions about the program, student teaching experiences, and many other topics that concern them. In the second semester, PSTs also enroll in the last physics teaching methods course, Multiple Representations in Physical Science, where they again work in groups during class meetings, work with a peer on a long project, and microteach together. The course culminates with a second oral exam. Preparation for the exam is a whole-class activity that lasts for three to four weeks.

3. *After-graduation experience*

By the time PSTs graduate from the program, they feel like members of a big family: They have bonded with peers from the same cohort, they have supported a younger generation, and they have met and worked with graduates of the program. In 2004, a PST cohort created a Web-based discussion group, and since then all new graduates join this group to stay in touch with each other. Since the fall of 2004, there has been an average of 70 messages per month on the discussion list, from a low of 15 in the summer to a high of 160 in some school-year months. Most of the questions are related to the teaching of specific physics topics, student difficulties and ideas, physics questions that are difficult for teachers, new technology, equipment sharing, interactions with students and parents, and the planning of meetings. The preservice teachers join the group during their student teaching, so that by the time they graduate they are well integrated into the community. The following quotation provides an example of how the teachers use the online community:

> I contribute or respond to the email group several times a week. Here is how I use the list to interact with our community:
>
> (1) Sharing resources (2-3 times per week): Usually I am sharing a found technology resource that might be interesting or useful, like an application or video. I also share resources I have created myself. Often, other community members will respond with improvements, alternatives, or related materials.
>
> (2) Sharing ideas (2-3 times per month): It's great to be able to write something like "hey, who had that lesson with the light bulb puzzle ..." and get a response that helps me recreate someone else's brainstorm. And when I share my own ideas, someone will often write back with improvements and new connections.
>
> (3) Logistical and emotional support (weekly or so): Soliciting suggestions and feedback on ideas to manage planning, assessment, grading, classroom management, etc. Sharing my own ideas and tools for these topics.

In addition to the online component, the community of graduates meets every two to three weeks at Rutgers. The meetings are organized by the program coordinator and last about two hours. The attendance varies from 15 to 20 people to sometimes as low as four. Several attend the meeting online (via Skype). In the early years of the community (2005–2010) the meetings' main purpose was to help new teachers with instructional challenges. Over the years, meetings evolved to include discussions of new, exciting, and challenging physics experiments or questions (called "physics challenges"), because we discovered that teachers missed the excitement of facing mental challenges and broadening their own understanding of physics. At first the physics challenges were posed by the program coordinator, but gradually the inservice teachers took the initiative of offering the challenges, and this new physics exploration has become the most attractive part of the meetings.

Over the years the meetings have changed days of the week. Recently we found that the best day and time for the meetings is Friday afternoon. The meetings start at 5:30 or 6:00 p.m. and last until 8:30 or 9:00 p.m. With this meeting schedule the teachers do not need to worry about teaching the following day, and after the meeting they can get together for dinner and more informal bonding.

One program graduate reflected:

> The Friday meetings are helpful in terms of providing hints and specific activities I can use in the classroom, but more importantly they provide a community where we feel free to share our difficulties and receive guidance and support in return. It is fun to spend time with a group of people who are as fascinated by physics as I am.

Maintaining the life of the community requires time and sacrifice of Friday nights every other week from the program coordinator. One might think that this is a heavy burden to carry. However, the meetings have such positive energy and provide such reinforcement for the program values that this emotional boost compensates for the invested time.

Two other issues are important to mention. The program has been in place for over 10 years, and most of the graduates teach in New Jersey after graduation. Many schools, after hiring one Rutgers graduate, wish to hire another. There are now five high schools that have two or more graduates of the program. This situation allows experienced teachers to really mentor the beginning teachers, creating a completely different first- and second-year teaching experience. As some of them comment:

> Having Jon [graduate of 2009 cohort] in the school with me, I think, is what helped make my first year such a positive experience. Knowing there is someone familiar with my teaching style and, on top of that, is aware of what parts of that teaching style generally work and don't work was very helpful in taking some of the stress of trial and error out of the job. For example, Jon told me that traditional homework assignments are very ineffective with the students at the level I was teaching. To accommodate for this learning style, I gave them smaller assignments that involved watching videos or using simulations to prepare for quizzes that I gave on most days. Another helpful aspect was the materials. Jon was able to provide me with plenty of material for almost every unit that, with a little adjustment for my teaching style, was in line with the way I wanted the students to learn and allowed me to spend more time to reflect on the teaching aspect and less on the formatting of activity sheets. I think the fact that I came out of my first year not haggard and, to the contrary, actually excited for the coming academic year is a testament to how helpful it is to have had someone from the program there to support and mentor me through my first year.

> Having another graduate of the program at my school helped my transition to teaching tremendously. I immediately had someone with whom I could collaborate, knowing that they graduated from the same teacher preparation program from which I had just graduated and would be familiar with the resources and pedagogy that I would be utilizing.

V. SUMMARY: DOES THE SYSTEM WORK?

The beginning of this chapter described three major challenges that exist in teacher preparation: (1) providing PSTs with enough experience of everyday instruction before they graduate from the program; (2) supporting them in implementing student-centered instruction; and (3) eliminating the feeling of isolation after they graduate to ensure that they remain in the profession.

Table III provides the summary of how the Rutgers physics teacher preparation program addresses these challenges through coursework, early teaching experiences, and community building.

A. Challenges of everyday instruction

While progressing through the Rutgers program, PSTs gain experience teaching the same physics content that they will be responsible for teaching in high school. The PSTs teach the content of first-semester high school physics in the Physics for the Sciences course (first eight weeks), and then teach this content again during the fall semester of student teaching. In addition, through the PSTs' microteaching and curriculum development activities in the methods courses, when they eventually secure a high school position they have

TABLE III. Efforts to address three issues in teacher preparation.

	Year 1		Year 2	
Issue in teacher preparation	Coursework	Early teaching experience	Coursework	Early teaching experience
Providing experience with everyday instruction	Coursework in physics-specific methods courses	Teaching in Physics for the Sciences	Course topics in physics-specific methods courses	Student teaching
Providing experience implementing student-centered instruction	Reflecting on professor's teaching and preparing and reflecting on their microteaching	Teaching in Physics for the Sciences, microteaching	Reflecting on student teaching experiences, planning lessons and units	Student teaching (Fall) and teaching in Physics for the Sciences (Spring)
Creating a community to combat the feeling of isolation	Group work on projects, microteaching, and oral exam	Reflecting on teaching using the Google discussion group	Sharing during student teaching, oral exam preparation	Cooperating teachers (graduates of the program), post-graduation community

access to two physics units that they developed on their own and another 10 units developed by their peers. As a result, there are very few course topics that program graduates need to teach in the first year that they have not already taught and planned. However, despite their strong preparation to teach physics content, there are other challenges that remain for first-year graduates, as illustrated by this comment:

> I felt thoroughly prepared in terms of knowing the subject matter and how to communicate it to the students, which amazed me given how little I knew just one year earlier! One of the things I felt most unprepared for was differentiated instruction, especially having students with special needs in my classroom, since during my student teaching I had only honors & AP students, and none with IEPs [individualized education plans]. Specifically I struggled with motivating some students to complete their homework, and in my regular class had difficulty balancing mathematical & conceptual as there were some students with considerably weaker math skills than others.

B. Implementing student-centered instruction

All instruction that PSTs experience from the start of the program (in both the roles of student and teacher) is purposefully student centered, and models the teaching we would like them to implement in a high school classroom. The repeated exposure to and implementation of this type of instruction creates new habits of teaching. Since in a typical year none of the PSTs have had any teaching experience before entering the program, but all have years of learning experiences in teacher-centered environments, many experience a shock at the beginning, and express it openly. They struggle though the first semester to conceptualize this new way of teaching, and even by the end of the semester many are not convinced that it would work in a high school setting. However, their student teaching experiences and observations of program graduates provide convincing evidence that such an approach works with high school students too. The program coordinator observes PSTs during student teaching, and in 100% of observed lessons she finds that the student teachers plan their lessons to be student centered, and that while implementing the lessons they attempt to follow through with this student-centered approach. The observed lessons involve students exploring physical phenomena in groups, discussing their observations, representing observations with motion and force diagrams or energy bar charts, presenting the results on a whiteboard, arguing their approaches, and solving problems in groups. Observations of program graduates carried out over the last seven years show consistently high scores on the Reformed Teaching Observation Protocol (RTOP) [26]. The scores are between 65 and 95, with an average of 75. (The score of 50 is the threshold between traditional and student-centered instruction.) As one of the graduates said after one year of teaching, "I felt the most prepared when it came to the content and my preparation of lessons that would lead students through the discovery process in learning physics."

C. Eliminating the feeling of isolation

The three-tier learning community (first-year cohort, second-year cohort, and graduates) helps alleviate the feeling of isolation during professional preparation and after graduation. Pairing PSTs with program graduates during student teaching also addresses issues (1) and (2). Many PSTs get hired in the schools where they did their student teaching, so that their former cooperating teachers, now colleagues, can provide additional support during their first years.

The online component of the post-graduation community is important because it allows participants to provide support when it is needed. As one of the teachers wrote:

> When I've had a discouraging day, the group is a great way to vent and share in a positive and supportive atmosphere. It's a great resource when I am feeling low. I always know that I am not the only one struggling with the daily challenges of a teacher, and I'm happy when I can extend help or at least empathy to my peers. It's also a great way to inspire myself to continual improvement and higher standards—I'm often amazed by what my colleagues understand and what they can accomplish in their lessons and in their relationships. I would otherwise have no idea of how many wonderful accomplishments are possible for physics students AND their teachers!

D. Other outcomes

The Rutgers physics teacher preparation program has been in place since 2002, with the first cohort finishing in 2004. Over 70 teachers have graduated from the program, and more than 80% of them are still teaching. (Average national teacher attrition rates are about 40% to 50% after five years.) Those who give their students pre- and post-Force Concept Inventory [27] tests achieve learning gains from 0.4 to 0.6 in the first year of teaching. Those who teach Advanced Placement courses enjoy excellent results, and in fact one of the program graduates consistently has the highest scores in New Jersey. Some graduates also teach selected courses in the program.

Can we say that the carefully crafted early teaching experiences and learning community contributed to this success? As we did not try to remove the program elements described in this chapter and study the effects on attrition rate, learning gains of the students, and so forth, such a claim is meaningless. The program is a complex organism, and all of its components contribute to the success of its graduates.

ACKNOWLEDGMENTS

I would like to thank the book editors and reviewers for their helpful suggestions and Caroline Coogan for assisting with the preparation of the manuscript.

[1] D. Boyd, P. Grossman, H. Lankford, S. Loeb, and J. Wyckoff, Teacher preparation and student achievement, Educ. Eval. Policy An. **31**, 416 (2009). doi:10.3102/0162373709353129

[2] P. E. Simmons, A. Emory, T. Carter, T. Coker, B. Finnegan, and D. Crockett, Beginning teachers: Beliefs and classroom actions, J. Res. Sci. Teach. **36**, 930 (1999). doi:10.1002/(SICI)1098-2736(199910)36:8<930::AID-TEA3>3.0.CO;2-N

[3] L. Darling-Hammond, The challenges of staffing our schools, Educ. Leadership **58**, 12 (2001). doi:10.1177/0022487105285962

[4] D. E. Meltzer, M. Plisch, and S. Vokos, eds., *Transforming the Preparation of Physics Teachers: A Call to Action. A Report by the Task Force on Teacher Education in Physics* (T-TEP) (American Physical Society, College Park, MD, 2012). http://www.phystec.org/webdocs/2013TTEP.pdf

[5] A. Tigchelaar and F. Korthagen, Deepening the exchange of student teaching experiences: Implications for the pedagogy of teacher education of recent insights into teacher behavior, Teach. Teacher Educ. **20**, 665 (2004). http://www.sciencedirect.com/science/article/pii/S0742051X04000812

[6] A. Arons, *Teaching Introductory Physics* (John Wiley & Sons, New York, NY, 1997).

[7] M. Brekelmans, T. Wubbels, and J. Levy, Student performance, attitudes, instructional strategies and teacher-communication style, in *Do You Know What You Look Like? Interpersonal Relationships in Education*, edited by T. Wubbles and J. Levy (The Falmer Press, Bristol, PA, 1993), pp. 56-63.

[8] L. S. Shulman, Knowledge and teaching: Foundations of the new reform, Harvard Educ. Rev. **57**, 1 (1987). http://hepg.metapress.com/content/J463W79R56455411

[9] J. H. van Driel, N. Verloop, and W. de Vos, Developing science teachers' pedagogical content knowledge, J. Res. Sci. Teach. **35** (6), 673 (1998). http://srvcnpbs.xtec.cat/cdec/images/stories/WEB_antiga/formacio/pdf/sfece/07-08/teachers.pdf

[10] V. Otero, N. Finkelstein, R. McCray, and S. Pollock, Who is responsible for preparing science teachers?, Science **313**, 445 (2006). doi:10.1126/science.1129648. For more information on the University of Colorado Boulder program, see http://stem.colorado.edu

[11] H. Borko and V. Mayfield, The roles of the cooperating teacher and university supervisor in learning to teach, Teach. Teacher Educ. **11**, 501 (1995). http://www.sciencedirect.com/science/article/pii/0742051X95000088

[12] L. Darling-Hammond, A. Pacheco, N. Michelli, P. LePage, and K. Hammerness, Implementing curriculum renewal in teacher education: Managing organizational and policy change, in *Preparing Teachers for a Changing World*, edited by L. Darling-Hammond and J. Bransford (John Wiley & Sons, San Francisco, CA, 2005), pp. 442-479.

[13] K. M. Zeichner and J. Gore, Teacher socialization, in *Handbook of Research on Teacher Education*, edited by R. Houston, M. Haberman, and J. Sikula (Macmillan, New York, NY, 1990), pp. 329-348.

[14] C. M. Evertson, Bridging knowledge and action through clinical experiences, in *What Teachers Need to Know*, edited by D. D. Dill (Jossey-Bass, San Francisco, CA, 1990), pp. 94-109.

[15] J. I. Goodlad, Studying the education of educators: From conception to findings, Phi Delta Kappan, **71** (9), 698 (1990). http://www.jstor.org/stable/20404256

[16] M. Corcoran-Smith, Reinventing student teaching, J. Teach. Educ. **42**, 104 (1991). doi:10.1177/002248719104200204

[17] A. Collins, J. S. Brown, and S. E. Newman, Cognitive apprenticeship: Teaching the crafts of reading, writing, and mathematics, in *Knowing, Learning, and Instruction: Essays in Honor of Robert Glaser*, edited by L.B. Resnick (LEA, Hillsdale, NJ, 1989), pp. 453-494.

[18] B. J. Reiser, Scaffolding complex learning: The mechanisms of structuring and problematizing student work, J. Learn. Sci. **13**, 273 (2004). doi:10.1207/s15327809jls1303_2

[19] L. Darling-Hammond, Keeping good teachers: Why it matters, what leaders can do, Educ. Leadership **60**, 6 (2003). http://eric.ed.gov/?id=EJ666108

[20] R. J. Murnane and R. J. Olsen, The effects of salaries and opportunity costs on length of stay in teaching: Evidence from North Carolina, J. Hum. Resour. **25**, 106 (1990). http://www.jstor.org/stable/145729

[21] Students in the post baccalaureate program have exactly the same coursework and clinical experiences, only they complete them after they finish their undergraduate degree.

[22] E. Etkina, Pedagogical content knowledge and preparation of high school physics teachers, Phys. Rev. ST Phys. Educ. Res. **2** (6), 020110 (2010). doi:10.1103/PhysRevSTPER.6.020110

[23] E. Etkina and A. Van Heuvelen, Investigative Science Learning Environment - A science process approach to learning physics, in *Research-Based Reform of University Physics*, edited by E. F. Redish and P. Cooney (American Association of Physics Teachers, College Park, MD, 2007). http://www.per-central.org/document/ServeFile.cfm?ID=4988

[24] E. Etkina, D. Brookes, and A. Van Heuvelen, *Instructor Guide to Accompany College Physics* (Pearson, San Francisco, CA, 2014).

[25] E. Etkina, A. Karelina, and M. Ruibal-Villasenor, How long does it take? A study of student acquisition of scientific abilities, Phys. Rev. ST Phys. Educ. Res. **2** (4), 020108 (2008). doi:10.1103/PhysRevSTPER.4.020108

[26] D. MacIsaac and K. Falconer, Reforming physics instruction via RTOP, Phys. Teach. **40**, 479 (2002). http://physicsed.buffalostate.edu/pubs/RTOP/TPTNov02RTOPcorrected.pdf

[27] D. Hestenes, M. Wells, and G. Swackhamer, Force concept inventory, Phys. Teach. **30**, 141 (1992). doi:10.1119/1.2343497

APPENDIX A

Excerpts from the Syllabus for the Teaching Internship Seminar Course
Teaching Internship Seminar (Physical Science Section)
15:255:536 (section)
3 Credits

Learning goals

The goals of the course are to learn how to plan, implement, and reflect on classroom instruction in physics/physical science that engages all students in productive and meaningful learning of physics content and practice. Achievement of those goals includes mastering time management, emotional control, physics experimental skills, listening to students, and, most importantly, communication skills (communication with students, cooperating teacher, parents and school administration). Additional goals include continued improvement of one's own physics understanding and acquisition of additional strategies that engage diverse learners in mastering physics (content and practices).

Grading and Activities

Your course final grade will be based on attendance, participation in the discussions, reflection on teaching, lesson plans, quizzes and exams that you will design, video analysis of your lesson, a research project, and teaching portfolio. Each assignment can be improved, as many corrections as needed are encouraged.

Activity	Total points
Attendance, participation	100
Reflection on teaching	100
Screencasts	100
Unit and lesson plan	100
Debate	100
Research on student learning	100
Teaching portfolio	100
Grand Total	**700**

Rubrics for self-assessment of teaching

Below is a list of abilities that you need to develop during student teaching. You can use the rubrics below to plan your lessons and self-assess them. Your cooperating teacher will have the rubrics too.

Ability	N/A	Well developed	Working towards it	Missed opportunity
		3	2	1
To start a lesson in an organized productive way		Students start working from the first second, everything is planned and no time is wasted.	The first seconds are spent unproductively but the lesson got on track within the first 3 min.	The beginning of the lesson did not lead to the organized, inspired work.

To create motivation for student learning		The content of the lesson is connected to student lives, or there is an interesting question, or motivation is created based on student success, students understand why they are doing what they are doing.	There is some attempt to motivate students but many do not know why they are doing what they are doing.	Motivation is based on "need for the test" or is absent.
To keep track of what every student is doing		The teacher scans the classroom often and notices subtle details of student learning activities and behavior; most students participate in the lesson and speak.	The teacher follows most of the students but misses a few, the omissions do not lead to the disruption of the lesson.	The teacher does not notice a crucial moment/s that leads to the disruption of the whole lesson; few students participate.
To help students develop study habits		A great deal of attention is given to building study habits: taking notes, planning learning, metacognition, drawing sketches and graphs, asking productive questions, time management.	Some attention is given to building study habits but it is not systematic.	No attention is given to study habits.
To use the board strategically		The board is a productive teaching tool that helps students organize their notes and follow the lesson, writing is clear, large letters, a ruler is used for the drawings and the whole lesson fits on one board.	The board is used but things are erased often, no ruler to draw graphs and other pictures, hard to follow.	The board is used randomly, it is clear that the teacher did not think it through.
To organize experimental work effectively		The experiments shown by the teacher are easy to see, students understand the point and either record and explain or predict, observe and reconcile. Experiments for the students are planned, who goes where and when is clear, no time is wasted, equipment is appropriate and works well.	Experiments done by the teacher work well but student participation is minimal, the purpose is not clear. Student experiments are planned but student work is not well thought through beforehand, time is wasted.	Teacher experiments are hard to see, the discussion is limited. Student experiments are not thought through – either time is wasted, students are disorganized or the physics point is lost.

To organize whole class discussion effectively		The teacher guides the discussion but does not dominate it, the summary is clear, lots of student-student talk, pauses for the students to take notes, main points are summarized on the board.	The discussion is two way mostly teacher-student-teacher, all summaries are done by the teacher, no time to take notes, the board is sketchy.	The teacher talks most of the time, students respond yes or no, the board is not used, no time or attention to notes.
To organize group work effectively		Students are used to working in groups, they arrange quickly, the teacher moves among the groups and group assignments are open-ended enough to promote fruitful discussions, white boards are used and all students participate; at the end there is a debriefing.	Students are used to working in groups but it takes some time to settle or group tasks are focused on one right answer, or white boards are not used productively, the teacher spends too much time with one group.	Students are not accustomed to working in groups, many do not participate, no debriefing, the teacher does not attend to all groups.
To manage time effectively		A productive **sense of urgency** is present, timing for activities is announced, the change of types of work occurs often but not too often.	The pace is either too slow or too fast.	The lesson drags.
To lead reflection effectively		All students participate, the reflection is focused on the important issues.	Few students participate, some comments are not useful.	Students reflect on non-important issues.
To assign homework effectively		The homework helps reinforce the past lesson or prepares for the future lesson, it is meaningful and instructions are clear.	The purpose of homework is unclear but the instructions are present.	No homework or no instructions.
To listen to the students		The teacher listens and responds to student comments productively.	The teacher listens but some responses are not productive.	Student comments are not noticed or ignored.
To use multiple representations		Multiple representations are used and are used productively.	Some representations are used productively.	Few representations are used and the purpose is unclear.

To use technology		Technology is used strategically.	Technology is used strategically sometimes	Technology is used but is not really needed to improve learning.
To pose productive questions and to respond to students' questions		The questions are high level, responses to student questions are done through reflective toss technique, they lead to deep thinking, no wrong physics answers on the teacher's part.	The questions are mixed, students questions are answered directly, the teacher's physics is correct.	The questions are mostly yes/no, students' questions are ignored, or teacher's responses have incorrect physics.
To encourage students to generate productive questions		There is a mechanism through which students learn to generate good questions, the teacher models how to ask good questions, the atmosphere in class is conducive to students asking questions.	Students questions are rare but are treated with respect	There are no students questions.
To generate explanations		Students are continuously encouraged to explain and devise mechanisms for evidence; students, not the teacher, evaluate provided explanations, students are encouraged to argue their point of view and multiple points of view are tolerated as long as the explanations are logical; the explanations provided by the teacher are correct from the physics point of view.	Students sometimes are pressed for explanations but not always, the teacher evaluates explanations by saying good or ok, instead of tossing them back to students, the explanations provided by the teacher are ok but not really deep.	The teacher does not press for explanations, argumentation is not encouraged, phenomena are analyzed macroscopically, mechanisms are missing, the explanations provided by the teacher have physics mistakes.
To build the lesson on students' ideas		The lesson plan takes into account student ideas documented in research and learned in course work and the lesson is continuously modified based on students' ideas emerging during the lesson	The lesson plan takes into account student ideas documented in research and learned in course work and but during the lesson students' ideas are largely go unnoticed	Students' ideas are not taken into account during the planning stage and are not used productively during the lesson.

APPENDIX B

Teaching Physical Science Exam Questions

The oral exam will take place during our regular class time on the 7th and at 2 pm if on the 14th and will take about 4 hours. To prepare for the exam you will need to prepare answers to the questions below, to solve all video puzzles in the section Surprising data and Puzzles on the video website, and to solve all problems from section 4 of the relevant chapters of the ALG (chapters 1-6, 9, 15, 16, 17, 18, and 19). You will also need to prepare a "cool experiment" to show during the exam, explain it and describe where in the learning sequence you will use it (as a hook, as an observational experiment, testing experiment, or application experiment). During the exam you will be given two questions from list below, a puzzle section on the video website and from the ALG. You will present the answer to the question at the board, so everyone can hear you and the other two items – on paper.

To prepare for the exam you need to work with your classmates and you need to start this preparation now as it is not a regular exam.

1. Show how to use NJ Science Standards while planning a lesson (use a specific lesson and specific standards, explain how the standards connect to the goal of the lesson).
2. For the following units: kinematics, linear dynamics, circular motion, momentum, energy, fluid statics, vibrations, electric fields, DC currents, magnetism and electromagnetic induction make a list of conceptual goals, quantitative goals, epistemological and metacognitive goals and explain why each goal is appropriate.
3. Use any of the verbatim students responses from your observations (or from your own students) to show how this particular response will affect your instruction (you can use a made-up scenario).
4. Use student work from physics 194, your observations or your high school teaching as evidence that students achieved a particular goal.
5. Use any lesson plan you wrote during this semester to show how you addresses all elements of the e-portfolio rubric for the lesson plan.
6. Use the unit plan you wrote for your project to show how you addresses all elements of the e-portfolio rubric for the unit plan.
7. Use an example of a lesson that you planned this semester. Show that the beginning of the lesson you designed builds on student ideas and engages them in meaningful exploration of physics ideas.
8. What difficulties do students have with kinematics concepts? Make a list and suggest on strategy per difficulty that helps alleviate it.
9. For the question above, provide examples of formative assessments that will tell you whether a student has a particular difficulty.
10. What are the difficulties that students have with kinematics graphs? Provide examples of the difficulties and sequences of instructional moves that are considered helpful for those difficulties.
11. Describe what students will do in class to construct the equation $\Delta x = v_{0x} t + ½ a_x t^2$.
12. Explain what concept of an index is and how you plan to use it to help your students construct the operational definitions of velocity, acceleration, density, $\vec{E}$ field and electric resistance.
13. Explain how an operational definition of a physical quantity is different from a cause-effect relationship for the same quantity. Provide at least two examples of such relationships and explain how your students will learn the difference.
14. What should students know about *any* physical quantity and how will they learn it? Use two quantities as an example.
15. What are science practices? Give examples and describe examples of lesson where students can employ them. Be very specific.
16. What should students know about friction force? How will they learn it? What questions will you ask to find out if they mastered those ideas?
17. Explain why one needs to draw a motion diagram first before drawing a force diagram while solving a dynamics problem. Provide an example of a problem to solve which both representations are essential. Explain where in the dynamics unit students will solve the problem.
18. Describe a difference between a cook-book and a design lab. Provide an example of a design lab for energy and show how students will use science practices in this lab.
19. What are scientific abilities rubrics? Give an example and show how you can use them to help students develop scientific abilities.
20. Explain the relationship between normal and friction force and demonstrate how to approach multiple-objects problems.

21. Explain why and under which assumption the energy of a bound system is negative. Show how to derive the expression for gravitational potential energy and electric potential energy of two objects. Use bar charts for your derivation.
22. Show how to help students develop mathematical expressions for 4 types of energy used in mechanics.
23. Describe student difficulties in the area of work-energy,
24. Describe curriculum sequence for teaching work-energy unit.
25. Describe how your students will learn what density is and the difference between the operational definition of density and cause-effect relationship.
26. Explain where the expressions for the fluid pressure and buoyant force come from.
27. Describe student difficulties with the concept of buoyant force and show specific questions that will assess whether your students have those difficulties.
28. Explain the difference between periodic motion and simple harmonic motion; describe and explain SHM; describe useful representations in this area.
29. Explain the difference between the concept of electric field and the physical quantities characterizing it.
30. Use multiple representations to explain the behavior of conductors and dielectrics in electric field.
31. Give an example of a DC circuit problem to solve which one needs to reason through potential difference not through current.
32. What should students know about power in electric circuits? How will they learn it? How will you know that they have learned it?
33. Give an example of a complete ISLE cycle for any of the concepts you choose.
34. What is the difference between two right hand rules in magnetism? Describe how your students will learn those and how you will assess them.
35. Outline the sequence of student learning of static electricity,
36. Outline the sequence of student learning of DC circuits,
37. Outline the sequence of student learning of magnetic fields
38. Outline the sequence of student learning of electromagnetic induction.

Joining hands to establish a teacher training program: An example from a major research university

Alice D. Churukian and Laurie E. McNeil
Department of Physics and Astronomy, University of North Carolina at Chapel Hill, Chapel Hill, NC 27599

At the University of North Carolina at Chapel Hill, the College of Arts and Sciences and the School of Education have formed a partnership to develop and implement a fast-track program to prepare science and mathematics majors to become licensed high school teachers in their chosen fields. At the time the program started, no baccalaureate program for secondary school teacher preparation existed at the university, so we were breaking new ground. We discuss how this program came into existence, describe the program and the collaboration among the School of Education and departments within the College of Arts and Sciences, and explain what we have learned that may be of use to other institutions.

I. GETTING STARTED

In 2005, the dean of the School of Education (SOE) at the University of North Carolina at Chapel Hill (UNC-CH) issued a call to action to the chairs of the science departments in the university's College of Arts and Sciences (CAS). He said, "I want there to be more science teachers, not more education students," meaning that he wanted more students majoring in science to become science teachers. This remarkable appeal for collaboration between the School of Education and the College of Arts and Sciences led to the establishment of a new program to prepare undergraduate science majors to teach in high schools, and to an unusual partnership between science and education faculty at an institution that has a greater focus on research that could change the world than on the public schools that prepare students to implement those changes.

Here we describe how our program, which came to be called the University of North Carolina Baccalaureate Education in Science and Teaching (UNC-BEST), came about, and what we believe similar institutions seeking to contribute to teacher education can learn from it. We discuss what motivated the various partners in the enterprise, how we secured the necessary resources, and the expertise we needed (and where we found it). We explain the tasks we had to complete in order to establish our program and get it approved by the state of North Carolina, and how the program fits into our physics curriculum. Finally, we describe how we have made the program sustainable (so far), the challenges that this involves, and the lessons we have learned. We hope that this account of what we did and why we did it will be instructive for physicists in similar institutions who wish to establish new programs for physics teacher preparation.

Physics faculty in major research universities ("research universities, very high research activity," or RU/VH in the Carnegie Classification), including those in public universities that are the flagship campuses of their state systems, typically see our primary missions to be the production of high-quality research and the mentoring of graduate students as they become physicists themselves. We also take pleasure in educating "the best and the brightest" undergraduate students in our institution, expecting that most of our physics majors will follow in our footsteps to graduate school in the subject we have made our lives' work. The preparation of undergraduates for immediate entry into the work force is often, at best, a secondary goal for physics departments in such institutions. Despite the well-documented shortage of high school physics teachers, major research universities have typically not regarded teacher education as being strong parts of our missions. Although institutions like ours may have schools of education, these schools are often focused on education research, education leadership development, or policy studies, rather than on preparing large numbers of teachers. Other campuses in public university systems (which in some cases began as normal schools) usually take a more prominent role in the preparation of future teachers. When faculty at research universities complain, as they often do, that students are ill-prepared for their physics courses, these faculty are ignoring the fact that by not preparing physics majors to become teachers, they are contributing to the problem.

In this chapter we will describe how the University of North Carolina at Chapel Hill decided to transcend this prevailing norm and establish a program to prepare undergraduate science majors to become high school science teachers. We chose to do this because increasing the number of teachers who have studied in depth the subject they will teach was "the right thing to do," but establishing our program has also proved to have significant benefits that we might not have anticipated, and that can be realized by institutions similar to ours. The challenges we faced are also likely to be representative of those that any major research university would encounter in such an effort, and our means of overcoming them may offer some lessons that could be useful to those who wish to establish a similar program. The need for more high school science teachers who majored in the subject they teach should be apparent to all readers of this book, and we provide here an example of how a major research university can help to meet that need.

A. Institutional and departmental context

The University of North Carolina at Chapel Hill is proud to be the first state university in the U.S. and the flagship campus of the 17-institution University of North Carolina system. With federally financed research and development expenditures in the top 10 among all U.S. universities (and in the top five among public universities), UNC-CH is unambiguously a research-intensive institution. Consistent with this focus on research, our School of Education pursues rigorous, practice-oriented research that is supported with substantial external funding. The SOE enrolls only about 200 undergraduate students (out of a total of about 18,500 at UNC-CH); these students major in early-childhood, elementary, or middle-grades education. At the time we established our program, UNC-CH offered *no* program in which undergraduate students could prepare for teacher licensure at the high school level in *any* subject. We truly started from scratch.

With the preparation of teachers being such a minor component of the mission of even the SOE (much less the institution as a whole), the science departments in the College of Arts and Sciences had virtually no interaction with the SOE. In order to contribute to the quality of science education at the secondary level, the science departments had to forge partnerships with one another and with the SOE to begin to graduate high school science teachers who know their science, know their discipline-specific pedagogy, and are in a position to make a real difference in the schools in which they teach. In the long term we expect our graduates to influence the students whom we admit to our campus and teach in our own courses. We hope by this account to inspire physics departments at other institutions to launch programs of their own and to share the lessons we have learned.

B. Genesis of the program

As described in the opening of this chapter, our journey began when the dean of the UNC-CH School of Education, Tom James (now provost and dean of Teachers College, Columbia University) addressed a meeting of the chairs of the science departments in the CAS (including one of the authors, L. E. McNeil) and offered to work with them to design a program in which science majors could prepare to become high school science teachers while completing their science degrees. Two of the chairs responded to that initial overture, with somewhat different motivations. Recognition of the extreme shortage of qualified physics teachers (in North Carolina and nationally) was one reason to consider such a move, while the needs of the large population of biology majors whose initial goal to attend medical school was not likely to be achieved (a group referred to as "post-premeds") was another. Neither of these incentives had been strong enough to provide an impetus to create a program before the dean proposed it, but they provided fertile ground for the program's germination and for the forging of a partnership between the two departments. Other departments (mathematics, geological sciences, and chemistry) later found good reasons to join the program—in some cases after a change of chair. The partnership among the science departments has proved to be important in our program's success in ways we will describe below.

Once the chairs of the physics and astronomy and biology departments made the decision to work together and establish a program, they immediately faced several problems. The largest of these was the fact that none of the existing faculty members in the two departments knew anything about secondary education. This is, of course, typical of the professoriate at a research university science department, within which there may perhaps be a junior member who participated in Teach for America, AmeriCorps, or the Peace Corps before going to graduate school, but whose experience in high schools is otherwise mostly limited to parent-teacher conferences or occasional outreach activities. Had our physics and astronomy department included a physics education research group, we would have had access to more expertise in educational matters, but not all such groups are well versed in the specifics of secondary schools. Beyond expertise in the science of teaching and learning, we also needed people with knowledge of the secondary school environment—a very different world from the ivory tower, and one that is governed by very different rules. None of our tenure-stream faculty members had an interest in devoting the time necessary to develop that specific expertise, nor would it have been wise for any of them to divert their efforts from the research activities that are a primary expectation for faculty at UNC-CH and institutions like it. The solution was to recruit a full-time, continuing, non-tenure-track faculty member whose duties would be directed exclusively toward the educational mission of the department. Although such positions carry fixed-term contracts (typically for three to five years at a time) rather than permanent tenure, the contracts are renewable and the positions can provide long-term career opportunities and a promotion ladder that runs parallel to the tenure track. Such faculty members carry different titles at different institutions. At UNC-CH they are called "lecturers"; other universities may call them "teaching professors," "academic professors," or "instructors."

C. Faculty

In order to recruit faculty members with the necessary expertise, we chose to appeal to the provost for support, since we sought to establish a program that would be a partnership between two of the units under her jurisdiction (CAS and SOE). It was clear from the outset that the preparation of highly qualified physics and biology teachers would require the creation of discipline-specific pedagogy courses rather than a generic science education methods course, and that we would therefore need a new faculty line in each department. However, judging that two positions might be more than the provost would be willing to risk on an untried venture, we made a strategic decision to request a position in biology, which has almost 20 times as many majors as does physics. The provost granted our request, and the biology department was fortunate to find a licensed high school teacher with undergraduate and master's degrees in biology and a Ph.D. in science education (from UNC-CH).

At the same time, we took steps to draw on external resources to establish the physics portion of the program. One of the authors (L. E. McNeil) wrote a proposal to the American Physical Society- and American Association of Physics Teachers-led Physics Teacher Education Coalition (PhysTEC) program and included in the budget funds for the partial support of a physics education specialist. The intention was that the remainder of the support could come from the department's instructional budget, which is used to pay part-time instructors, in exchange for the specialist's teaching one or more existing physics courses. When the grant was awarded, we were able to use it as external validation of our ideas and as evidence that the program was worthy of support, and we thereby persuaded the provost to grant a faculty line to the physics and astronomy department. This allowed us to devote the PhysTEC grant funds to a Teacher-in-Residence (TIR) instead (see Section IV below). Drawing on the resources of the physics education research community, we were able to hire a physicist with a Ph.D. in curriculum and instruction and with expertise in physics education at all levels (one of the authors, A. D. Churukian). With knowledgeable new colleagues in our departments and the backing of the dean of the SOE, we were poised to begin the task of creating a teacher education program for our biology and physics majors.

II. COLLABORATION WITH THE SCHOOL OF EDUCATION

From the beginning we sought to develop a program in which a student could complete a science major for either a for either a Bachelor of Arts (B.A.) or a Bachelor of Science (B.S.) degree, meet the requirements of the North Carolina Department of Public Instruction (NCDPI) to become eligible for teacher licensure in his or her chosen field, and graduate in four years. The initial team comprised SOE faculty members and administrators with thorough knowledge of NCDPI licensure rules, the two discipline-based science education specialists hired for the program, and the two department chairs from the CAS side.

A. Teacher licensure

Because there was no existing baccalaureate program at UNC-CH to prepare high school teachers, it was necessary to seek permission from the North Carolina Department of Public Instruction to establish a program in which teachers could be licensed. The two science education specialists, one of the SOE faculty members, and one of the SOE administrators worked together to create the necessary documentation. In North Carolina, science teachers usually do not have subject-specific licensure such as physics for grades nine through 12, but rather a comprehensive science license. Any teacher with a comprehensive science license is considered to be highly qualified to teach any science at the ninth through 12th grade level. We were seeking permission to have a program for subject-specific licensure. While the education requirements were fairly straightforward and standard to all programs, the content knowledge requirements proved to be a moving target. Fortunately, the SOE administrator was adept not only at interpreting the NCDPI documentation but also at addressing the right questions to the correct people. Because of this, once we submitted our proposal to the NCDPI, very little clarification and/or modification was needed for acceptance.

B. Streamlining the program

In order to establish a program that would allow science majors to become eligible for teacher licensure with the smallest possible number of extra courses, we needed to streamline existing education courses to the minimum needed to meet NCDPI requirements. Instead of addressing the requirements in terms of courses, we looked at each requirement individually and determined what courses could be created to meet them. By doing so, we were able to reduce the number of program-required courses to a total of three three-credit education courses encompassing everything from educational psychology to classroom management, a four-credit discipline-based pedagogy course, and a 12-credit student teaching internship. For the education courses, the SOE team members were able to build on their earlier efforts to streamline the pre-licensure program to support entry into high school teaching by professionals in other fields (such as engineering). This measure, known as "lateral entry," was adopted to address the critical shortage of science (and other) teachers in the state. The science education faculty members in each of the participating science departments developed the corresponding pedagogy courses.

In this streamlined program there is very little room for redundancy. Every requirement is met in one—and only one—of the four courses or the internship, or by standard exams. For example, in the pedagogy course students meet the requirement of experience in a high school classroom before beginning the student teaching internship. Showing breadth of content knowledge, on the other hand, is achieved by passing the comprehensive science Praxis exam. The Praxis Series is a set of tests [1] developed and administered by the Educational Testing Service to measure the knowledge and skills of teacher candidates. Each state has different requirements for which exams are needed and what constitutes a passing score. To further streamline the process, the education courses were designed to also count toward general education graduation requirements and/or requirements for the respective majors. The partnership between the science departments and the SOE was instrumental in crafting the courses such that they met both the NCDPI requirements and the requirements of the general education curriculum as overseen by the Office of Undergraduate Curricula in the CAS.

III. CREATING A MULTIDISCIPLINARY PROGRAM

A. Departmental specialists

The initial partners with the SOE in establishing the teacher preparation program were the Department of Physics and Astronomy and the Department of Biology. The program has since expanded to include the Departments of Mathematics, Chemistry, and Geological Sciences. Each department has a specialist who teaches the department's pedagogy course and recruits and advises the students. The specialist also takes on other duties, such as teaching other courses, training TAs, and consulting with faculty on educational matters. In physics and astronomy, for example, the specialist is a member of the cadre of faculty who coordinate and teach the large-enrollment introductory physics courses. In addition, she prepares the teaching assistants (both graduate and undergraduate) for their duties each week, provides training for all new teaching assistants each semester, collaborates with other faculty members in course redesign, and acts as a faculty consultant on educational matters when requested.

The biology specialist teaches an undergraduate research course and a graduate seminar, Teaching College Science. In addition, she teaches the biology pedagogy course twice per year, supervises biology student teachers in the field, and organizes and runs the annual UNC-BEST alumni retreat. The mathematics specialist is the director of the Math Help Center, coordinates with the Physics Tutorial Center director (the help and tutorial centers share a common room), teaches a TA training seminar each fall and a math course each semester, and supervises mathematics (and sometimes other) student teachers during their internships. In geological sciences the specialist is responsible for teaching the large introductory geology courses each semester; the geological sciences specialist also manages the coordinating laboratories, writes laboratory activities, and trains and supervises TAs. The chemistry specialist was the General Chemistry Laboratory Manager prior to the start of the UNC-BEST program and was asked to add the program to her responsibilities. This variety of ways in which the specialists contribute to their departments' missions highlights how a partnership among the science departments can be forged by appealing to each department's specific needs.

B. Program coordinator

In addition to the subject specialists, our program also has an overall coordinator. This person is a clinical faculty member within the School of Education tasked with providing primarily practical instruction and application of practical knowledge; the coordinator is also responsible for the day-to-day aspects of the program within the SOE. These aspects include teaching the UNC-BEST seminar (see below), acting as the liaison between the UNC-BEST steering committee (consisting of department chairs and deans) and the operations committee (consisting of teaching faculty), managing grants associated with the program (these include the Burroughs Wellcome Fund, a U.S. Department of Education Teachers for a Competitive Tomorrow grant, and a grant from the Howard Hughes Medical Institute), managing the process of admission into the program, managing student teacher placement, assisting students with completing and submitting licensure paperwork, and supervising student teachers in the field. In addition to coordinating the program, the coordinator is also a member of the Master of Arts in Teaching (MAT) faculty and teaches two courses for the MAT program each semester. The involvement of a faculty member who has primary duties in teacher preparation (as opposed to education research) has proven to be important to the success of the SOE/CAS partnership.

C. Cohort seminar course

As we developed the program, we knew that its multidisciplinary character would have students scattered across the campus and rarely, if ever, crossing paths. We wanted to manifest the partnership among the science departments by finding a way to bring the students together in a cohort, so they could have a support group of like-minded colleagues with similar experiences as they went out into their first teaching positions and beyond. We added a one-credit seminar to the course requirements to serve this purpose, with a secondary goal of giving us a venue to address topics that may otherwise slip through the cracks of the streamlined program. The seminar meets five times each semester. In its original format, the seminar began as a social gathering and then transitioned to a more serious topical discussion. The topics ranged from taking and discussing the Myers-Briggs Type Indicator personality assessment to a question-and-answer session with a panel of local high school science teachers. The seminar now also covers technical details such as taking the Praxis exams and registering for the online system preservice teachers use to submit materials for their applications for licensure.

Originally, we encouraged students to attend the seminar as soon as they were accepted into the program, and we required them to enroll for one semester before graduation (usually the semester preceding their student teaching semester). This format worked well in the first few years of the program, but as our numbers grew it became unwieldy. Students at various stages in the program were intermixed, so that some attended the seminar three or four semesters in a row whereas others attended for only one semester. This meant some students were seeing the same topics repeatedly, whereas others were missing important information. The seminar now has a more formal structure with required readings and other assignments, and students are expected to take it in the semester prior to their student teaching. The technical details are handled concurrently with their student teaching internship. While the overall sense of community may have been lost, students can now form cohorts that are more cohesive because the students are at the same stage in the program and are having similar experiences.

IV. MODIFYING THE PHYSICS CURRICULUM

At UNC-CH, as elsewhere, curricula for science majors are jam-packed and rigorous. Each of the departments participating in our teacher preparation program offers both a Bachelor of Arts and a Bachelor of Science track. The B.S. track is designed for students who intend to pursue graduate study or seek technical employment in the field after graduation. The B.A. track is designed to offer more flexibility for students to pursue broad interests in addition to their primary science discipline. It requires fewer courses in the major (beyond the core courses) but has more extensive general education requirements outside the sciences than does the B.S. track.

In order to attract the largest possible number of students and allow them (especially those pursuing the B.S. track) to complete the teacher preparation program within four years, it was our goal to have a "fast-track" curriculum that would allow science majors to become eligible for teacher licensure with the fewest additional courses. This goal was made even more salient by our administration's increased emphasis on students' graduating after no more than eight semesters in residence.

A. Degree requirements

To accomplish the goal of graduating teachers in eight semesters, we arranged for each course in our program to fulfill at least one other requirement (either in the major or in general education), so that the total number of courses taken by preservice teachers equaled the number of courses taken by other students within their majors. This meant that the discipline-specific pedagogy course taught by each department had to count as an elective within the major. We were able to persuade our colleagues within the physics and astronomy department that the Teaching and Learning Physics course would help our students deepen and solidify their knowledge of the fundamental concepts taught in the first two semesters of our introductory physics sequence (which is a prerequisite for it). We further argued that Teaching and Learning Physics would play a role in the physics curriculum similar to that of other elective courses in which students use the knowledge they had acquired in core courses to study specialized topics. Just as students had the option of using their knowledge of electromagnetism to study optics in another elective course, they could use their knowledge of basic physics to learn how to teach it to others. The fact that the content of the course is firmly grounded in research conducted by physicists in physics departments (although not in our department, which has no physics education research group) was also a factor in our colleagues' willingness to agree that the course be treated like any other physics elective.

The extensive fieldwork in high schools that is incorporated into the pedagogy course allowed it to also fulfill a general education requirement in "experiential education." Courses that meet this requirement allow students to, in alignment with the UNC-CH Undergraduate Bulletin, "connect academic inquiry with a structured, active learning experience in which [they] exercise initiative and apply academic knowledge in various real-world contexts" [2]. Embedding the pedagogy course within the physics curriculum has the further advantage that some students who take it as a physics elective catch the "teaching bug" and opt to join the teacher preparation program or to, upon graduation, pursue teaching high school as a career through some other path.

The remaining courses in our preservice teacher program are taught by School of Education faculty and are considered by the CAS to be in the realm of the social and behavioral sciences. As noted above, these courses satisfy general education requirements in that realm, as well as UNC-CH's "U.S. diversity" requirement to "help students develop a greater understanding of diverse peoples and cultures within the United States" [2]. In this way, each of the required courses in the teacher preparation program also fulfills another graduation requirement, and no student has to take extra courses in order to complete both the science major and the teacher licensure program simultaneously.

In addition to the required courses, all students in our program must complete a student teaching internship. This is a full-time obligation for one semester, and can be undertaken only after the student has completed all other requirements for licensure. Science majors who enter UNC-CH with a significant number of credits (e.g., from Advanced Placement or International Baccalaureate examinations based on work done in high school) are sometimes able to complete all their other graduation requirements in seven semesters, making it possible for them to do their student teaching in the eighth semester. Students for whom this is not possible may delay graduation for a semester to complete the internship.

When we began the program, an alternative track to licensure was for the student to graduate after eight semesters and then complete the student teaching internship as a post-baccalaureate student. This route was made possible by the "lateral entry" program that North Carolina (among other states) has adopted to provide a means for professionals in other fields such as engineering to become licensed teachers. In this program a school hires an eligible person to teach and recommends him or her for a provisional teaching license. The teacher affiliates with an approved teacher education program, and upon completion of its requirements the teacher is granted a license. We have discouraged students from planning to take this route, however, since success depended upon finding a school willing to hire a teacher with no teaching experience.

B. Pedagogy course

The Teaching and Learning Physics pedagogy course is at the heart of our physics teacher preparation program. We designed the course to be co-taught by the specialist and an experienced high school physics teacher who would serve as our Teacher-in-Residence. We structured the course in three main parts: theory, application, and practice. The course meets on campus twice per week and has an associated "laboratory" component in the field. Instruction in topics such as constructivism, epistemology, types of questions, assessment, and lesson planning is provided by the specialist. Drawing on his or her vast experience in the physics high school classroom, the TIR provides instruction in application of the pedagogical theory by leading the students in performing and discussing laboratory activities and problem solving on the topics covered in the physics portion of the North Carolina Standard Course of Study. Overlaid on top of these classroom activities is the "practice" component of the course: Students spend a minimum of 30 hours during the semester in a local high school physics classroom. Here they observe the mentor teacher, assist in the classroom as deemed appropriate, interview various members of the school staff to get a feel for the school environment, and teach two micro-lessons.

C. Teacher-in-Residence

One aspect of our program that we have found to be essential to our success is the TIR. He or she brings to the program vast knowledge based on experience in the classroom, for which no amount of theoretical understanding can substitute. Each of our TIRs has had at least 30 years of classroom experience. All have been retired (although some have been teaching part-time after retirement), and were therefore willing to add this role to their activities. In addition to their experience and expertise, the TIRs bring a different perspective to the classroom and add authenticity to theoretical discussions by sharing personal stories relating to particular situations or events. They can also excite young physicists to be dedicated educators by being steadfast role models who demonstrate that enthusiasm for teaching can be sustained over a lifetime.

V. SUSTAINABILITY

Once we had created the program, obtained approval for it, recruited our first students in biology and physics, seen them through to graduation, and placed them in teaching positions (a process that took a rather breathless two years), we were immediately faced with the challenge of sustaining the program over the long term amidst the vagaries of the university's budget. It was clear that our university administration was pleased with our new program, and our first graduates were highlighted in a news release and in the printed program for the university's commencement ceremony. But sustaining the program after its initial success would require broadening our partnership beyond physics and astronomy, biology, and education, as well as seeking additional external and internal funds to support the expansion.

The mathematics and geological sciences departments had begun to express interest in participating in the program, so we turned to the U.S. Department of Education's Teachers for a Competitive Tomorrow program [3] to seek funds to expand the collaboration among our science departments. Our successful proposal included partial support for education specialists in math and geology and for the coordinator of the program in the SOE. It also included funds for partnerships with high-need school districts (where we have since been able to place some of our graduates) and for activities in support of our new teachers, including an annual alumni retreat. This expansion of our collaboration among science departments helped put our teacher preparation program on a broader disciplinary footing and attach it more firmly to the central mission of the College of Arts and Sciences. As the grant funds become exhausted, we have further embedded our program in the CAS by making clear the links between it and a new initiative wherein the college seeks to improve the teaching of introductory science courses by introducing active-engagement, evidence-based pedagogy.

Our education specialists are especially well placed to contribute to this initiative, so we have made sure that they are deeply involved in the planning and execution of the pedagogical changes as well as in the preparation of proposals to the National Science Foundation and the Association of American Universities to support the course transformations across multiple science departments. All of this activity makes manifest the value of these individuals and the expertise they bring to the science departments involved in this collaboration, which helps maintain the internal support for UNC-BEST.

We have also cultivated a partnership with the higher administration of our campus. From the advent of our program, it has been important to emphasize the contribution it makes to the mission of our state university, which is to enhance the wellbeing of the citizens who support it with their tax dollars. Four years after we established UNC-BEST, our SOE discontinued its Master of Arts in Teaching program in high school science teaching because the school did not have sufficient resources to maintain both programs (and the MAT program had very low enrollment even prior to the inception of UNC-BEST). As a result, the SOE's partnership with CAS through UNC-BEST is now the *only* way in which a UNC-CH student can prepare to serve the people of North Carolina by obtaining licensure as a high school science teacher. We remind our campus administration of this fact at every opportunity. This is an ongoing task, as administrators come and go, but they are always happy to have examples of ways in which our campus serves the external constituencies they frequently find themselves addressing.

VI. LESSONS LEARNED

Our first lesson was that it would be very difficult to establish a physics teacher preparation program from scratch without partnering with at least one other science department. At a typical research university, the number of undergraduate students who receive degrees in physics is significantly smaller than the number of students who receive degrees in biology. Physics majors constituted only 3.3% of the bachelor's degree recipients in the five disciplines participating in UNC-BEST (physics, biology, mathematics, geology, and chemistry) from May 2009 to May 2013, whereas biology graduates constituted 64.8%. In order to avoid having the physics teacher program labeled "low performing" because of its small number of graduates (fewer than one per year on average), and thus subject to budget cuts, it was vital for us to partner with a department with a much larger number of graduates. The fact that the need for qualified teachers is much greater in physics than in biology is not necessarily a compelling argument for the allocation of resources when administrators are faced with a realistic expectation of graduating less than one new teacher per year on average. This fact, coupled with the fact that the resources needed do not scale linearly with the number of students enrolled (e.g., the instructional personnel costs are approximately the same for a pedagogy class with 10 students as for one with 20), made it very important to join forces with a much larger department. There is safety in numbers.

The second lesson was that despite our recruiting efforts, the number of physics majors we can expect to join the physics teacher preparation program is small. The fraction of science graduates who have gone through UNC-CH's teacher preparation program has been similar among the various disciplines: 2.5% in biology, 3.2% in physics, 4.6% in mathematics, 4.3% in geology, and 1.4% in chemistry (which came late to the program). Although we hope to raise those percentages over time, the reality is that students who choose to attend major research universities rarely do so with the intention of becoming high school teachers. The number of physics teachers such a program will graduate will always be limited by the arithmetic of taking a small percentage of an already small number. This does not mean that the enterprise is not worthwhile, as the number of qualified physics teachers prepared nationally is so small compared to the need that every new teacher makes a difference. However, it reinforces the wisdom of partnering with a department with a much larger number of majors, so that the small percentage of a larger number becomes a respectable figure (i.e., 46 new biology teachers in five years compared to three new physics teachers).

A third lesson concerns the beneficial effect of the physics teacher preparation program on our department's instruction. Partly as a result of hiring a member of the physics education research community with expertise in curriculum and instruction, we have been able to redesign our introductory physics courses to incorporate active-engagement pedagogy, resulting in better learning outcomes and happier students. The guidance and coordination that this specialist provides has helped us transform our large-enrollment introductory offerings from multiple courses taught independently (with varying degrees of success) by faculty who neither communicated with one another nor used research-validated pedagogy, into unified courses taught in multiple sections by a team of faculty members who agree upon a single vision that incorporates active learning. The synergy afforded by simultaneously transforming our introductory courses and establishing a physics teacher preparation program was unexpected (though perhaps obvious in hindsight), and it has helped our department take a leadership position as we partner with other science departments to improve science education at our institution.

A fourth lesson was in the value of partnership with our School of Education. There is far more involved in becoming an effective physics teacher than learning physics and how to teach it, important as these factors are. The SOE's capacity to help our preservice teachers learn about the physical, cognitive, language, and socio-emotional development of children, and how these factors affect the design of appropriate learning experiences for the high school classroom, has been invaluable. Beyond the classroom, our students need to learn about key legal issues related to education and to develop strategies to communicate effectively with students and parents, work with exceptional and at-risk youth, monitor and communicate student progress, and manage student behavior. Given the diverse communities in which our graduates can expect to teach, they need to explore the impact of students' lives outside school on their learning, including the impact of culture, family, and personal values; our graduates also need to understand how history, identity, and sociocultural issues affect academic opportunities and achievement. All of the necessary expertise to prepare our students in this way resides in our School of Education, and certainly *not* in the physics faculty. Furthermore, the regulations involved in the licensing of teachers and the state bureaucracy that administers those regulations are quite opaque to physics faculty members but are part of the normal routine of a professional school within a large university. We therefore value our amicable working relationship with the SOE, in which each partner does what it can do best while working toward a common goal.

A fifth lesson is perhaps more of a cautionary tale. We are confident that the graduates of our physics teacher preparation program are highly qualified to teach high school physics, and to do so more effectively than someone who did not study physics as an undergraduate. However, since a typical high school in North Carolina has (at most) one physics teacher, when UNC-CH graduates are first hired they often find themselves teaching subjects such as physical science, chemistry, or biology if the school already has someone teaching physics. Our students who have a minor or a second major in mathematics are encouraged to seek a mathematics teaching license in addition to the physics license, which makes them eligible to teach mathematics classes if no physics classes are available to them. As noted above, our graduates are considered "highly qualified" to teach *any* science subject, because they receive a comprehensive science license in addition to their physics certification (as a result of taking the comprehensive science Praxis exam to satisfy the NCDPI requirement for "breadth of knowledge"). They receive this license despite the fact that their content knowledge and pedagogical content knowledge in other sciences may be limited. We have thus found ourselves, in a sense, contributing to the problem of teachers who are teaching outside of their subjects, despite our intention to address it. It is our expectation—one beginning to be realized—that our graduates will take on the teaching of additional physics classes as the value of their superior preparation to teach the subject gains recognition in their schools. We know that the students in those schools will greatly benefit when that happens, and that the partnership at UNC-CH that made it possible will have been worthwhile.

VII. CONCLUSION

At the University of North Carolina at Chapel Hill we have developed and implemented a fast-track program to prepare science and mathematics majors to become licensed high school teachers in their chosen fields. Our program is a collaboration among the various science and mathematics departments and the School of Education. Without this collaboration we would not have been successful at preparing highly qualified physics teachers. We hope the lessons we have learned are beneficial to others at institutions similar to ours who wish to establish programs of their own. The success of our program demonstrates that institutions like UNC-CH can contribute to the goal of increasing the number of highly qualified physics teachers for the next generation and beyond.

---

[1] See http://www.ets.org/praxis

[2] University of North Carolina at Chapel Hill, *2013-2014 Undergraduate Bulletin* (2013). http://www.unc.edu/ugradbulletin/pdf/2013-14.pdf

[3] See http://www2.ed.gov/programs/tct/index.html